The Handbook of Environmental Chemistry

Founded by Otto Hutzinger

Editors-in-Chief: Damià Barceló • Andrey G. Kostianoy

Volume 18

The Handbook of Environmental Chemistry
Recently Published and Forthcoming Volumes

Global Risk-Based Management of Chemical Additives I: Production, Usage and Environmental Occurrence
Volume Editors: B. Bilitewski, R.M. Darbra, and D. Barceló
Vol. 18, 2012

Polyfluorinated Chemicals and Transformation Products
Volume Editors: T.P. Knepper and F.T. Lange
Vol. 17, 2012

Brominated Flame Retardants
Volume Editors: E. Eljarrat and D. Barceló
Vol. 16, 2011

Effect-Directed Analysis of Complex Environmental Contamination
Volume Editor: W. Brack
Vol. 15, 2011

Waste Water Treatment and Reuse in the Mediterranean Region
Volume Editors: D. Barceló and M. Petrovic
Vol. 14, 2011

The Ebro River Basin
Volume Editors: D. Barceló and M. Petrovic
Vol. 13, 2011

Polymers – Opportunities and Risks II: Sustainability, Product Design and Processing
Volume Editors: P. Eyerer, M. Weller, and C. Hübner
Vol. 12, 2010

Polymers – Opportunities and Risks I: General and Environmental Aspects
Volume Editor: P. Eyerer
Vol. 11, 2010

Chlorinated Paraffins
Volume Editor: J. de Boer
Vol. 10, 2010

Biodegradation of Azo Dyes
Volume Editor: H. Atacag Erkurt
Vol. 9, 2010

Water Scarcity in the Mediterranean: Perspectives Under Global Change
Volume Editors: S. Sabater and D. Barceló
Vol. 8, 2010

The Aral Sea Environment
Volume Editors: A.G. Kostianoy and A.N. Kosarev
Vol. 7, 2010

Alpine Waters
Volume Editor: U. Bundi
Vol. 6, 2010

Transformation Products of Synthetic Chemicals in the Environment
Volume Editor: A.B.A. Boxall
Vol. 2/P, 2009

Contaminated Sediments
Volume Editors: T.A. Kassim and D. Barceló
Vol. 5/T, 2009

Biosensors for the Environmental Monitoring of Aquatic Systems
Bioanalytical and Chemical Methods for Endocrine Disruptors
Volume Editors: D. Barceló and P.-D. Hansen
Vol. 5/J, 2009

Environmental Consequences of War and Aftermath
Volume Editors: T.A. Kassim and D. Barceló
Vol. 3/U, 2009

The Black Sea Environment
Volume Editors: A. Kostianoy and A. Kosarev
Vol. 5/Q, 2008

The Handbook of Environmental Chemistry Also Available Electronically

The Handbook of Environmental Chemistry is included in Springer's eBook package *Earth and Environmental Science*. If a library does not opt for the whole package, the book series may be bought on a subscription basis.

For all customers who have a standing order to the print version of *The Handbook of Environmental Chemistry*, we offer free access to the electronic volumes of the Series published in the current year via SpringerLink. If you do not have access, you can still view the table of contents of each volume and the abstract of each article on SpringerLink (www.springerlink.com/content/110354/).

You will find information about the

- Editorial Board
- Aims and Scope
- Instructions for Authors
- Sample Contribution

at springer.com (www.springer.com/series/698).

All figures submitted in color are published in full color in the electronic version on SpringerLink.

Aims and Scope

Since 1980, *The Handbook of Environmental Chemistry* has provided sound and solid knowledge about environmental topics from a chemical perspective. Presenting a wide spectrum of viewpoints and approaches, the series now covers topics such as local and global changes of natural environment and climate; anthropogenic impact on the environment; water, air and soil pollution; remediation and waste characterization; environmental contaminants; biogeochemistry; geoecology; chemical reactions and processes; chemical and biological transformations as well as physical transport of chemicals in the environment; or environmental modeling. A particular focus of the series lies on methodological advances in environmental analytical chemistry.

Series Preface

With remarkable vision, Prof. Otto Hutzinger initiated *The Handbook of Environmental Chemistry* in 1980 and became the founding Editor-in-Chief. At that time, environmental chemistry was an emerging field, aiming at a complete description of the Earth's environment, encompassing the physical, chemical, biological, and geological transformations of chemical substances occurring on a local as well as a global scale. Environmental chemistry was intended to provide an account of the impact of man's activities on the natural environment by describing observed changes.

While a considerable amount of knowledge has been accumulated over the last three decades, as reflected in the more than 70 volumes of *The Handbook of Environmental Chemistry,* there are still many scientific and policy challenges ahead due to the complexity and interdisciplinary nature of the field. The series will therefore continue to provide compilations of current knowledge. Contributions are written by leading experts with practical experience in their fields. *The Handbook of Environmental Chemistry* grows with the increases in our scientific understanding, and provides a valuable source not only for scientists but also for environmental managers and decision-makers. Today, the series covers a broad range of environmental topics from a chemical perspective, including methodological advances in environmental analytical chemistry.

In recent years, there has been a growing tendency to include subject matter of societal relevance in the broad view of environmental chemistry. Topics include life cycle analysis, environmental management, sustainable development, and socio-economic, legal and even political problems, among others. While these topics are of great importance for the development and acceptance of *The Handbook of Environmental Chemistry,* the publisher and Editors-in-Chief have decided to keep the handbook essentially a source of information on "hard sciences" with a particular emphasis on chemistry, but also covering biology, geology, hydrology and engineering as applied to environmental sciences.

The volumes of the series are written at an advanced level, addressing the needs of both researchers and graduate students, as well as of people outside the field of "pure" chemistry, including those in industry, business, government, research establishments, and public interest groups. It would be very satisfying to see these volumes used as a basis for graduate courses in environmental chemistry. With its high standards of scientific quality and clarity, *The Handbook of*

Environmental Chemistry provides a solid basis from which scientists can share their knowledge on the different aspects of environmental problems, presenting a wide spectrum of viewpoints and approaches.

The Handbook of Environmental Chemistry is available both in print and online via www.springerlink.com/content/110354/. Articles are published online as soon as they have been approved for publication. Authors, Volume Editors and Editors-in-Chief are rewarded by the broad acceptance of *The Handbook of Environmental Chemistry* by the scientific community, from whom suggestions for new topics to the Editors-in-Chief are always very welcome.

Damià Barceló
Andrey G. Kostianoy
Editors-in-Chief

Volume Preface

This first volume of the RISKCYCLE book analyzes the chemical additives used in the production of important industrial sectors. Additives are substances that are used to improve the characteristics of the final product.

In the first part, a review of the additives used in the sectors of textiles, electronics, lubricants, plastics, paper and leather is carried out with an emphasis on the function of each compound inside the product. For example, flame retardants reduce the potential ignition of the products; others are water repellent or biocides. These additives due to their chemical structure may have a potential impact on the environment or human health when released from the product.

The potential effects of the mainly used additives are presented in this volume in detail together with data on the world production and trade of these compounds. The aim of the RISKCYCLE project is to assess the risk of chemical additives at global scale, and it is therefore essential to know more about the world distribution of these compounds and their potential impact.

In the second part, a detailed perspective of the chemicals used in the aforementioned sectors in different countries is presented in case studies. One very interesting case concerning the recycling of paper in Vietnam is thoroughly analyzed. The emissions of additives from plastics in Sweden give a new insight on the subject, and the recycling of electronic waste in Brazil is also a case to highlight. These different case studies show the magnitude of the trade of these chemicals all over the world and their potential impact on human health and the environment if they are not correctly treated.

We hope that this book is of interest for the scientific community and will help to increase knowledge and awareness on issues raised by the different chemical additives used in major industrial sectors.

Finally, we would like to thank all the authors who have contributed to this book, for their effort in gathering the information and elaborating the different chapters.

Brussels, Belgium — Georges Deschamps, EU Executive Office

Pirna, Germany — Bernd Bilitewski
Barcelona, Spain — Rosa Mari Darbra, Damià Barceló

Contents

Introduction

B. Bilitewski, R.M. Darbra, and D. Barceló

Abstract Many potential hazardous compounds are traded worldwide as chemicals or incorporated as additives in consumer and industrial products. Their release to the environment has been a concern of the European Commission, UNO, WHO and OECD. The discussion of the assessment and management of chemicals and products led to the creation of the OECD programme Globally Harmonised System of Classification and Labelling of Chemicals (GHS). The World Summit encouraged countries to implement GHS with a view of having the System operating by 2008. The need to form GHS on a global scale is part of the EU policy. GHS aims to have the same criteria worldwide to classify and harmonise the responsible trade and handling of chemicals and products and at the same time protect human health and the environment. The EU will ensure transition from the current EU Classification and Labelling (C + L) system to the implementation of GHS, which harmonises with REACH registration. However, a complete picture on the global state of implementation is not available. With the growing level of worldwide trade, we however face unsafe consumer products on the market. These examples show that on a global perspective the trade of chemicals and products in a circular economy is not acceptable without globally agreed assessment methods and harmonised C + L (GHS).

B. Bilitewski (✉)
Institut für Abfallwirtschaft und Altlasten, TU Dresden, Pratzschwitzer Str. 15, 01796 Pirna, Germany
e-mail: abfall@mail.zih.tu-dresden.de

R.M. Darbra
CERTEC, Department of Chemical Engineering, Universitat Politècnica de Catalunya, ETSEIB, Diagonal 647, pav. G, 08028 Barcelona, Spain

D. Barceló
Department of Environmental Chemistry, IDAEA-CSIC, Jordi Girona, 18-26, 08034 Barcelona, Spain

B. Bilitewski et al. (eds.), *Global Risk-Based Management of Chemical Additives I: Production, Usage and Environmental Occurrence*, Hdb Env Chem (2012) 18: 1–8, DOI 10.1007/698_2011_117, © Springer-Verlag Berlin Heidelberg 2011, Published online: 20 October 2011

The overall objective of the Coordinating Action RISKCYCLE is to define together with international, European and national experts from different programmes future needs of R + D contributions for innovations in the field of risk-based management of chemicals and products in a global perspective using alternative testing strategies to minimise animal tests.

Keywords Chemicals, Circular economy, Globally Harmonised System, Risk assessment

The global trade of chemicals and products containing chemical additives such as paint, cosmetics, household cleaners, paper and cardboard, plastic toys, textiles, electronic appliances, petrol and lubricants has resulted in a substantial release of harmful substances to the environment with risk to man and nature on a worldwide scale.

The discussion of the assessment and management of chemicals and products at the *1992 Earth Summit* in Rio de Janeiro led to the creation of the OECD programme *Globally Harmonised System of Classification and Labelling of Chemicals (GHS)*. The *World Summit on Sustainable Development in Johannesburg 2002* encouraged countries to implement the GHS (adopted by UN ECOSOC in July 2003) as soon as possible, with a view of having the system operating by 2008.

In spite of some common efforts to harmonise the safety assessment of chemicals and products, a new problem with Recovered Material (as illustrated in Fig. 1) additionally appeared. The figure shows a simplified material flow in a circular economy at global scale with its risks for health and the environment due to the worldwide trade of chemicals and products. Circular Economy is a concept that is transforming traditional patterns of economic growth and production. The conventional perception of economic systems is that they are linear. This linear system is converted to a circular system when the relationship between resource use and waste residuals is taken into consideration.

Although the practices of circular economy have been present throughout history, the modern concept of circular economy (now under discussion in Asia and very seriously in China) was only introduced in Germany in 1998. The new threat comes from closing the loop in a global scale. Plastic, paper and cardboard, lubricants and other products undergo a recycling process and make their ways into a recovered material with unpredictable and not foreseen health and safety problems. In this way, unsafe consumer and industrial products get onto the global market. Latest reports in the news on tooth paste, toys and drinking cups from Asia releasing hazardous components prove this new problem.

Viet Nam News December 22, 2008 wrote that more than 75% of the country's empty plastic containers carrying all kinds of chemicals are recycled and sold back to the market to produce new plastic products of different type. Bangkok Post on February 8, 2009 noted that the use of substandard fertilisers and agricultural chemicals of unknown origin has badly affected rubber production in the region.

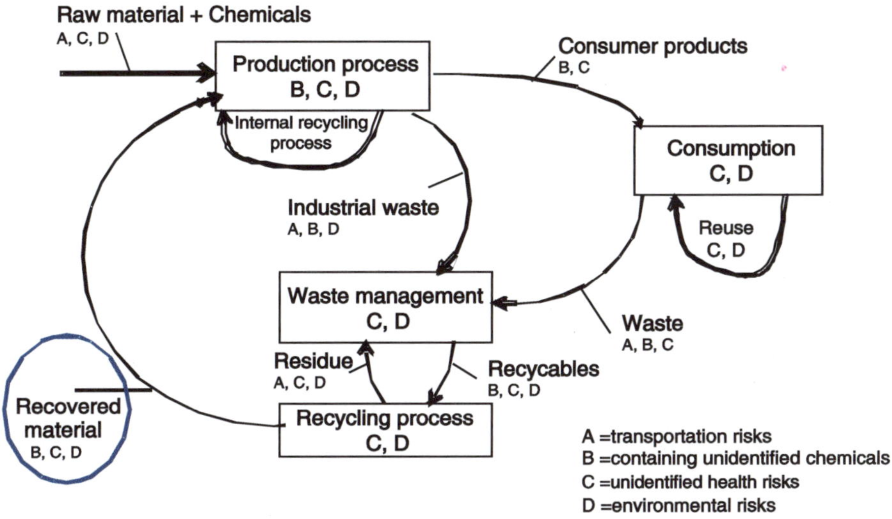

Fig. 1 Simplified material flow of a circular economy in a global scale with health and environmental risks

One compound with estrogenic activity that has been studied extensively as an intermediate in the production of polycarbonate and epoxy resin is Bisphenol A (BPA). The migration of BPA from polycarbonate flask used for baby food is enhanced when sterilisation of flask is carried out at temperatures over 80°C [1]. A recent study carried out in Germany demonstrated that estrogenic activity was detected as a result of migration of endocrine disruptors form bottled mineral water in plastic containers made of polyethylene terephthalate (PET). In contrast to polycarbonate, PET should be free form BPA. Different authors detected several phthalates in PET plastic bottles. So, it is most probably that a migration of DEHP (Di(2-ethylhexyl) phthalate) (see Table 1) occurred from PET to the bottled mineral water and afterwards the estrogenic activity was detected [2].

Waste electrical and electronic equipment, also known as e-waste, refers to the end-of-life products encompassing consumer electronics, information communication and household devices. Many of them have a short lifetime like computers, television, printers and cell phones. Therefore, e-waste is generated in large quantities with an annual volume about 20–50 million tonnes worldwide with an estimated increase of 3–5% per year [3]. Toxic substances present in e-waste are indicated in the enclosed table (Table 1). Among them, it can be found heavy metals such as lead, mercury and cadmium and persistent organ halogen compounds such as polychlorinated biphenyls (PCBs) and brominated flame retardants (BFRs). Uncontrolled e-waste recycling has become a topic of serious concern in recent years. It is estimated that up to 80% of e-waste from industrialised countries is exported to Asian developing countries, such as Vietnam for recycling and

Table 1 List of the main chemicals present in the different industrial products that will be studied in the following chapters of this book

Industrial sector	Chemical	Synonym/Substance/Acronym	CAS No.
Lubricants	Perfluoro octane sulfonate	PFOS	2795-39-3
	Perfluoro octanoic acid	PFOA	335-67-1
	Nonylphenoxy acetic acid	NPAA	3115-49-9
Textiles	Hexabromocyclododecane	HBCDD	25637-99-4
	5-Chloro-2-(2,4-dichloro-phenoxy)-phenol	Triclosan	3380-34-5
Plastics	Di(2-ethylhexyl) phthalate	DEHP	117-81-7
	Lead	Pb(II)	–
	Tri-*n*-butyltin hydride	Tributylstannane	688-73-3
Electronics	Tetrabromodiphenyl ethers	2,2',4,4'-Tetrabromodiphenyl ether (BDE 47)	5436-43-1
	Pentabromodiphenyl ethers	2,2',4,4',5-Pentabromodiphenyl ether (BDE 99)	60348-60-9
		2,2',4,4',6-Pentabromodiphenyl ether (BDE 100)	189084-64-8
	Decabromodiphenyl ether	BDE 209	1163-19-5
	Lead	Pb(II)	–
	Triphenyl phosphate	TPP	115-86-6
	Mercury	Hg(II)	–
Leather	Pentachlorophenol	PCP	87-86-5
	(Benzothiazol-2-ylthio) methyl thiocyanate	TCMTB	21564-17-0
Paper	Bisphenol A	4,4'-Isopropylidenediphenol (BPA)	80-05-7
	Dibutyl phthalate	DBP	84-74-2
	Isothiazolinones	5-Chloro-2-methyl-isothiazolin-3-one (CMI)	26172-55-4
		2-Methyl-2-isothiazolin-3-one (MI)	2682-20-4

exploiting the inexpensive labour costs and weak enforcement of environmental laws.

The paper chain is a very good example for successful recycling in Europe. However, if a deeper analysis is made, the recycling of paper and cardboard, especially for graphical paper, can introduce chemicals (from the original paper) into recovered material with unpredictable and not foreseen health and safety problems. This is the case for the thermal paper, used in cash machines and as copy paper. It has a colour developing layer with the chemical BPA. BPA is introduced into the paper cycle through the recovery of used thermal paper. BPA is found in the wastewater and detected in the next paper product. Toilet paper has a high concentration of BPA, which can be found in the wastewater after use. Printing ink used in newspaper is contaminating the cardboard for packaging and entering into the packed food exceeding the threshold values for Polycyclic Aromatics in the food by up to more than ten times [4]. All these examples show that in a circular economy the trade in a global dimension is not acceptable without a globally agreed risk assessment for existing and newly developed chemicals and products without using additional test animals.

Within this situation, the overall objective of the introduced co-ordination action RISKCYCLE aims to establish and co-ordinate a global network of European and international experts and stakeholders from different programmes and countries of the EU, USA, Japan, China, India, Brazil, Vietnam etc. to explore the synergies of the research carried out within different programmes and countries, and to facilitate the communication among researchers, institutions and industries and make the information about the risks of hazardous chemicals and additives in products and the risk reduction measures for substances widely available. As a result of this, RISKCYCLE has to define together future needs of R + D contributions for innovations in the field of risk-based management of chemicals and products of a circular economy in a global perspective making use of alternative strategies to animals test.

When addressing how this objective will be achieved, it is relevant to consider what information on present activities in this area is available and what is still unknown.

The key pieces of information that will be required and collected are:

- Where are the critical points throughout the product's life cycle for the release of chemical substances?
- How potent is the material set free? Has an evaluation and control of the risk of the substances taken place?
- Has a development of strategies for limiting the risks of these substances been done? If yes, for which substances?
- Do the effects caused by the chemicals have a global or only local meaning?
- Is the release of specific substances in the circular economy an actual risk or a perceived risk?
- Is the development of new "3R" methods (based on the principles of Refinement, Reduction and Replacement) as a modern alternative approach to the use of animals in safety assessment on a global scale, known and supported by regulators?

The specific objectives of RISKCYCLE are:

- To exploit complementary elements needed with regard to the research objectives, methodologies and data of ongoing as well as recently completed EU and international projects.
- To specify demands for tools for ecological design of consumer products, production, use and reuse of products and waste recycled to secondary material and products. Methods such as LCA (Life Cycle Assessment), risk assessment and risk reduction strategies, environmental impact analysis, material flow analysis and economics-related tools are considered to achieve socio-eco-efficient solutions.
- To create a powerful platform enabling discussion among all stakeholders on different topics: usage, risks and chemical properties of consumer products; labelling; fate of certain chemicals in products traded, used and recycled in a

global scale. The identification of problems and solutions is also an interesting aim of this platform.

- To contribute to the UN Globally Harmonised System (GHS) for chemical substances and mixtures.
- To start with a conceptual development of a global strategy for a risk-based management of chemicals and additives in recycling and trade products.
- To identify alternative testing strategies and methods to avoid the enlargement and the outsource of animal tests to East and Southeast Asia.
- To identify knowledge and research gaps for future research activities.
- To consider the most effective way of ensuring continuing progress in this field involving EU and other partners at global scale including also international organisations.

The RISKCYCLE network closely collaborates with related projects, EU and international bodies and authorities to communicate and agree on standards and to avoid duplication and redundant work.

Chemicals and additives in products being produced and marketed globally is making essential an international harmonised assessment and management. The European Commission has a vital role in stimulating global awareness throughout all sectors and in encouraging harmonisation of a system of classification and labelling (GHS) of chemicals and products. Especially recycling products with an unknown origin must be carefully considered and evaluated.

Being aware of the fact that humans and the environment are subject to the risk of chemical impacts on a day-to-day basis, management of chemicals including reduction of risks has led to the rise of several policy instruments such as different directives as well as to co-operations both at the global and European levels.

The *inter-Organization Programme for the Sound Management of Chemicals (IOMC)* was established in 1995 by UNEP, ILO, FAO, UNIDO, UNITAR and the OECD (the participating organisations) following recommendations made by the 1992 UN conference on Environment and Development to strengthen co-operation and increase international co-ordination in the field of chemical safety. The purpose of the IOMC is to promote co-ordination of the policies and activities pursued by the Participation Organisations, jointly or separately, to achieve the sound management of chemicals in relation to human health and the environment. It is clear that the contamination caused by the recycling activities, for instance e-waste should be mitigated to reduce human exposure to BFR and other toxic chemicals reported in Table 1. During e-waste recycling activities, dust ingestion by the children working at the e-waste sites should be carefully considered and investigated in the next coming years as a pathway of exposure.

At the level of the European Community, the need to protect the Community's environment and to create common standards to protect consumers in order to ensure the free circulation of goods among the Member States has been recognised leading to *REACH* (2006), the new European Community Regulation on chemicals and their safe use (EC 1907/2006).

It deals with the Registration, Evaluation, Authorisation and Restriction of Chemical substances with the aim to improve the protection of human health and the environment through the better and earlier identification of the intrinsic properties of chemical substances. At the same time, innovative capability and competitiveness of the EU chemicals industry should be enhanced. The benefits of the REACH system will come gradually, as more and more substances are phased into REACH.

At the 2002 Earth Summit in Johannesburg, an international agreement was reached that governments should adopt the *Globally Harmonised System of Classification and labelling of Chemicals (GHS)* by 2008 to simplify the global trade and to ensure a sound management of chemicals. These international criteria were agreed by the United Nations Economic and Social Council (UN ECOSOPC). In July 2007, the European Commission adopted the United Nations GHS for chemical substances and mixtures. The EU expects to be able to substitute the current EU classification and labelling system for chemicals (Directive 67/548/ EEC, Directive 1999/45/EC and Directive 91/155/EEC in 2010 for simple chemicals and is expected 2015 for chemical mixtures).

Also in accordance with the Johannesburg implementation plan and Agenda 21, the *Dubai Declaration on International Chemical Management* (February 2006) promotes sound management of chemicals and hazardous wastes throughout their life cycle. The Dubai convention is determined to protect children and the unborn child from chemical exposures that impair their future lives.

Related to the children protection and as a specific example of the risk of chemicals at global level and in a circular economy, the toys market can be approached. Very recently, the European Commission has come forward with measures to improve toy safety in Europe. The Commission wants to strengthen EU-rules, especially those relating to the use of chemical substances in toys. The proposal aims at enhancing the safety of toys replacing and modernising the 20 years old Toys Directive 88/378/EEC of 3rd May 1988.

The revision has a threefold objective: first and foremost there will be new and higher safety requirements to cope with recently identified hazards, second it will strengthen manufacturers' and importers' responsibility for the marketing of toys and finally it enhances the market surveillance obligations of Member States. It will prohibit the use of chemical substances that are susceptible to provoke cancer, prohibit allergenic fragrances or oblige toy manufacturers to issue appropriate warnings to improve the prevention of accidents.

The RISKCYCLE project will influence policy issues at a global scale, not only in developing countries but also in developed ones and will create awareness and enhance state-of-the-art on risk-based management of chemicals and products among stakeholders.

The primary aim of RISKCYCLE is to identify future R&D needs required to establish a risk-based assessment methodology for chemicals and products that will help minimise animal testing while ensuring the dual aim of allowing the development of new chemicals and minimising risks for health and the environment. In order to achieve this goal, the first step will be to assemble and evaluate existing information on the chemicals and especially the additives used in consumer and industrial products. Many potential hazardous compounds are traded worldwide as additives in different products.

RISKCYCLE will focus on the fate and behaviour of these additives in six sectors: textile, electronics, plastics, leather, paper and lubricants. Table 1 lists the main chemicals present in the different industrial sectors that will be studied in the following chapters of this book.

References

1. Nam S-H, Seo Y-M, Kim M-G (2010) Bisphenol A migration from polycarbonate baby bottle with repeated use. Chemosphere 79:949–952
2. Wagner M, Oehlmann J (2009) Endocrine disruptors in bottled mineral water: total estrogenic burden and migration from plastic bottles. Environ Sci Pollut Res 16:278–286
3. Tue NM, Sudaryanto A, Minh TB, Isobe T, Takahashi S, Viet PH, Tanabe S (2010) Accumulation of polychlorinated biphenyls and brominated flame retardants in breast milk from women living in Vietnamese e-waste recycling sites. Sci Tot Envrion 408:2155–2162
4. Kersten A, Hamm U, Putz H-J, Schnabel S (2011) Wochenblatt für Papierfabrikation 1:14–21

Part I
Chemical Additives in a Global Context

Additives in the Paper Industry

Antoni Ginebreda, Daniel Guillén, Damiá Barceló, and Rosa Mari Darbra

Abstract An overview of the paper sector from the point of view of additives use is presented. In the first section, the general trends of the sector concerned in terms of world production and trade flows are reviewed together with an overall description of the main characteristics of the production process, including pulp production and paper manufacturing and finishing. The second part is focused on the description of the different additive classes used to provide paper with the required functional characteristics and to facilitate and improve the production process as well, namely, retention aids, sizing agents, binders, wet strength agents, coating agents, optical brightener agents, biocides and dyes. Main compounds or compound families within each additive class are listed, and some related information regarding their function, world consumption and environmental potential impacts is also included.

Keywords Additives, Binders, Biocides, Optical brighteners, Paper

Contents

A. Ginebreda (✉), D. Guillén, and D. Barceló
Department Environmental Chemistry, Institute of Environmental Diagnostic and Water Studies (IDAEA), Spanish Council of Scientific Research (CSIC), Jordi Girona 18–26, 08034 Barcelona, Spain
e-mail: agmqam@cid.csic.es

R.M. Darbra
CERTEC, Department of Chemical Engineering, Universitat Politècnica de Catalunya, ETSEIB, Diagonal 647, pav. G, 08028 Barcelona, Spain

B. Bilitewski et al. (eds.), *Global Risk-Based Management of Chemical Additives I: Production, Usage and Environmental Occurrence*, Hdb Env Chem (2012) 18: 11–34,
DOI 10.1007/698_2011_109, © Springer-Verlag Berlin Heidelberg 2011,
Published online: 21 June 2011

Abbreviations

AKD	Alkyl ketene dimer
AOX	Adsorbable organic halides
ASA	Alkenyl succinic anhydride
CEPI	Confederation of European paper industries
CPD	3-Chloro-1,2-propanediol
DADMAC	Diallyldimethylammonium chloride
DCP	1,3-Dichloro-2-propanol
ECH	Epichlorohydrin
FWA	Fluorescent whitening agent
MF	Melamine formaldehyde
MUF	Melamine–urea formaldehyde
OBA	Optical brightener agent
PAC	Polyaluminium chloride
PAE	Polyamine-epichlorohydrin resin
PAM	Polyacrylamide
PEI	Polyethylenimine
TCMTB	2-(Thiocyanomethylthio)benzothiazole
UF	Urea formaldehyde

1 Introduction

Although the flattened stalks of papyrus reeds were used by the Egyptians as a writing surface some 5,500 years ago, paper as we know it today was developed in China in about 105 AD [1]. Since then, its role in many key aspects of human society, such as printing material, art and decoration, packaging, etc., has been undisputed. Nowadays though partially substituted by modern electronic information media and advanced plastic and composite packaging materials, this good is still competing successfully with them. Table 1 shows the more representative typical uses of paper today.

It is known that the pulp and papermaking industry consumes a large amount of resources, both raw materials (wood and water) and energy (fossil fuel and

Table 1 Functional use of paper and board [2]

Functional use	Typical grade	Typical end product	Important trends
Information (collection, distribution and storing)	– Newsprint – Coated and uncoated magazine (SC and LWC) – Coated and uncoated wood-free printing and writing	– Newspapers – Journals – Books – Computer printouts – Xerographic copies – Inserts – Illustrations	Increased use of multicolour printing and copying Electronic media taking over banking and trading documents Increased recycling as raw material Increased use of additives
Packaging (transportation, distribution and protection)	– Liner – Sack – Corrugating medium – Folding box board – Liquid packaging board – Wrapping	– Bags – Boxes – Wrappings – Containers	Increased use for distribution of food General increase in recycling of packaging materials Increased use of composites
Hygienic (personal care, cleanliness and disease prevention)	Tissue – Dry crepe – Wet crepe	– Toilet tissue – Kitchen towels – Facial tissue – Napkins – Hand towels – Hospital clothing – Wipers	Use increases with general living standard End of the chain for the recycling of fibres Use of virgin fibre for top-end products
Speciality (great variety of uses)	– Official papers – Filter paper – Fire resistant papers	– Notes – Stamps – Air filters – Coffee filters – Baking paper	An ever-increasing number of new applications

electricity). For instance, papermaking fresh water consumption ranks third in the world, after the primary metals and chemical industries [3].

Pulp and paper mill discharges may contain many compounds that can be hazardous for living forms, due to its high toxicity or because they can disturb the hormonal behaviour in some species [4, 5] (these compounds are known as endocrine disruptors). Moreover, oxygen depletion is a common effect produced by the discharges of such effluents, potentially causing anoxia to fish and other aquatic organisms [6]. The most relevant paper mill emissions to water, air and soil are shown in Table 2.

Since the end of the 1970s until recently, the main emphasis was put on the role of chlorinated substances formed in the bleaching process [7]. Dioxins and furans had been detected [8] in some effluents of pulp mills and the public discussion focused on the harmful effects of chlorine bleaching [9]. The public concern about the potential environmental hazard imposed using chlorine in the bleach plants has brought about a drastic decrease in the use of molecular chlorine as a bleaching chemical during the last decade. The environmental control authorities in many

Table 2 Kraft pulp mill most relevant emissions to water, air and soil [2]

Wastewater	Air emissions	Solid waste
– Organic substances (COD, BOD) – Extractive compounds like resin acids, etc. – Chlorinated organics (AOX), chlorate – Nitrogen, phosphorus – Suspended solids – Metals, salts – Coloured substances	NOx, SO_2, CO, CO_2, dust malodorous reduced sulphur compounds (TRS), e.g. methyl mercaptan, dimethyl sulphide, hydrogen sulphide, VOC, chlorinated compounds (AOX), visible plume	– Boiler ashes – Dregs, lime mud – Sand and stones – Green liquor sludge – Wood waste, bark, rejects – Primary and biosludge – Cleaning and mixed household type waste – Others

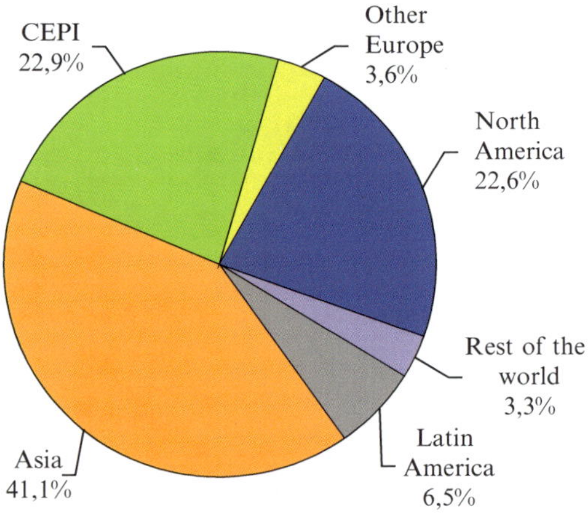

Fig. 1 Paper consumption by region in 2008 [11, 12], total: 391.6 million tonnes

countries have set severe restrictions on the discharges of chlorinated organics measured as adsorbable organic halides (AOX) into the aquatic environment. A substantial reduction of AOX emissions was achieved through the replacement of molecular chlorine by chlorine dioxide or the introduction of other oxygen-containing chemicals such as molecular oxygen, peroxide and ozone [10]. Hence, the enforcement of the increasingly strict environmental regulations has doubtlessly contributed to a reduction of the pollution associated to the paper industry.

2 Overall Worldwide Consumption and Trading

There is a considerable variation in the worldwide consumption of paper per capita; even within a given region, there are large differences among the various membership countries. Europe, North America and Asia are the regions where consumption is highest (Fig. 1), thus showing the importance of paper as resource for developed societies. Table 3 shows the countries that forms The Confederation of European Paper Industries CEPI.

In the following tables and figures, the import and export rates for paper (Table 4 and Fig. 2) and recycled paper (Table 5 and Fig. 3) trade at world scale from 2007 to 2009 are reported, thus showing its relevance.

Table 3 CEPI Members in 2009 [11]

Austria	Portugal
Belgium	Romania
Czech Republic	Slovak Republic
Finland	Spain
France	Sweden
Germany	Switzerland
Hungary, Italy	The Netherlands
Norway	UK
Poland	

Table 4 CEPI exports and imports of paper to other regions 2007–2009 (tonnes), *E* export, *I* import [11, 12]

	2007		2008		2009		Share of total % (2009)	
	E	I	E	I	E	I	E	I
Other Europe	6,349	1,553	6,203	1,621	5,314	1,491	35.8	31.8
North America	2,613	2,219	2,098	2,062	1,501	1,769	10.1	37.7
Latin America	1,475	421	1,378	561	1,075	454	7.2	9.7
Asia	4,686	692	4,688	524	4,744	651	31.9	13.9
Rest of the world	2,540	533	2,563	391	2,228	330	15.0	7.0
Total	17,663	418	16,930	5,160	14,862	4,694	100.0	100.0

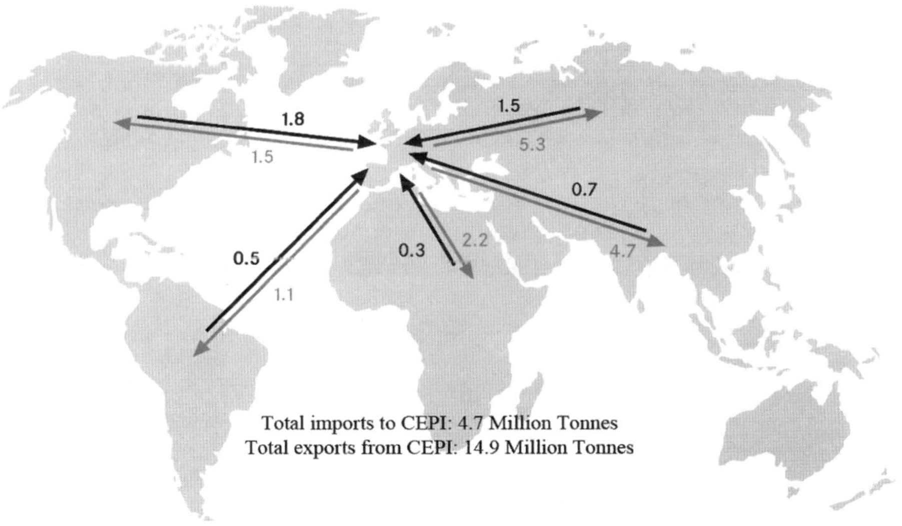

Fig. 2 Trade flow paper (units: million tonnes) [11]

Table 5 CEPI exports and imports of recovered paper to other regions 2007–2009 (tonnes), *E* export, *I* import [11, 12]

	2007		2008		2009		Share of total % (2009)	
	E	I	E	I	E	I	E	I
Other Europe	372	950	353	938	352	954	2.7	84.1
North America	60	208	96	161	29	158	0.2	0.1
Latin America	29	7	45	6	12	1	0.1	0.1
Asia	8,897	3	10,616	2	12,337	12	96.3	1.1
Rest of the world	147	4	81	7	76	9	0.6	0.8
Total	9,504	1,173	11,192	1,114	12,805	1,134	100.0	100.0

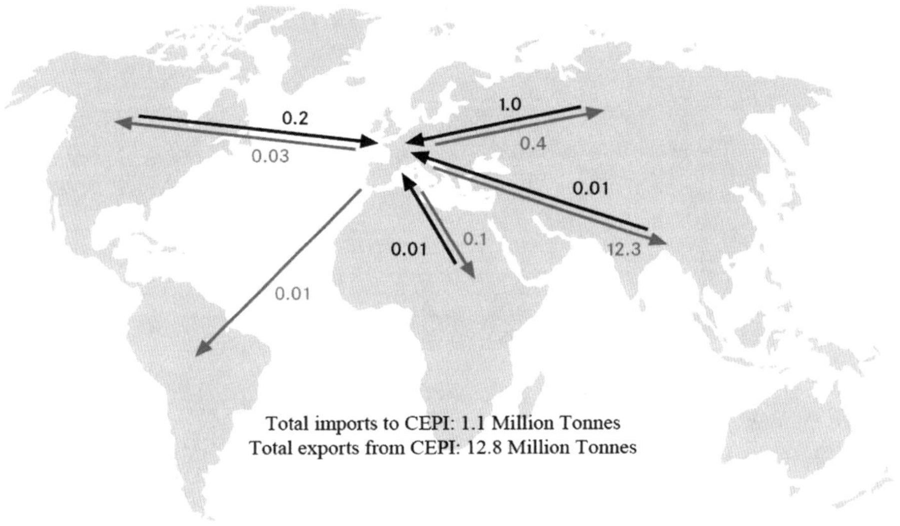

Fig. 3 Recycled paper flow (units: million tonnes) [11]

With landfills reaching capacity and communities opting for more sustainable systems, recycling is the most natural choice. As shown in Table 5, in the last years Asia is the only region that has increased the trade of recycled paper, specially China buys ships of waste paper from Europe and USA for processing in its recycling factories. In waste buyers countries, future efforts should focus on improving its waste separation management to recycle its own paper waste.

3 Production Process

3.1 Pulp

Paper structure is based mainly on a network of cellulose fibres and a series of compounds that are added to improve product properties. In papermaking process, the main raw material used is wood, but straw, hemp, grass, cotton and other cellulose-bearing material can be also used.

Wood contains around 50% of water and the rest is composed of 25% hemicellulose, 25% lignin and 5% other organic and inorganic compounds.

The wood supplied to the paper mill has to be debarked before it can be used to produce one of the several varieties of pulp (the base material for the subsequent production of paper). The debarked trunks are either pulped to fibres (chemical pulp) or processed to chips (mechanical pulp).

Another way of paper production is using recovered paper as fibre source, in Europe there is an average utilization rate of recovered paper of 43% [2]. But in spite of the economic value, recycled fibres have long been known to have inferior strength properties compared to those of virgin fibres [13]. Therefore, it is necessary to add primary fibres (from wood) to ensure the strength and other properties of the final product. One problem that arises when recycled fibres are used is the presence of contaminants; so these "raw" fibres will require the application of large quantities of cleaning and bleaching agents before their use.

Two different technological processes are distinguished on chemical pulping, namely sulphate (also termed Kraft process) and sulphite pulping. The major differences between them from an environmental point of view are to be found in the chemistry of the so-called cooking process, the chemical preparation and recovery system, and the reduced bleaching required because of better initial brightness of sulphite pulp.

Kraft process is, however, the dominating pulping process worldwide for the manufacture of chemical pulp due to the superior strength properties exhibited by the obtained pulp and its applicability to almost all wood species. Present-day bleaching Kraft process uses large amounts of chlorine and chlorine chemicals.

3.2 Paper Manufacturing (Papermaking)

Papermaking (Fig. 4) involves five basic steps that will provide the pulp and subsequently the finished product. As regards its environmental impact to the aquatic environment, the final effluents are a combination of wastewaters from each of the five different unit processes and the method used.

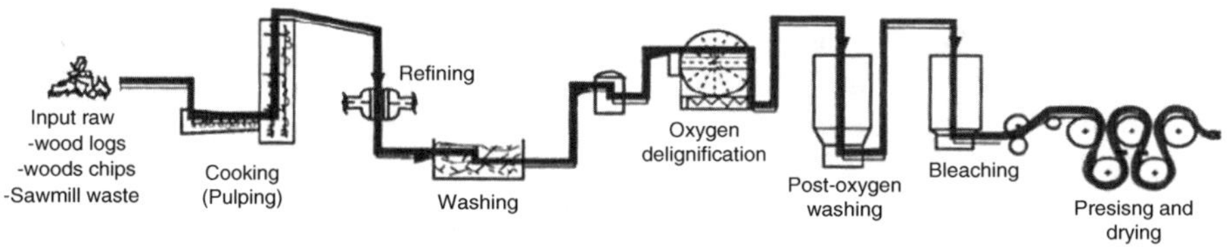

Fig. 4 Mass stream overview of a sulphate pulp mill [3]

Debarking: Before wood can be pulped the bark must be removed, as bark does not contain the types of fibres desirable for papermaking. This process is followed by wood chipping which converts the plant fibre into smaller pieces called chips. Debarked logs are washed with water, and due to the nature of the raw material used, i.e. hard wood, softwood, agro-residues, results in the transfer of tannins, resin acids, etc. to the water effluent.

Pulping releases the cellulosic fibres from the chips by breaking down the wood's intercellular lignin glue. After pulping, a pulp with lignin and hemicellulose low content is obtained, which is ideal for use in the paper manufacturing. Pulping can be carried out using mechanical, chemical or hybrid process. Chemical pulping is actually the most used in paper industry, which is of two types: sulphate pulping (Kraft pulping) and sulphite pulping.

Bleaching is used on the brown pulp (due to the presence of residual lignin) obtained after pulping to meet the desired colour, especially for printable paper. Chlorine and hypochlorite were commonly used until the 1990s [14], when it was realized that they were the origin of organochlorine harmful by-products (including dioxins and furans). While it is banned in the developed world, in the majority of mills in the developing countries elemental chlorine is still used for bleaching. Undesirable by-products resulting from using these chemicals are chlorinated organic substances, some of which are toxic, mutagenic, persistent and bioaccumulable and are known to cause numerous harmful disturbances in biological systems [15, 16]. In recent years, efforts are focused on reducing the use of chlorine being replaced by chlorine dioxide, hydrogen peroxide and ozone. The use of enzymes has also been tested [17], and they can provide a very simple and cost-effective way to reduce the use of chlorine, chlorine compounds and other bleaching chemicals.

Washing removes the bleaching agents from the pulp. Generally an alkali (caustic soda) is used to extract any lignin remains and bleaching agents from the pulp, and hence this process is also known as the alkali extraction stage.

Paper finishing is a process in which the pulp is mixed with chemical additives to obtain a final product with the desired properties. In the next part of this chapter, we will extensively discuss the different types of additives used for this purpose and the environmental concern due to its use in the paper sector.

4 Additives Used in Paper Finishing

From their origins, paper industry has been using chemical additives to modify the properties of this good or to improve the production process. Therefore, modern paper industry is largely indebted to chemical additives. Without them, paper would be a low resistant material with a brownish colour, unable to serve as printing media or for decorative packaging purposes. At the same time, it can be truly said that the paper industry has served as driver for the design, development and production of many chemical additives.

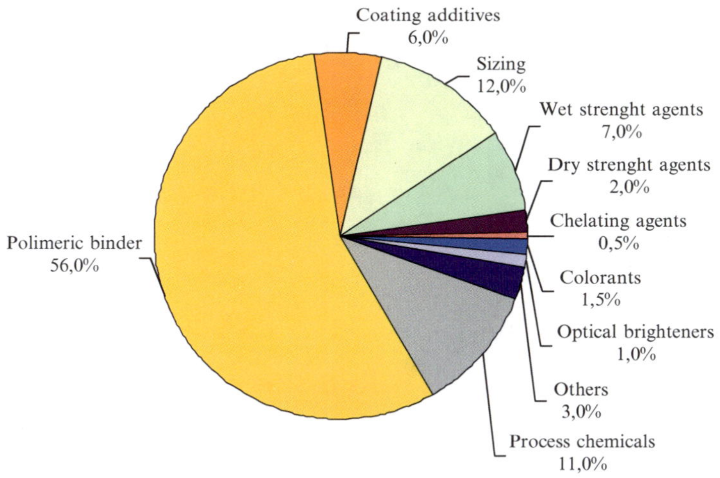

Fig. 5 Market shares of different synthetic chemical additives related to the global consumption [2]

A great variety of the different chemical additives used in paper finishing can be found in the market, which include binders, sizing agents, coating aids, strength agents (wet and dry), biocides, optical brighteners and colourants among others. An overview of their respective market shares is presented in Fig. 5.

The consumption of chemical additives, including starch and alum, as a proportion of paper production is only 3%, and restricting the attention to the synthetic chemical additives, they account for only 1%. A clear distinction needs to be done between process chemicals, which are used to assist in the production process and products whose function is to modify one or more particular properties of the paper manufactured.

The use of chemical additives especially contributes to reduce the consumption of water and energy and allows increasing the use of waste paper as well as to save raw materials by decreasing the grammage of the obtained paper. Without chemical additives in some cases this would be not possible at all.

The aim of this chapter is therefore to discuss the role of chemical additives, understood as a substance that remains in the final product modifying any undesirable characteristic or conferring some desirable property. The present text describes the different classes of product aids used in the paper sector including an assessment about their potential impacts on the environment.

4.1 Retention Aids

A large amount of the added substances in paper manufacturing is not retained on the fibres during processing. Retention aids are thus used to increase retention in the cellulosic fibres avoiding losses of fillers by decreasing the fibre polarity and the specific surface area. Retention aids are normally used in almost all paper grades, but especially in printing papers, since they enhance the opacity and brightness of the cellulosic pulp.

These additives not only play an important role in the paper quality but also provide other operational and process advantages, such as higher production speeds, the use of larger proportion of recycled paper and mineral fillers, and decreasing the amount of fresh water used [18]. Retention aids act synergistically with other chemicals additives – such as deaerators – which improve the overall performance of other wet end chemicals and reduce their consumption. Finally, retention aids are a major factor in the reduction of the paper grammage [19] and contribute to the improvement of effluent quality [20].

Main categories of retention aids are listed in Table 6. They can be classified in several classes, namely inorganic salts (such as aluminium sulphate and polyaluminium chloride, PAC); modified organic natural products (such as cationic starch); and high molecular polymers such as polyacrylamide and polyethylenimine. Figure 6 shows the world consumption distribution of the aforementioned classes.

Table 6 Main classes of retention aids additives

Inorganic compounds	Aluminium sulphate (alum)
	Polyaluminium chloride (PAC)
	Colloidal silicic acid/silica gel
	Bentonite
Modified natural materials	Modified starch (cationic)
	Galactomannane (guar)
	Carboxymethylcellulose (CMC)
Synthetic organic polymers	Polyacrylamides (PAM)
	Polyethylenimine (PEI)
	Polyamidomine (PAAm)
	Poly-DADMAC
	Polyvinylamine (PVAm)
	Polyethylene oxide (PEO)

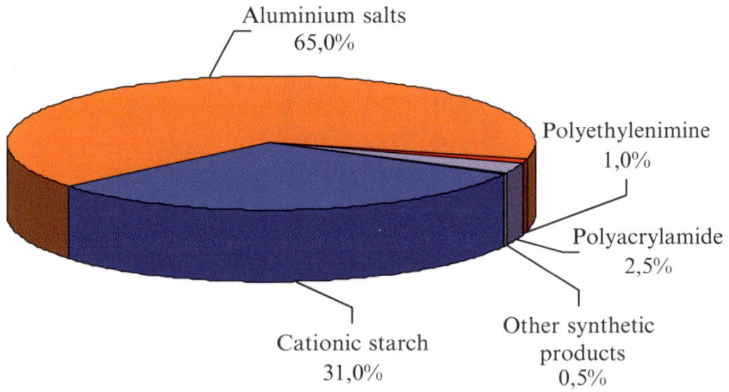

Fig. 6 Retention aids worldwide consumption, total 2.3 million tonnes/year [18]

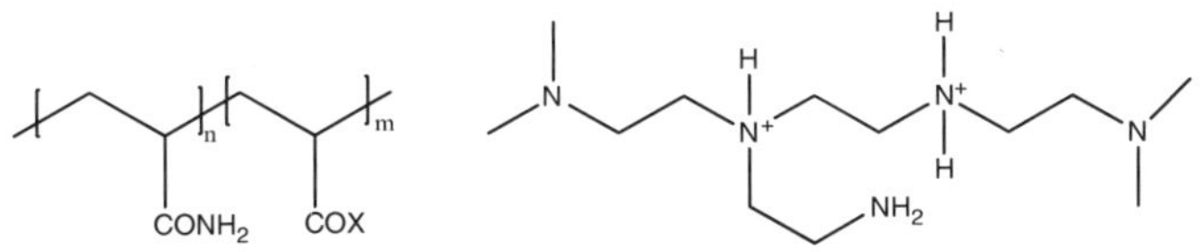

Polyacrylamides (PAM) Polyethylenimines (PEI)

Fig. 7 Chemical structures of two polymers used as retention aids

Even though global consumption of polymeric retention aids appears to be low, these compounds are widely used. This is due to its high efficiency at relatively low amounts (typically of the order of 0.5 kg retention aid agent per 1 ton of dry weight papermaking fibre) [20].

Polyacrylamides (PAM) and polyethylenimine (PEI) (Fig. 7) are the retention aids most used in papermaking. Other less used groups of products are polyamines and poly-DADMAC compounds. Paper industry is developing new and more efficient retention aid products that offer improved properties to paper, as is the case of the synthetic polymer polyvinylamine. Pelton and Hong found that polyvinylamine and their copolymers increase the dry and wet strength of newsprint [21].

4.2 *Sizing Agents*

The term "internal sizing" is used in paper technology to describe the practice of adding chemicals to aqueous slurries that contain cellulosic fibres, so that the resulting fibre network becomes hydrophobic and the final product is thus capable to resist the penetration of water or other fluids.

Fluid penetration resistance is a key property for products designed to act as containers, such as packaging for food products, paper cups, paper bags and items that can be exposed to the rain [22]. Currently, some sizing additives can be found on the market, which are able to overcome the inherent wettability of the cellulose and hemicellulose (two of the main chemical components of ordinary paper). The world consumption for these additives in the paper industry is reported in Fig. 8.

Sizing agents can be classified into two main categories (Table 7), namely internal sizing agents (wet end) which are added to the pulp slurry and surface sizing agents which are applied to the paper surface after the paper sheet is formed [23].

Rosin is one of the most common and traditional internal sizing agents used. From the 1950s onwards, the rosin sizing process underwent many changes, for example, the development of fortified sizes, which are natural rosins modified with maleic anhydride or fumaric acid via Diels–Alder reaction [24].

Internal sizing chemicals used in papermaking at the wet end are mainly alkyl ketene dimer (AKD), alkenyl succinic anhydride (ASA), rosin and polymeric compounds. By making the paper web more hydrophobic, these additives influence

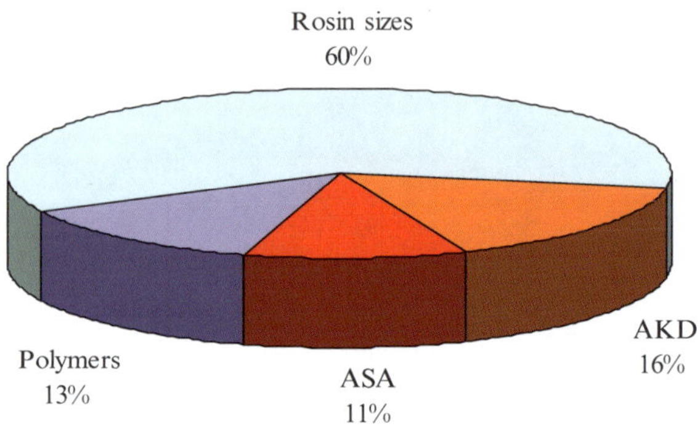

Fig. 8 Sizing agents worldwide consumption [18]

Table 7 Sizing agents

Wet end sizing agents	Rosin size and wax emulsions
	Alkyl ketene dimer (AKD)
	Alkenyl succinic anhydride (ASA)
	Polymeric sizing agents
Surface sizing agents	Styrene acrylate
	Styrene maleinate

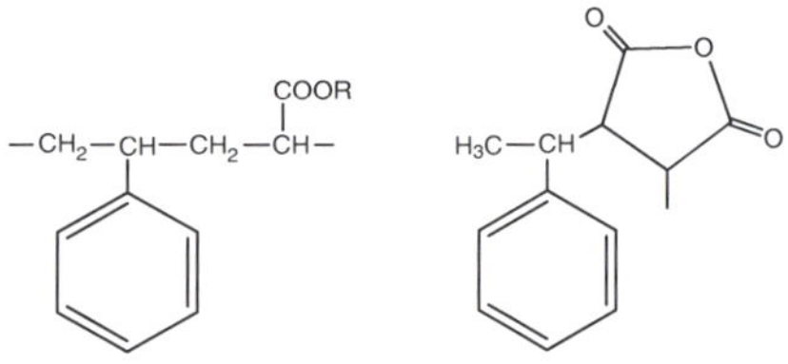

Fig. 9 Examples of some styrene copolymers used as sizing agents

dewatering and retention of fillers and fibres in the paper sheet. In addition to paper quality, the main effect of internal sizing agents is on runnability of the paper machine. Papermaking is a fast industrial chemical process with production velocities well over 25 m/s [25], hereby internal sizing will be a leading issue to optimize the manufacturing performance.

Synthetic copolymers usually used for surface sizing (Fig. 9) are often based on styrene copolymers. These compounds enhance the longevity of paper because they are retained in the cellulose network without the need of fixing agents like alum, which is added as fixing agent for rosin.

The acidic nature of alum ($Al_2(SO_4)_3 \cdot 18H_2O$) sometimes makes it difficult to preserve paper for long periods of time [23].

Furthermore, it has been reported that synthetic copolymer sizing agents provide the additional advantage of enhancing paper strength [23].

Nowadays, the most commercially successful of the foregoing sizing agents are AKDs and ASAs, whose chemical structures are reported in Fig. 10. They require lower levels of addition, compared to the classical rosin-type sizing agents.

Fig. 10 Chemical structures of AKD and ASA

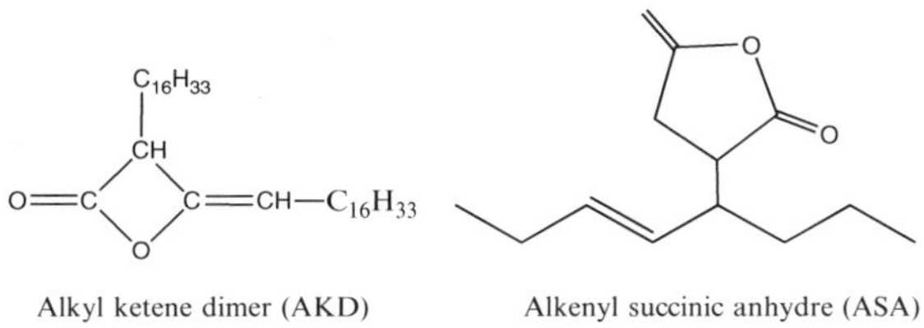

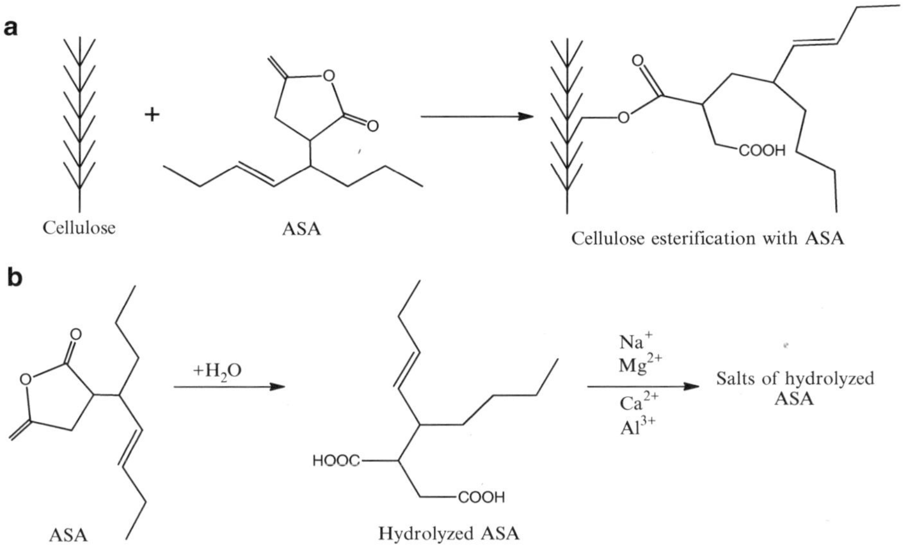

Fig. 11 (**a**) ASA reaction with cellulosic fibre; (**b**) ASA hydrolysis [26]

ASA that is obtained by reaction of an isomerized olefin with maleic anhydride is more reactive to cellulose than AKD, usually prepared by dimerization of stearoyl chloride [22]. ASA reacts with hydroxyl groups (Fig. 11a) of cellulose or hemicellulose components by formation of an ester linkage. There are some factors that may decrease the sizing agent performance; one can be the ASA hydrolysis (Fig. 11b) that is the undesired competing reaction to ASA–paper interaction.

4.3 *Wet Strength Agents*

Some types of paper can only fulfil their purpose if they have adequate wet strength. Such papers include, for example, filter papers, hygienic papers, papers for bags, label papers, wallpaper base papers, laminate base papers and packaging papers. These products are subjected to strain in the wet state, and for this reason the use of strength additives is required to obtain better physicomechanical properties.

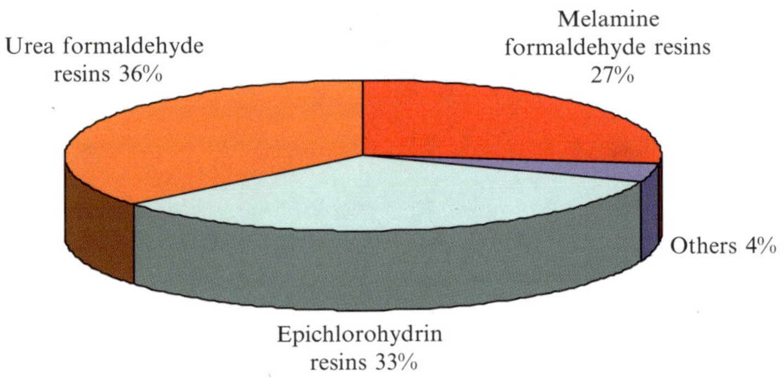

Fig. 12 Wet strength agents worldwide consumption, total consumption 220.000 tonnes (1996) [18]

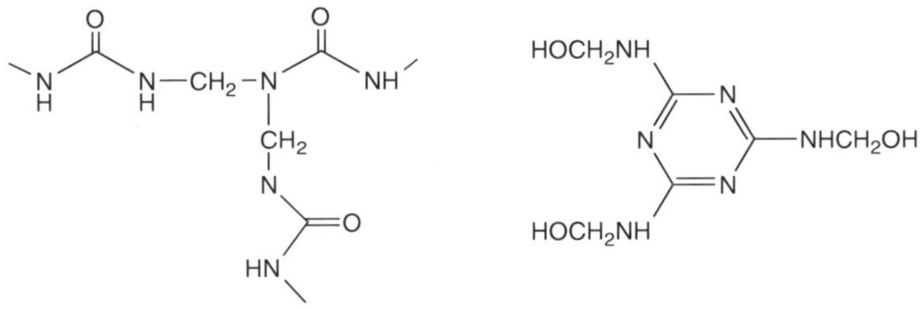

Fig. 13 Urea formaldehyde resin (*left*) and melamine formaldehyde resin (*right*)

Wet strength additives are water-soluble compounds with high reactivity with the paper fibres and are mainly cationic polymeric compounds.

Wet strength additives include synthetic resins and other compounds suitable for a wide range of paper products. Worldwide consumption of these additives is depicted in Fig. 12. Synthetic resins are often based on formaldehyde-derived polymers, the most common being urea formaldehyde (UF) and melamine formaldehyde (MF) (Fig. 13). Due mainly to cost reasons UF resin is often preferred to the more expensive MF resins. Cost savings by substituting a portion of melamine in MF by urea would be highly desirable and could be achieved either by mixing MF and UF or by preparing a suitable melamine–urea formaldehyde (MUF) copolymer [27].

There are two proposed action mechanisms to explain development of wet strength into the paper fibres [28]: (a) protection mechanism, based on the assumption that the additive is distributed on the fibre and cross-links with itself to form an insoluble network on the surface and inside the fibre; (b) reinforcement mechanism, which assumes that the additive reacts with the cellulose (or hemicellulose) ether bonds thus cross-linking cellulose molecules.

Efforts are still directed to develop technologies that avoid formaldehyde releases during the manufacturing process, which is a compound of high concern due to its toxicity and relationship to allergic diseases [29]. These alternatives

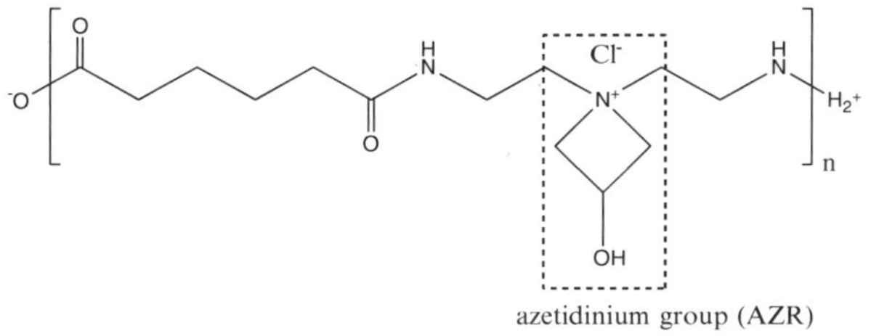

Fig. 14 Polyamideamine-epichlorohydrin (PAE)

include the use of epichlorohydrin (ECH)-derived compounds, mechanical modification of the pulp to promote auto-adhesion of the fibres and the use of enzymes to enhance the bonding of wood fibres [30].

Aqueous solutions of polyamine-epichlorohydrin resin (PAE) (Fig. 14) have been used as wet strength agents for making tissue paper, paper towels and base paper for liquid packaging. Typically, 0.1–1% (based on dry weight pulp) PAE is added to pulp slurries as a wet end additive in papermaking [31].

PAE synthesis is performed by a two step process: first polyamideamine pre-polymer is prepared by reaction of dimethyladipate with diethylenetriamine, and then PAE is finally obtained by subsequent addition of epichlorohydrin. During that reaction two by-products are formed: 1-3-dichloro-2-propanol (DCP) (formed by reaction of the ECH epoxide group with chloride ions) and 3-chloro-1,2-propane-diol (CPD) (formed by hydrolysis of ECH in water) [32]. These two chlorinated compounds are of particular concern due to their known toxicity: CPD is a male anti-fertility agent and mutagen in bacterial assays [33], whereas DCP is a suspected carcinogen [34].

Owing to this fact, many environmental authorities have issued amendments to regulate the releases of these compounds. This action has directly affected the use of epichlorohydrin resins. As mentioned before, organic halogen compounds are globally monitored by means of the parameter “adsorbable organic halides” (AOX). Among the chemical additives, only epichlorohydrin-based wet strength resins make any significant contribution to the AOX load from paper mills effluent. Compelled by regulatory pressures, industry has put a great deal of effort into developing “low-AOX” and recently “free-AOX” epichlorohydrin resins with no detectable amounts of undesirable organochlorine by-products, being the most promising ones are those based on glyoxal acrylamide, polyisocyanate and polyvinylamine [19].

4.4 Paper Coating

Pigments coating are the major industrial operation used to improve the appearance and printability of paper accounting for 50% of the total amount of chemical additives used worldwide by the paper industry [19] (calculated on dry substance).

Pigmented paper coating is usually performed by means of a high solid aqueous suspension of inorganic pigments and binders. Inorganic pigment consists in a finely divided inorganic substance that not only improves the paper opacity and colour but also contributes to the control of flow during application. The original and more classical pigments used for surface coating were kaolin clay and $CaCO_3$, but modern practice has broadened the field to include many others, such as blanc fixe satin white, precipitated chalk, lithopone, titanium dioxide, alumina and talc [35], each one having certain desirable characteristics.

On the other hand, binders are additives aiming to improve properties such as surface strength, glossiness, water resistance, printing adaptability [36] and speed in paper machines.

The most common coating binders for papermaking are polymer dispersions (Fig. 15). Among them, styrene-butadiene latex occupies the dominant position for paper coatings. Some representative chemical structures are depicted in Fig. 16. Binders are used either alone or in combination with others such as starch.

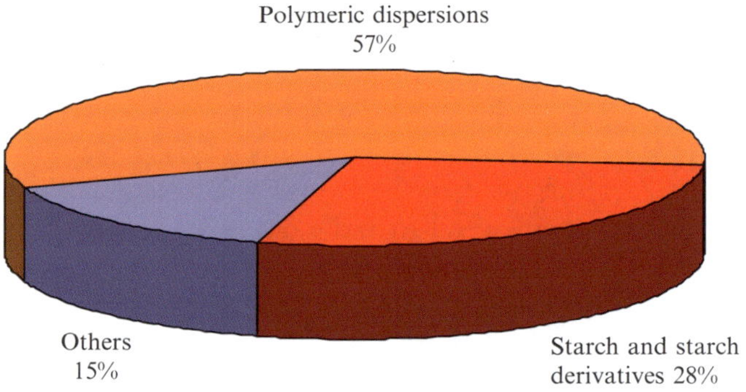

Fig. 15 Worldwide consumption of binders in the paper industry [18]

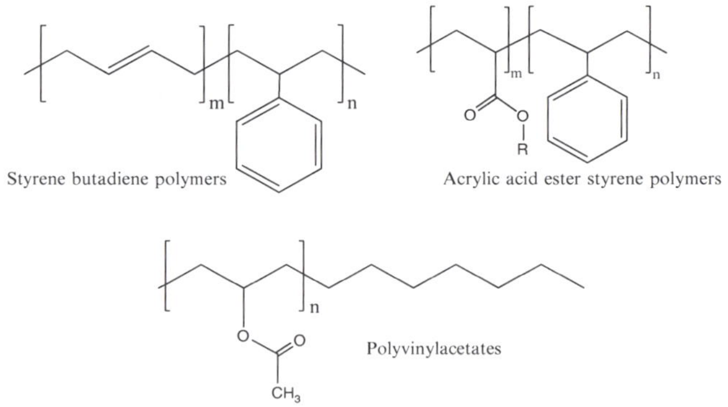

Fig. 16 Chemical structure of main additives used as binders

4.5 Optical Brighteners Agents

Most of the raw materials used in pulp and papermaking are treated with chemical oxidants to remove their colours (bleaching process), but chemical bleaching alone is not sufficient to remove colour completely and usually a certain yellowness still remains. Optical brightener agents (OBAs) are therefore used as functional pigment in the process of the chemical bleaching to the fibre, paper and pulp to make them appear brighter. Thus, it has become an established practice for paper producers to use high brightness pigments, such as calcium carbonate and titanium dioxide, and to incorporate fluorescent agents as components of paper coating formulations.

These additives (also known as "fluorescent whitening agents", FWAs) influence the optical properties of the paper by directly absorbing ultraviolet light (300 nm) and re-radiating in the blue-violet range of spectrum (under 550 nm), conferring to paper the feeling of bright whiteness [37]. In this way, the degree of whiteness can be increased tremendously, especially if the FWAs are combined with shading dyestuffs [19].

FWAs are water-soluble dyes due to the presence of a sulphonic acid functional group (SO_3H) in the molecule. Derivatives of 4,4′-diaminostilbene-2,2′-disulphonic acid (Fig. 17) are the most popular in paper industry because of their fastness properties. Tetrasulphonated derivatives with medium substantivity account for about 80% of the market [19], thus making them the top of this product group.

The contamination of aquatic bodies by potential harmful organic chemicals such as FWAs is one of their associated great environmental problems: tonnes per year of these compounds are discharged into sewage treatment plants or directly into the aquatic environment worldwide [38]. The FWAs are not easily degradable by the wastewater treatment, and some researchers [39, 40] have identified in rivers and in bank filtrates some aromatic sulphonates derivatives. It has been proved [38] that the major sink in the environment for such compounds is possibly direct photodegradation by sunlight in surface waters.

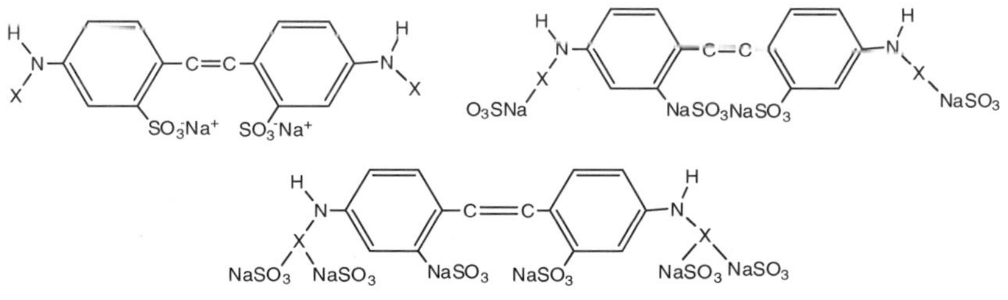

Fig. 17 Some representative types of optical brighteners agents (OBAs) used for paper manufacturing

4.6 Biocides

Microbial activity can cause severe quality and runnability problems on paper machines as well as impair the quality of the finished product [41]. Since paper machines work on an aquatic environment, a temperature of 30–60°C and a pH range of 4.5–9.0, they offer an ideal media for microbes to grow and reproduce. Moreover, cellulose and various degradable additives present in paper machine waters offer a good source of nutrients for microbes. To face these problems, it is a common practice to dose biocides during papermaking.

Biocides are products used to control the growth of micro-organisms that encompass algicides, bactericides, fungicides and virucides. They act either by killing micro-organisms (the biocidal effect) or by inhibiting their growth (biostatic effect). An ideal biocide should have the following characteristics [42]:

- Be applicable over a wide range of operating conditions, such as pH and temperature
- Not interfere with other paper mill additives
- Have a broad spectrum of activity towards microbes
- Be efficient and fast acting
- Be environmentally friendly and non-toxic to non-target organisms (no organic solvents, heavy metals, dioxins or furans) and also safe for the operator
- Be easy-to-handle
- Be cost-competitive

Currently, a wide range of compounds useful to treat many different types of micro-organisms is available on the market. Figure 18 shows the market shares for the main types of biocides.

Fifty years ago, chlorophenol compounds (penta- and tetrachlorophenol) (Fig. 19) were widely used in the wood preservation industry (by pressure treatment) and as additives to inhibit microbial growth in a wide array of products (paper, leather, textile, etc.), but nowadays chlorophenol compounds have been

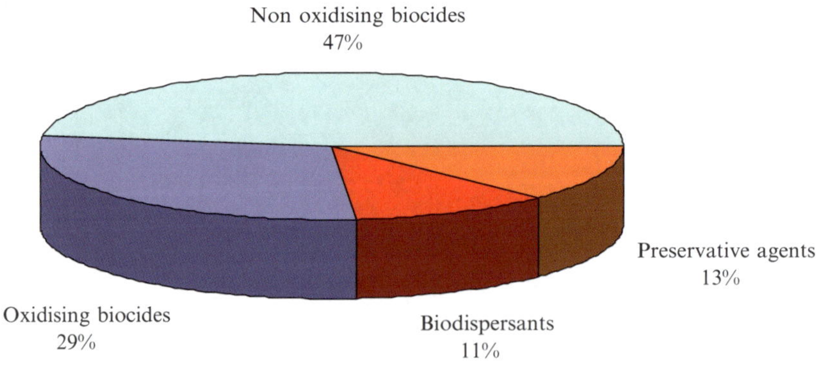

Fig. 18 Papermaking biocide sales in Europe (estimated in 2005) [18]

banned because of concerns about their toxicity, persistence, occupational health impacts and hazardous impurities (potential precursor of dioxins and furans) [43].

In replacement of traditional chlorophenols, the paper industry has a wide range of compounds to fight against micro-organisms. Broadly speaking, two classes of biocides can be differentiated: oxidizing and non-oxidizing biocides (Tables 8 and 9).

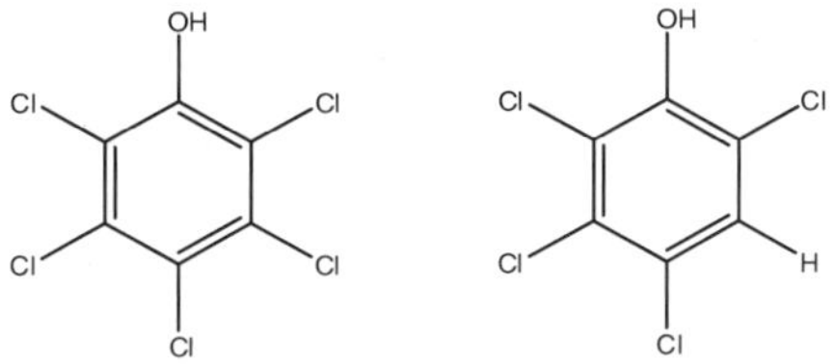

Pentachlorophenol Tetrachlorophenol

Fig. 19 Chlorophenol-based compounds' chemical structure

Table 8 Oxidizing biocides used in the paper industry [19]

Class		Synonym	Chemical base
Non-halogenated		Peroxyacetic acid	Peroxyacetic acid
		Hydrogen peroxide	H_2O_2
Halogenated	Stabilized	Halogenated alkylhydantoin, e.g. BCDMH, MCDMH; DCDMH	Bromochlorodimethylhydantoin; Dichlorodimethylhydantoin; Monochlorohydantoin
		SDIC, organoamines	Chloroisocyanurate
		Ammonium bromide	NH_4Br
		Bromosulphide	Bromosulphamic acid
	Non-stabilized	Sodium hypochlorite	NaOCl
		Sodium bromide	NaBr
		Chlorine gas	Cl_2
		Chlorine dioxide	ClO_2

Table 9 Non-oxidizing biocides commonly used in the paper industry [19]

Class	Name	Chemical base
Quaternary ammonium compounds	QAC	*N,N*-dimethyl-*N,N*-didecyl ammonium chloride
		Benzalkonium chloride
Amides	DBNPA	2,2-Dibromo-3-nitrilopropionamide
Nitro compounds	Bronopol	2-Bromo-2-nitropropane-1,3-diol
Cyanates	MBT	Methylene bisthiocyanate
Thiazoles	CMIT	5-Chloro-2-methyl-4-isothiazolin-3-on
	MIT	2-Methyl-4-isothiazolin-3-on
	BIT	1,2-Benzisothiazolin-3-on
	TCMTB	2-(Thiocyanomethylthio)benzothiazole
Thiocarbamides	Carbamate	Dimethyldithiocarbamate
Aldehydes	GDA	Glutardialdehyde
	EGHF	Ethylene glycol hemiformal
	THPS	Bis[tetrakis(hydroxymethyl) phosphonium]sulphate

The fate of biocides in the environment can follow several pathways: while a part can be degraded (chemically or biologically) another may remain in circulating waters. Ultimately, a fraction can be present in the discharged effluent or remain in the solid matter [6]. Variable amounts of biocides (and their degradation products as well) may be retained in the paper when it is dried, being the proportion retained in the paper related to the ratio of adsorption to desorption on the fibres and fillers. In turn, this also depends on the chemical structure, physicochemical properties of each compound, electrostatic charge, pH and zeta potential of the solids in the water [19].

TCMTB-based compounds and formulations have been used globally for approximately 20 years as an alternative to chlorophenol compounds. TCMTB and its transformation products (Fig. 12) have a high acute toxicity to algae, fish and daphnia. Benzothiazole compounds may also be potentially harmful to the environment.

4.7 Dyes

From an historic perspective, colourants are probably the oldest additives used in papermaking. In contrast with other additive classes, the main role of colouring is for aesthetic or appearance purposes. In any case, the colouring step is one of the most important since it is directly related to the final value of the product. Consequently, there is a large market for products useful for colouring (see Fig. 20), which is common to other industrial sectors such as textile or leather. Readers are thus referred to the respective chapters in this book in which industrial dyes have been extensively discussed.

Given below is a comprehensive list of the various grades of paper that may require dyeing:

- Writing and printing paper
- Tissue – facial, toilet, towel, napkin
- Copier papers
- Boards/cover papers

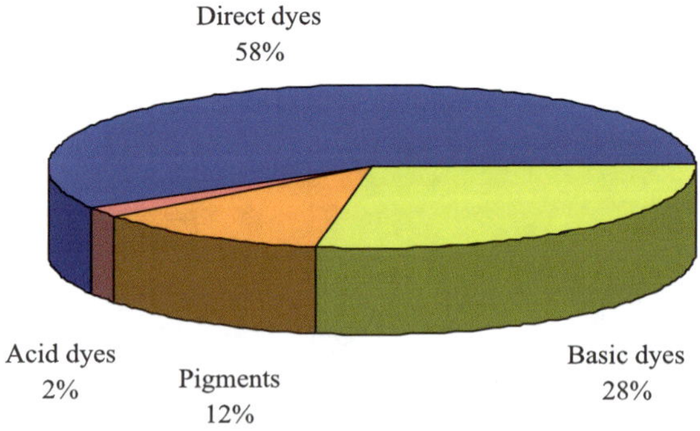

Fig. 20 Chemical classes of paper colourants (value shares of worldwide consumption 2001, total 500 million €) [18]

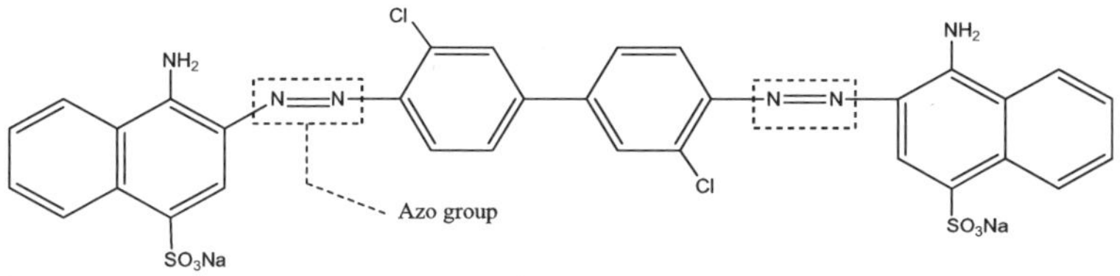

Fig. 21 Direct Red 061 structure may release the hazardous amine 3,3′-dichlorobenzidine

- Decorative laminated paper
- Packaging grades – corrugated case stuff
- Envelope grades
- Specialty papers such as label, laundry tag, posters

While inorganic colour pigments and natural organic dyes were among the first additives to be used in papermaking, present industrial practice is dominated by synthetic dyes.

As discussed in the leather chapter, azo dyes are widely used for colouring products due to its high efficiency and the wide range of colours available in the market. Nevertheless, there is nowadays a great concern about their use since some of these compounds may release hazardous aromatic amines by breaking the azo group (Fig. 21).

According to the directive 2002/61/EC [44], dyes are classed according to the aromatic amines that may be generated from the azoic dyes by reductive cleavage of the N=N diazo bond. If the potentially formed aromatic amines are listed in this regulation (additives in the leather industry chapter, Table 9), the use of the parent colourant is forbidden.

5 Conclusions

Paper industry makes extensive use of additives with a double purpose: (1) first of all, additives are used throughout the manufacturing process to enable a better runnability and higher productivity using less water and energy and thus contributing to minimize the environmental impact of the paper and board production. (2) Second, a wide range of additives is used to confer the desired quality properties to the final product (otherwise lacking in the natural paper fibres).

Paper is nowadays a good present in many aspects of our daily life. It is thus not strange that its production and trade have reached a global scale. Therefore, the concurrent spreading of paper additives is an issue of concern, since it is seldom considered in a life cycle perspective. Many uncertainties remain open, thus requiring its actual assessment further attention. This is the main objective of the RISKCYCLE project (a coordination action funded under the 7th Framework Program of the European Community).

Historically, from an environmental perspective the paper industry has attracted particular interest due to the known large amounts of organic compounds released to the environment by this sector. However, additives have deserved much less attention. Hence, within the framework of the aforementioned project, we have reviewed and summarized in this chapter the current state of the art of the additives currently used in the paper and board industry, particularly focusing on their environmental impacts. Additives considered include retention aids, sizing agents, fillers, wet strength agents, coatings, optical brighteners, biocides and dyes. Special emphasis has been put on those substances that may pose a risk to the environment. The main environmental issues identified are emissions to water and air, sludge build-up and energy consumption. It is the pollution of water bodies, however, which is of major concern due to the large volumes of wastewaters generated during the paper manufacturing, depending on both the raw material and the type of paper produced.

References

1. Thompson G, Swain J, Kay M, Forster CF (2001) The treatment of pulp and paper mill effluent: a review. Bioresour Technol 77(3):275–286
2. European Commission (2001) Reference document on best available techniques in the pulp and paper industry. Integrated Pollution Prevention and Control (IPPC)
3. Avsar E, Demirer GN (2008) Cleaner production opportunity assessment study in SEKA Balikesir pulp and paper mill. J Clean Prod 16(4):422–431
4. Fukazawa H, Hoshino K, Shiozawa T, Matsushita H, Terao Y (2001) Identification and quantification of chlorinated bisphenol A in wastewater from wastepaper recycling plants. Chemosphere 44(5):973–979
5. Soares A, Guieysse B, Jefferson B, Cartmell E, Lester JN (2008) Nonylphenol in the environment: a critical review on occurrence, fate, toxicity and treatment in wastewaters. Environ Int 34(7):1033–1049
6. Lacorte S, Latorre A, Barcelo D, Rigol A, Malmiqvist A, Welander T (2003) Organic compounds in paper-mill process waters and effluents. Trac-Trends Anal Chem 22 (10):725–737
7. Shimp RJ, Owens JW (1993) Pulp and paper technologies and improvements in environmental emissions to aquatic environments. Toxicol Environ Chem 40(1–4):213–233
8. Hackett GAR, Easton CA, Duff SJB (1999) Composting of pulp and paper mill fly ash with wastewater treatment sludge. Bioresour Technol 70(3):217–224
9. Thacker NP, Nitnaware VC, Das SK, Devotta S (2007) Dioxin formation in pulp and paper mills of India. Environ Sci Pollut Res 14(4):225–226
10. Bijan L, Mohseni M (2004) Using ozone to reduce recalcitrant compounds and to enhance biodegradability of pulp and paper effluents. Water Sci Technol 50(3):173–182
11. Confederation of European Paper Industries, CEPI. www.cepi.org
12. RISI, The leading information provider for the global forest products industry. http://www.risiinfo.com/
13. Wistara N, Young RA (1999) Properties and treatments of pulps from recycled paper. Part I. Physical and chemical properties of pulps. Cellulose 6(4):291–324
14. Lankinen VP, Inkeroinen MM, Pellinen J, Hatakka AI (1991) The onset of lignin-modifying enzymes, decrease of AOX and color removal by white-rot fungi grown on bleach plant effluents. Water Sci Technol 24(3–4):189–198

15. Ali M, Sreekrishnan TR (2001) Aquatic toxicity from pulp and paper mill effluents: a review. Adv Environ Res 5(2):175–196
16. Schnell A, Steel P, Melcer H, Hodson PV, Carey JH (2000) Enhanced biological treatment of bleached kraft mill effluents – I. Removal of chlorinated organic compounds and toxicity. Water Res 34(2):493–500
17. Bajpai P, Anand A, Sharma N, Mishra SP, Bajpai PK, Lachenal D (2006) Enzymes improve ECF bleaching of pulp. BioResources 1(1):34–44
18. Hubbe MA, Nanko H, McNeal MR (2009) Retention aid polymer interactions with cellulosic surfaces and suspensions: a review. BioResources 4(2):850–906
19. Committe ZT (2008) Chemical additives for the production of pulp & paper. Deutscher Fachverlag GmbH. Frankfurt am Main
20. Howard GJ, Hudson FL, West J (1977) Water-soluble polymers as retention aids in a model papermaking system. 1. Polyacrylamides. J Appl Polym Sci 21(1):1–16
21. Pelton R, Hong J (2002) Some properties of newsprint impregnated with polyvinylamine. Tappi J 1(10):21–26
22. Hubbe MA (2007) Paper's resistance to wetting – a review of internal sizing chemicals and their effects. BioResources 2(1):106–145
23. Roberts JC (1996) Paper chemistry, 2nd edn. Blackie Academic & Professional, London, New York, p xiii, 267 p
24. Garnier G, Bertin M, Smrckova M (1999) Wetting dynamics of alkyl ketene dimer on cellulosic model surfaces. Langmuir 15(22):7863–7869
25. Ishida Y, Ohtani H, Tsuge S, Yano T (1994) Determination of styrene copolymer sizing agents in paper by pyrolysis-gas chromatography. Anal Chem 66(9):1444–1447
26. Gess JM, Rende DS (2005) Alkenyl succinic anhydride (ASA). Tappi J 4(9):25–30
27. Kandelbauer A, Petek P, Medved S, Pizzi A, Teischinger A (2010) On the performance of a melamine-urea-formaldehyde resin for decorative paper coatings. Eur J Wood Wood Prod 68 (1):63–75
28. Espy HH (1995) The mechanism of wet-strength development in paper – a review. Tappi J 78 (4):90–99
29. Cockayne SE, McDonagh AJG, Gawkrodger DJ (2001) Occupational allergic contact dermatitis from formaldehyde resin in clothing. Contact Dermatitis 44(2):109–110
30. Lund M, Felby C (2001) Wet strength improvement of unbleached kraft pulp through laccase catalyzed oxidation. Enzyme Microb Technol 28(9–10):760–765
31. Obokata T, Yanagisawa M, Isogai A (2005) Characterization of polyamideamine-epichlorohydrin (PAE) resin: roles of azetidinium groups and molecular mass of PAE in wet strength development of paper prepared with PAE. J Appl Polym Sci 97(6):2249–2255
32. Boden L, Lundgren M, Stensio KE, Gorzynski M (1997) Determination of 1,3-dichloro-2-propanol and 3-chloro-1,2-propanediol in papers treated with polyamidoamine-epichlorohydrin wet-strength resins by gas chromatography mass spectrometry using selective ion monitoring. J Chromatogr A 788(1–2):195–203
33. Velisek J, Davidek T, Davidek J, Hamburg A (1991) 3-Chloro-1,2-propanediol derived amino alcohol in protein hydrolysates. J Food Sci 56(1):136–138
34. Eder E, Weinfurtner E (1994) Mutagenic and carcinogenic risk of oxygen-containing chlorinated C-3 hydrocarbons – putative secondary products of C-3 chlorohydrocarbons and chlorination of water. Chemosphere 29(9–11):2455–2466
35. Mansour OY, Sefain MZ, Ibrahim MM, El-Zawawy WK (2000) Paper coating mixture: preparation, application, and study of their rheological properties. J Appl Polym Sci 77 (8):1666–1678
36. Duan H, Zhao C, Wu Y, Zhang Q, Wang S (1999) Performance in paper coating of styrene/acrylate copolymer latex. Polym Adv Technol 10(1–2):78–81
37. Veleva S, Valcheva E, Valchev I, Radeva G (2001) Application of an exponential kinetic equation to the interaction of optical brighteners with pulp. React Kinet Catal Lett 72(2):355–364

38. Wong-Wah-Chung P, Mailhot G, Bolte M (2001) 4,4′-Diaminostilbene-2,2′-disulfonate (DSD) behaviour: under irradiation in water. Decrease of its activity as a fluorescent whitening agent. J Photochem Photobiol A Chem 138(3):275–280
39. Poiger T, Field JA, Field TM, Giger W (1996) Occurrence of fluorescent whitening agents in sewage and river water determined by solid-phase extraction and high-performance liquid chromatography. Environ Sci Technol 30(7):2220–2226
40. Stoll JMA, Giger W (1998) Mass balance for detergent-derived fluorescent whitening agents in surface waters of Switzerland. Water Res 32(7):2041–2050
41. Oppong D, King VM, Zhou X, Bowen JA (2000) Cultural and biochemical diversity of pink-pigmented bacteria isolated from paper mill slimes. J Ind Microbiol Biotechnol 25(2):74–80
42. Kiuru J, Tsitko I, Sievanen J, Wathen R (2010) Optimization of biocide strategies on fine paper machines. BioResources 5(2):514–524
43. Ward JEH (1989) The environmental impacts and behaviour of TCMTB. Information memorandum of the BC Ministry of Environment, Waste Management Program, Toxicology Unit
44. Directive 2002/61/EC of the European Parliament and of the Council of 19 July 2002 amending for the nineteenth time Council Directive 76/769/EEC relating to restrictions on the marketing and use of certain dangerous substances and preparations (azocolourants). http://eur-lex.europa.eu/LexUriServ/LexUriServ.do?uri=OJ:L:2002:243:0015:0018:en:PDF

Additives in the Leather Industry

Daniel Guillén, Antoni Ginebreda, Mira Petrovic, Damià Barceló, Rosa Mari Darbra, and Stefan Rydin

Abstract An overview of the leather sector from the point of view of additives use is presented. In the first section, the general trends of the sector under concern in terms of world production and trade flows are reviewed, together with an overall description of the main characteristics of the production process including beamhouse (i.e., salting, soaking, unhairing, liming, deliming, pickling), tanning, post-tanning (i.e., neutralization, bleaching, re-tanning, fatliquoring), and finishing operations. The second part is focused on the description of the different additive classes used to provide leather with the required functional characteristics and to facilitate and improve the production process as well, namely, fatliquoring agents, surfactants, coating agents (binders, lacquers), waterproofing agents, flame retardant agents, biocides, and dyes. Main compounds or compound families within each additive class are listed, and some related information regarding their function and environmental potential impacts is also included.

Keywords Additives, Coating, Dyes, Leather, Waterproof

Contents

D. Guillén (✉), A. Ginebreda, M. Petrovic, and D. Barceló
Department Environmental Chemistry, Institute of Environmental Diagnostic and Water Studies (IDAEA), Spanish Council of Scientific Research (CSIC), Jordi Girona 18–26, 08034 Barcelona, Spain
e-mail: dggqam@cid.csic.es

R.M. Darbra
Department of Chemical Engineering, CERTEC, ETSEIB, Universitat Politècnica de Catalunya, Diagonal 647, pav. G, 08028 Barcelona, Spain

S. Rydin
Nordeconsult Sweden AB, Sofiaparken 3C, 222 41 Lund, Sweden

B. Bilitewski et al. (eds.), *Global Risk-Based Management of Chemical Additives I: Production, Usage and Environmental Occurrence*, Hdb Env Chem (2012) 18: 35–56,
DOI 10.1007/698_2011_105, © Springer-Verlag Berlin Heidelberg 2011,
Published online: 27 April 2011

Abbreviations

BA	Butyl acrylate
BFR	Brominated flame retardant
DDT	Dichlorodiphenyltrichloroethane
EA	Ethyl acrylate
EU	European Union
HBCD	Hexabromocyclododecane
NP	Nonylphenol
NPE	Nonylphenol ethoxylate
octaBDE	Octabromodiphenyl ether
PBDE	Polybrominated diphenyl ethers
PCP	Pentachlorophenol
PU	Polyurethane
TCMTB	2-(Thiocyanomethylthio)benzothiazole
TeCP	Tetrachlorophenol
USSR	Union of Soviet Socialist Republics
VOC	Volatile organic compounds

1 Introduction to Leather Sector

The use of leather goes back to prehistoric times, when hides and skins of animals killed for their meat started being used to protect the body from adverse climate conditions.

Leather raw materials are animal hides and skins (large animals such as cattle have hides whereas small animals such as sheep have skins). Skins or hides are largely constituted by collagen, so it is this fibrous protein and the main responsible of the properties conferred to the skin. Hence, its chemistry is the matter with which the tanner is mostly concerned.

The tannery operation consists of converting the raw skin, a highly putrescible collagenous matrix, into leather, a stable material which can be used in the manufacture of a wide range of products. Leather is widely used to make items such as clothing, shoes, car furniture, boat and airplane seats, and many more items of daily use. Each item requires the use of a leather type with certain properties. Here comes into play the use of chemical compounds that are able to change the properties of these raw and hides.

Environmental concerns have been growing since many years ago around leather industry; this sector has always had an unfriendly image for the environment since it uses large amounts of water and generates enormous amounts of solid waste and liquid effluents, characterized by high organic and salt contents as well as toxic compounds. The aim of this chapter is discussing the use of finishing additives for leather, i.e., chemicals that improve leather specific properties: stability, appearance, water resistance, temperature resistance, elasticity, flame retardance, etc.

1.1 Worldwide Production and Trade Flows

Leather processing is an important activity in many developing countries, which are dependent on the rural economy. The increasing requirement of leather and their related products led to a global output that has risen over the past 30 years. It has been estimated that around 18 billion square feet of leather is made annually around the world [1] with a trade value estimated to be approximately US$ 70 billion.

Production of raw hides and skins depends on animal population. On a word scale, significant cattle populations are to be found in China, USA, Brazil, Argentina, India, former USSR, and the EU. Sheep and lambskins originate predominantly in China, New Zealand, Australia, the Near East, and the EU [2]. In the tables and figures below (Table 1, Figs. 1, 2) are showed the main leather producer countries worldwide (Table 2) and values of international trade (Table 3).

For countries like China or USA, leather is a valuable economic source. Thus, for instance, the USA is the Europe's main supplier of raw hides and skins. Italian leather industry represents around 15% of world leather production and about 60% of Europe's total.

Table 1 Leather production for some countries (units: million square feet) [3]

Country	2003	2004	2005	2006
China	4,893.5	6,124.1	6,382.5	6,599.3
Italy	1,934.8	2,218.3	2,219.1	2,146.9
India	1,458.5	1,233.8	1,248.2	1,738.6
Brazil	883.1	1,326.8	1,433.4	1,647.3
Argentina	508.3	653.2	690.5	712.7
Mexico	780.9	535.1	555.9	591.8
USA	813.6	563.2	475.0	477.0
Spain	521.9	484.4	464.0	442.0
World total	*21,922.6*	*22,036.8*	*22,306.6*	*22,700.5*

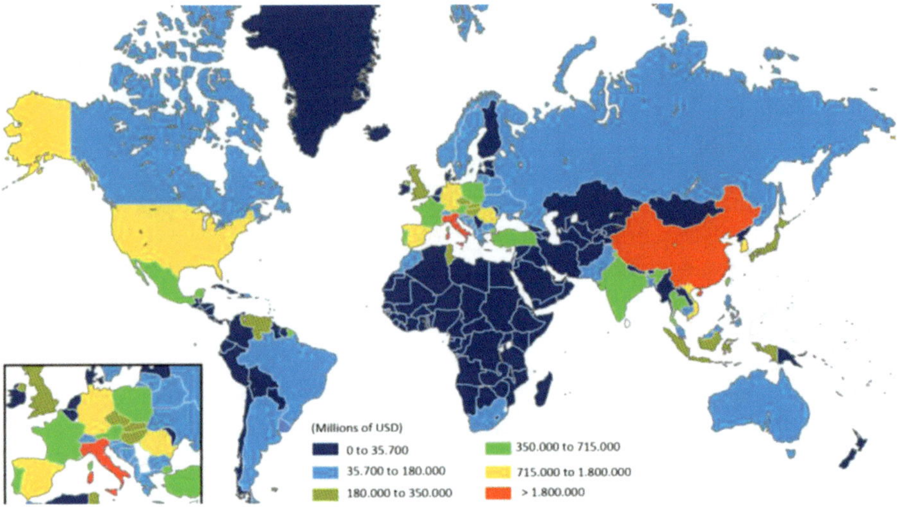

Fig. 1 List of importers for raw hides, skins, and leather in 2008 [4]

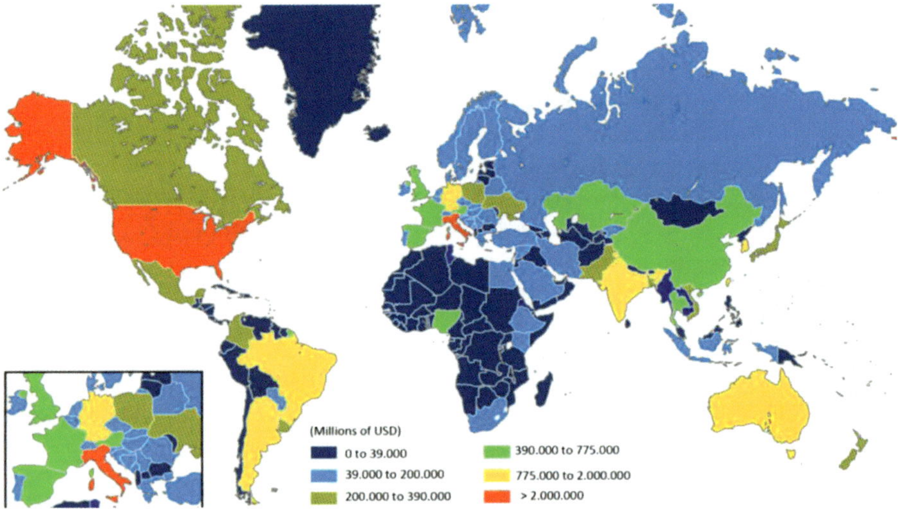

Fig. 2 List of exporters for raw hides, skins, and leather in 2008 [4]

2 Leather Manufacturing from Raw Hides

Traditionally, leather processing involves four steps (Fig. 3), namely, beamhouse, tanning, post-tanning, and finishing processes. While beamhouse operations are designed to clean the skin collagen from unwanted materials and preparing it for subsequent tanning, tanning itself serves for stabilizing the protein against putrefaction.

Table 2 The value of international trade in hides, skins, and leather footwear average 2003–2005, compared with other commodities (units: million US Dollars) [3]

	Developing countries	Developed countries	World
Raw hides and skins	598.4	4,410.1	5,008.5
Wet blue, crust, and finished leather	9,340.0	8,137.7	17,477.7
Leather footwear	13,990.9	17,347.7	31,338.6
Total	*23,929.3*	*29,989.5*	*53,824.8*
Meat from cattle, sheep and goats	5,264	18,841	24,105
Cotton	3,086	6,122	9,208
Coffee	6,738	2,869	9,607
Tea	2,567	691	3,258
Rice	5,779	1,231	7,010
Sugar	6,930	4,222	12,281

Table 3 Main operations/processes usually involved in leather manufacturing

Salting	Function	Historically the most common way of preserving food and hides, salt inhibits growth and activity of bacteria
	Chemicals	Usually NaCl
Soaking	Function	Allows hides and skins reabsorbing any water lost after curing. Removes dung, blood, and dirt, and discards interfibrillary material
	Chemicals	Water, alkali, and surfactants. Enzymes can be added to improve the efficiency of the process
Unhairing and liming	Function	The aim of this process is removing hair, epidermis, interfibrillary components, and opening up the fiber structure
	Chemicals	Sulfides (NaHS or Na_2S), thiourea dioxide, and sodium dithionite. Alternatives include organic sulfur compounds (mercaptans or sodium thioglycolate) in combination with strong alkali. Enzymatic preparations are sometimes added to improve the performance of the process
Deliming	Function	Skin structure is further opened by enzyme treatment and unwanted material is removed
	Chemicals	Hydrochloric, sulfuric, formic, lactic, and boric acids
Pickling	Function	Preserving skins for up to many years. The skins are stirred in a solution until they are at or near equilibrium pH value of 3.0–3.5 optimal conditions for tanning
	Chemicals	Sulfuric, formic, hydrochloric, and boric acids
Tanning	Function	Stabilizing the structure by replacing some of the collagen links with complexed chromium ions or another tanning agent
	Chemicals	– Mineral tannages: aluminum, chromium, and zirconium. Currently chromium III is the most used agent for tanning – Vegetal tannins obtained from wood, barks, leaves, roots, etc. – Alternative tanning agents such as syntans, oil tannage, and aldehydes

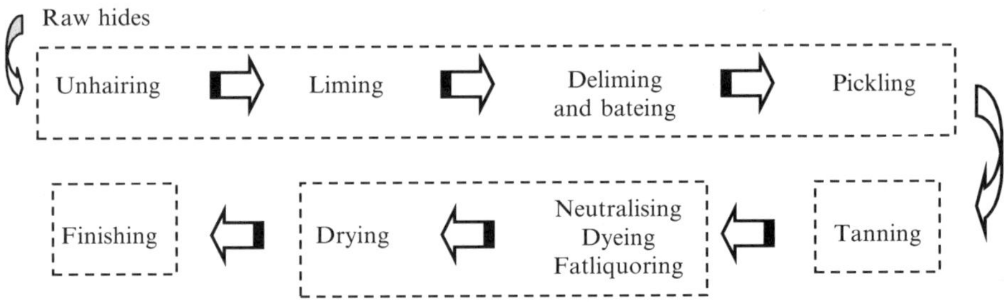

Fig. 3 Main four steps for leather manufacturing

Processes used in each of these categories may change depending on the raw materials used and the final manufactured products. In Table 3 are summarized the main stages involved in the tanning process [2, 5].

Incorporation to leather of some properties that otherwise are lacking, like thermal and mechanical resistance, water and fire resistance, flexibility, gloss or color as well as improving its aesthetical appeal are the objectives of post-tanning and finishing steps. These operations are achieved by extensive use of chemical additives.

3 Use of Chemical Additives

3.1 Coating

Topcoats technology traditionally requires large amounts of organic solvents. Lacquers and lacquers emulsions (with water or surfactant) were used for years to improve leather surface properties. Lacquers were obtained mainly by dissolving nitrocellulose in solvents (e.g., ethyl acetate, methoxypropanol, or methyl isobutyl ketone).

Nowadays, there is an increasing interest on using aqueous topcoats in the leather sector. Waterborne topcoats offer several well-recognized advantages over conventional lacquers or lacquers emulsions such as: handling and storage of inflammable solvents are eliminated, environmental and working conditions become healthier, and atmosphere emissions of volatile organic compounds (VOC) are reduced [2, 36].

Polymeric binders are products capable of forming on drying a film on the surface where they are applied and they represent the basic element in an aqueous leather finishing. Three types of polymeric binders with a different chemical base are mainly used [6]: acrylic, polyurethane (PU), and butadiene, being the first two the most largely used. Each one of these binders provides a characteristic finish to the final product.

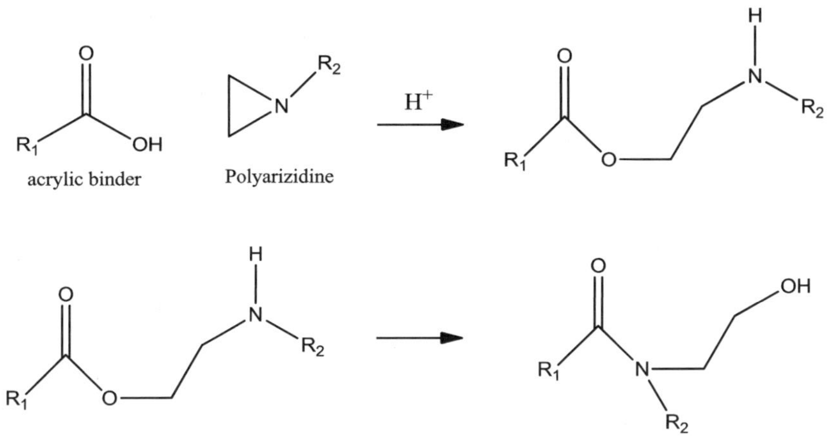

Fig. 4 Polyaziridine reaction with carboxylic groups of the acrylic binder

To achieve the physical properties required for most leather goods, it is necessary to use cross-linkers that react with polymers [7]. Cross-linkers are chemical compounds that act on reactive groups in the polymeric chains joining them, forming lattices or a highly branched molecular mesh. Suitable cross-linkers for use in leather finishing ideally must react at ambient temperature in a relatively short time. The most commonly used are polyaziridine, polysane, polyisocyanate, epoxy resin, and polycarbodiimide. A comparative study [8] of different cross-linkers indicates that polysane presents a better cross-linking reaction rate than the others aforementioned. Cross-linking reaction (Fig. 4) improves the wet physical properties such as wet rub fastness, float resistance, and wet flexing. The use of aziridines requires special handling care since they are corrosive materials and can cause skin irritation [9].

3.1.1 Acrylic Binders

Acrylic polymers have found extensive use in leather finishing. These binders are formed by means of the polymerization of monomers containing carboxylic groups, amide groups, or amine groups (Table 4).

The main properties of acrylic binders are: softness, heat and light stability, thermoplasticity, good adherence, and low VOC content. However, in some cases they present low chemical resistance and low mechanical properties [10].

Elasticity is a precious property in leather products; leathers with good elasticity give a comfortable and soft feeling when touched. Recent research [11] showed that leathers treated with butyl acrylate (BA) and ethyl acrylate (EA) exhibited better elasticity properties than other selected comonomers.

Table 4 Monomers most commonly used to produce acrylic binders

Carboxylic groups	Acrylic acid	$CH_2=CH-COOH$
	Methyl acrylic acid	$CH_2=C(CH_3)-COOH$
	Vinyl acetate	$CH_2=CH-O-C(=O)-CH_3$
	Methyl acrylate	$CH_2=CH-C(=O)-O-CH_3$
	Ethyl acrylate	$CH_2=CH-C(=O)-O-CH_2-CH_3$
	Butyl acrylate	$CH_2=CH-C(=O)-O-CH_2-CH_2-CH_2-CH_3$
	Itaconic acid	$HOOC-CH_2-C(=CH_2)-CO-OH$
	Hydroxyethyl acrylate	$CH_2=CH-CO-O-CH_2-CH_2-OH$
Amide groups	Acrylamide	$CH_2=CH-CONH_2$
	Methylolacrylamide	$CH_2=CH-CO-NH-CH_2OH$
Amine groups	2-(*tert*-butylamino)ethyl methacrylate	$CH_2=C(CH_3)-CO-O-CH_2-CH(CH_3)-NH-C(CH_3)_3$

3.1.2 Polyurethane Binders

PU dispersions (Fig. 5) are gaining popularity in leather finishing due to their excellent properties such as good abrasion resistance, high chemical resistance, high strength, and low temperature flexibility. PU binders contain carbamate or urethane groups in their structural unity that are formed by the reaction of an isocyanate group (R–N=C=O) with a hydroxyl group.

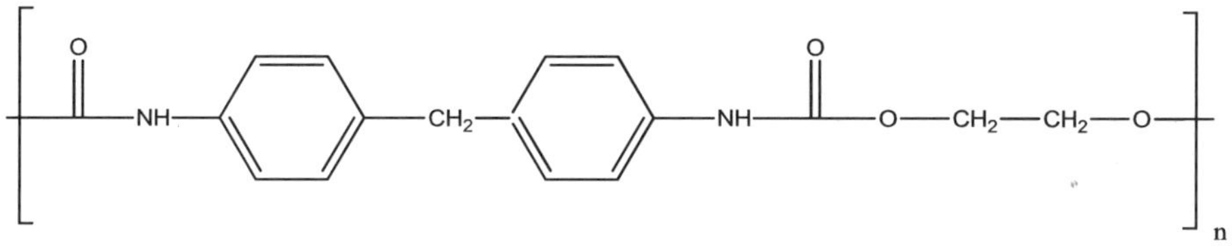

Fig. 5 Polyurethane binder chemical structure

Table 5 Most widely used fatliquoring substances

Biological fatty substance	Vegetable oils	Linseed oil, castor oil, nut oil, Word oil, sunflower oil, and soybean oil
	Vegetable fats	Coconut fat, palm oil fat, and palm kernel fat
	Animal oils	Seal oil, whale oil, sardine oil, shark liver oil, neatsfoot oil, and lard oil
	Animal fats	Butter fat, bone fat, and horse grease
	Waxes	Carnabua wax, beeswax, and wool grease
Nonbiological fatty substances	Parafin waxes, mineral oils, olefins, processed hydrocarbons, synthetic fatty acid esters and waxes, fatty alcohols, and alkyl benzenes	

3.2 Fatliquoring

The removal of natural fats on leather has disadvantages: fibers stick together, which makes the leather become rigid. To avoid this, after tanning, the leather is subjected to a fatliquoring step, in which fibrils are lubricated with oil (Table 5) and thus producing soft leather with a pleasant touch sensation. Fatliquoring also influences physical characteristics of leather, such as tensile strength, extensibility, wetting properties, waterproofness, and permeability against water vapor and air.

To allow a small amount of oil to be spread uniformly over the large surface area of the leather, fatliquors are used in emulsion form with water. Oil is dispersed in microscopical drops capable of penetrating down of the leather structure [12].

Currently [13] the reuse of natural fats removed during the tanning process is being developed for the fatliquoring treatment. This process integration can reduce environmental impacts, which are of major concern in the leather industry.

3.3 Waterproofing Leather

There are many polar groups on the protein fibers, such as –OH, $-NH_2$, –COOH, CO, or NH, which make possible their interaction with water. These functional

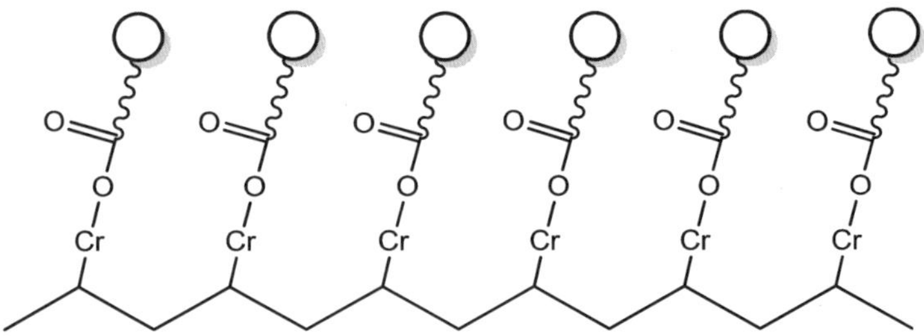

Fig. 6 Model structure of chromium tanned leather with fatliquoring agent combination

groups provide leather with the property of hydrophilicity, which is achieved by enhancing its surface energy. Therefore, the essential of waterproofness is decreasing the surface-free energy of leather.

For hydrophobic leather manufacturing, fatliquoring is the responsible step. It basically consists of applying to the product surface some emulsifying oil which has surfactant properties, thus preventing leather to become hydrophilic (Fig. 6). Typical materials used as waterproofing agents are hydrocarbon oil or waxes, fatty acid esters of glycerin or other polyols, silicones, fluorocarbons, and others [14].

When leather surface is sprayed with a fluorine-containing polymer layer, the surface-free energy decreases notably and conversely the waterproofness is increased [15]. However, there is evidence that creating a hydrophobic coating on leather is insufficient because it has a greater effect on the grain surface than in the rest of structure, and when the leather is flexed the hydrophobic film can be damaged or deformed resulting in a reduction of waterproofness.

The surface tension of waterproof leathers makes them difficult to finish without the help of surfactant flow aids and solvents to reach good flow out and penetration. The so-obtained products often exhibit a loss of waterproofness, as it is evidenced by lower Maeser flexes (standardized test for evaluating water penetration).

3.3.1 Use of Surfactants in the Leather Sector

Surfactants are used in many different processes throughout the tannery, e.g., soaking, liming, degreasing, tanning, and dyeing. Environmental concern about surfactants is mainly related to their biodegradability, toxicity of metabolites, and parent products [16].

Alkylphenol ethoxylates work as effective surfactants. More specifically, nonylphenol ethoxylates (NPEs) exhibit excellent wetting and emulsification properties. In anaerobic conditions [17], NPEs can suffer degradation releasing nonylphenol (NP) (Fig. 7) that is persistent in the aquatic environment, moderately bioaccumulative, and extremely toxic to aquatic organisms. Moreover, its endocrine-disrupting properties have been repeatedly proved, meaning that they interfere adversely with animal-based endocrine systems (hormones) which can lead to detrimental effects on human and animal health.

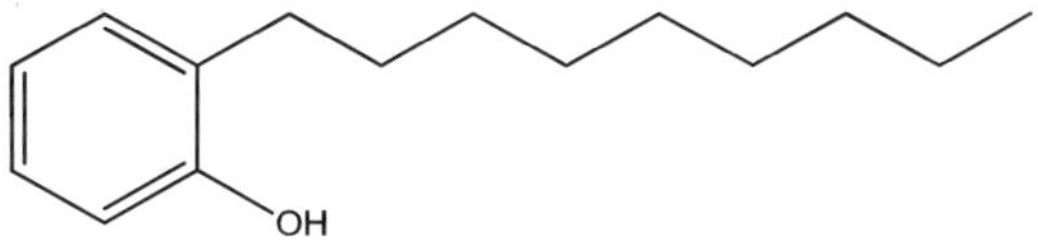

Fig. 7 Nonylphenol chemical structure

Owing to this fact, the use of NP and NPEs has been banned in Europe [18]. The main alternatives in the leather industry are alcohol ethoxylates, both linear and branched (with different chain lengths and ethoxylation degrees), and glucose-based carbohydrate derivatives such as alkylpolyglucoside, glucamides, and glucamine oxides. These compounds show lower toxicities than NPE and can be degraded to nontoxic products. The efficiency of linear alcohol ethoxylates as degreasing agents is comparable to that of NPE.

3.4 Biocides

Storing pickled and chrome-tanned leather under humid conditions and high temperature and the low pH value and high fat level of these leathers are factors causing the growth of microorganisms. For this reason, the use of fungicides against the proliferation of microorganisms is often necessary.

In tannery operations, there are three major types of microorganisms that may cause leather degradation, namely, bacteria, yeasts, and molds (sometimes known as "fungi"). All three types are capable of spoiling the raw material.

Since these organisms prefer an acidic environment, fungi are usually encountered in the tannery after the pickling process. In the stages following pickling, the sites most conducive to fungal growth are:

1. Pickled hides and skins
2. Chrome-tanned stock in the blue
3. Colored skins
4. Vegetable-tanned leathers, until dried
5. Vegetable tan liquors

Before 1970, the chemical components of the most effective fungicides available for tanning industry were predominantly halogenated phenols such as pentachlorophenol (PCP) (Fig. 8) and tetrachlorophenol (TeCP) [19]; in less extent organomercurial compounds were also used. The banning of the use of organomercurial compounds by more strict environmental regulations, and the restriction in the use of chlorophenolic compounds, has driven the development of new fungicides able to minimize the leather degradation caused by microorganisms' growth, but with more acceptable environmental behavior (Table 6).

Other chlorinated compounds such as dichlorodiphenyltrichloroethane (DDT) and lindane were used for protecting raw and hides from the insect damage, especially in areas with dry and hot climates, as in many third world countries Africa, compounds (Fig. 8). The organochlorine pesticide dieldrin was introduced

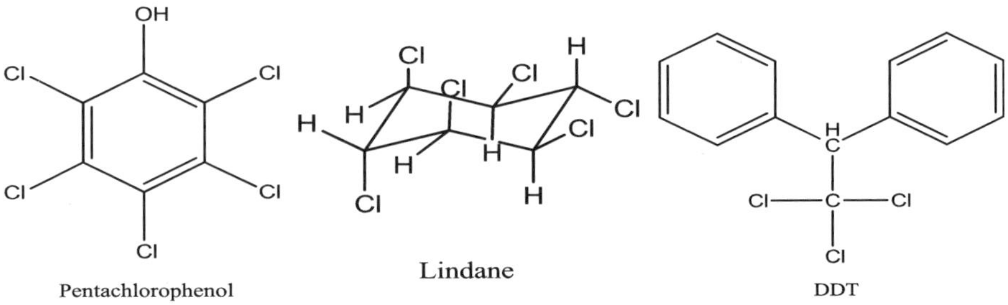

Fig. 8 Formerly used biocide compounds (nowadays banned) for leather preservation

Table 6 The most common biocides used in the leather industry

Chemicals added to salt to prevent halophilic bacterial growth "red heat"	Inorganic compounds	Sodium carbonate
		Boric acid
		Sodium bisulphate
		Sodium metabisulphite
		Sodium silico fluoride
		Sodium fluoride
		Zinc compounds
	Organic compounds	Dichlorophen(5,5′-dichloro-2,2′-dihydroxy diphenyl methane)
		Substituted benzothiazole
		Mixed phenolics
Short-term preservation chemicals	Inorganic compounds	Boric acid
		Hypochlorites
		Chlorites
		Sodium metabisulphite
		Sodium silico fluoride
		Sodium fluoride
		Zinc compounds
	Organic compounds	Quaternary ammonium compounds
		1,2-Benzisothiazolin-3-one (BIT)
Bactericides added to soak liquors	Organic compounds	Methylene bis-thiocyanate (MBT)
		2-Bromo-2-nitropropane-1,3-diol (bronopol)
		Sodium or potassium dimethyldithiocarbamate
		Thiadiazine
Fungicides used during pickling, tanning and post-tanning operations	Organic compounds	Mixed phenolics + pyrithion (3-methyl-4-chlophenol,2-phenylphenol)
		Isothiazolone
		2-Thio-cyano-methyl-thiobenzothiazole (TCMTB)
		Methylene bis-cyanate (MBT)
		TCMTB/MBT mixture
		Paranitrophenol

to provide protection against ectoparasites (such as fleas and lice). The usage of DDT and lindane was banned long ago, and nowadays both are included in the Stockholm Convention on Persistent Organic Pollutants [20].

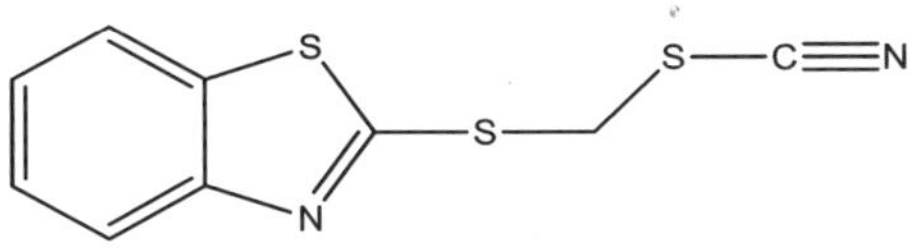

Fig. 9 TCMTB chemical structure

Currently other types of compounds that are less persistent and dangerous for the environment are being used. Among them, TCMTB is one of the most widely used (Table 6, Fig. 9), being its effectiveness as biocide against microbial growth well reported in different studies [21, 22].

3.5 Flame Resistance

During the last decade, fire retardance has become an ever increasingly important issue [2]. The degree to which leather is flammable must be identified to satisfy government regulations in automobile, aircraft, upholstery, clothing, and furniture applications. It is impossible to make leather completely resistant to charring and decomposition when exposed against potential ignition sources (fire, sparks, and even cigarettes), but a degree of flame resistance can be achieved. Flame retardant treatments for leather merely increase the amount of oxygen or heat required for combustion [23], thus diminishing the spread of flame. These additives can be applied to the leather by dipping (immersion) in a solution of the dissolved agent or by spraying the leather surface [24].

Brominated flame retardants (BFRs) have been added to many consumer goods, including leather, for several decades. The global demand for polybrominated diphenyl ethers (PBDEs) was estimated at 70,000 tonnes in 2001, of which penta-, octa-, and deca-formulation represented approximately 24%, 4%, and 72%, respectively [25, 26]. BFRs use in Europe has been reduced during the latest years due to legislation. Hence, some of these compounds, such as pentabromocyclododecane (PentaBDE) and octabromocyclododecane (octaBDE), have been banned [18]. Hexabromocyclododecane (HBCD) (Fig. 10) is the second most used BFR in Europe [27], and due to its persistence, toxicity, and bioaccumulation, a global ban is currently being considered under the framework of the Stockholm Convention on Persistent Organic Pollutants [20].

Current alternatives to brominated flame retardants for leather manufacturing are listed in Table 7.

3.6 Dyes

Dyeing is one of the most important steps in leather making as it is usually the most evident property first assessed by the customer. In post-tanning, it is possible to combine several finishing steps on a single operation giving rise to the so-called

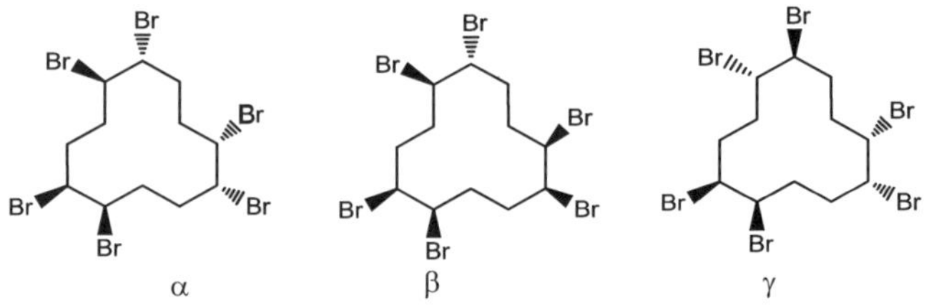

Fig. 10 Technical HBCD consisting on a mixture of three diastereoisomers, α- (10%), β- (9%), and γ-HBCD (81%)

Table 7 Nonbrominated flame retardants used to improve leather fire resistance

Group and subgroup		Chemicals
Inorganic salts (water soluble)		Ammonium borate
		Ammonium molybdate
		Ammonium polyphosphate (APP)
		Ammonium sulfamate
		Borax
		Boric acid
		Diammonium phosphate
		Sodium phosphate
		Sodium phosphate dodecahydrate
		Sodium stannate
		Sodium tungstate
		Zinc chloride
Phosphorus compounds	Butylated phenyl phosphates	Bis-(*tert*-Butylphenyl)phenylphosphate (BBDP)
		Tert-Butylphenyl diphenyl phosphate (BDP)
		Tris-(*tert*-Butylphenyl)phosphate (TBDP)
	Propylated phenyl phosphates	Bis-(isopropylphenyl) phenylphosphate (BIPP)
		Isopropylphenyl diphenyl phosphate (IPP)
		Tris-(isopropylphenyl)phosphate (TIPP)
	Others	2-Ethylhexyldiphenyl phosphate
		Cresyl diphenyl phosphate (CDP)
		Isodecylphosphate (IDP)
		Tricresylphosphate (TCP)
		Triphenylphosphate (TPP)
		Trixylylphosphate (TXP)

compact processing. In the case of dyeing, this is a common practice and coloring is incorporated into the final product through a unique dye-fatliquoring operation.

Organic dyestuffs used in tanneries can be classified in anionic and cationic dyes. Around 98% of leather dyes on the market are dyes which fix using ionic interaction between the anionic sulphonate group of the dye and protonated (cationic) amine groups present in collagen [28].

The molecular structural characteristics responsible for color (chromophoric groups) typically involve a system of π-delocalized electrons. This is usually achieved by synthesizing molecules that contain extended conjugated double bond systems and other groups alike (i.e., diazo groups), such as aromatic cycles, etc. Dyes commonly used in tanning industry can be classified into four classes [29], namely, acid dyes, basic dyes, direct dyes, and mordant dyes.

3.6.1 Acid Dyes

This class of dyes (i.e., acid orange 8, Fig. 11) is the most used in the leather industry, particularly for chromium tanned leather. Acid dyes consist of relative small and typically hydrophilic molecules that react, under acidic conditions, predominantly through electrostatic reaction between their sulfonate groups and the protonated amino groups of lysine. These compounds provide good fastness properties and a wide range of colors.

3.6.2 Basic Dyes

Basic dyes structure (i.e., basic green 4, Fig. 12) is essentially the same as acid dyes, except that basic dyes carry a net positive charge from the cationically charged amino substituents, even though they may also have anionic sites.

Because their structures contain quaternary ammonium groups derived from large aromatic tertiary amines, they tend to be relatively hydrophobic; consequently, they are often soluble in oils and nonaqueous solvents. They are characterized by strong and brilliant colors.

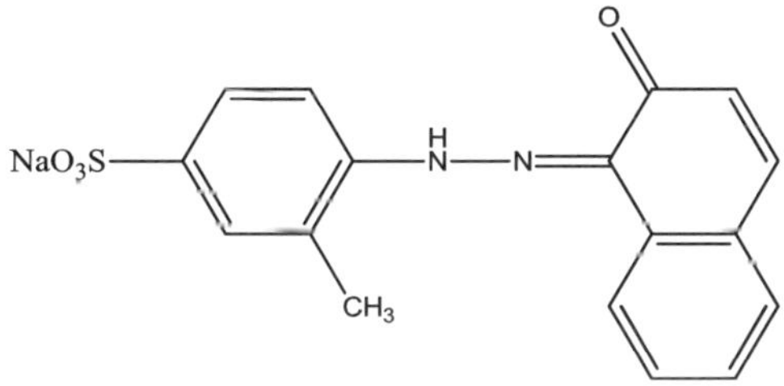

Fig. 11 Acid orange 8

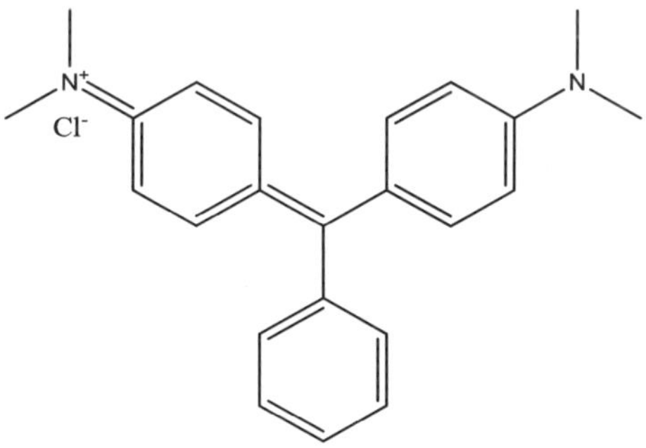

Fig. 12 Basic green 4

3.6.3 Direct Dyes

Direct dyes (i.e., direct violet 51, Fig. 13) have the same sort of structural features as acid and basic dyes, but with higher molecular weight. In these compounds, pH adjustment is not required for fixation as they have a higher number of reactive sites. They present fastness properties average to good.

3.6.4 Mordant Dyes

The original mordant dyes (i.e., mordant blue 13, Fig. 14) were typically composed of plant extracts, which produced relatively dull and pale shades and fixed poorly to leather or textiles when used alone: to properly fix the coloring agent with the substrate, it was necessary to provide an additional fixing mechanism, by applying metal salt either before or with the dye.

The formation of metal complexes means that the final color struck depends on the mordant metal used. Table 8 reports the different metal ions used together with their respective final colors.

3.6.5 Concern About Azo Dyes

Azo dyes have been widely used as colorants in a wide variety of products such as textiles, paper, and leather, since the early 1900s [30]. Although the dye molecule is biologically inactive, their synthetic precursors (i.e., aromatic amines) are not.

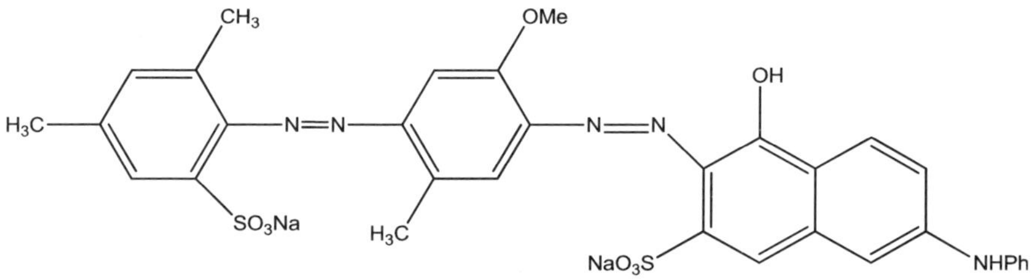

Fig. 13 Direct violet 51

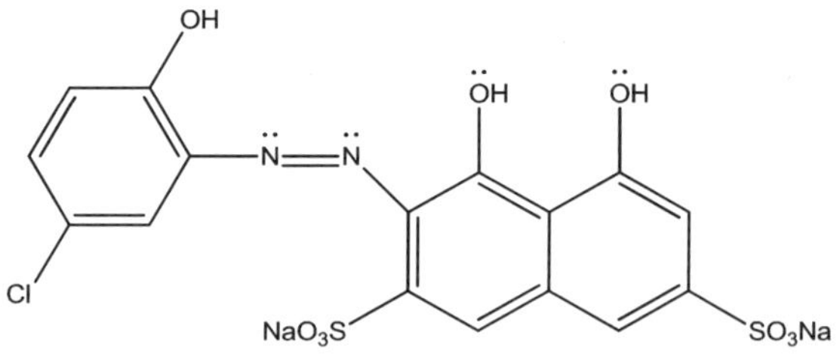

Fig. 14 Mordant blue 13

Table 8 Dye metal ions with their respective color

Metal ion	Color
Al (III)	Red
Sn (IV)	Pink
Fe (III)	Brown
Cr (III)	Puce-brown
Cu (II)	Yellow-brown

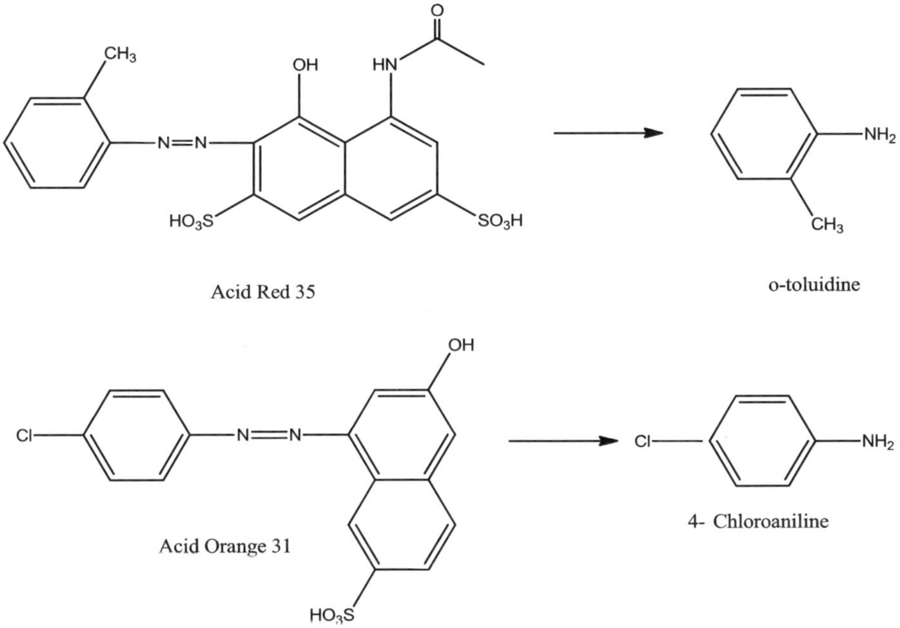

Fig. 15 Azo dyes and released carcinogenic amines

Furthermore, it has been found [31, 32] that living organisms are able to cleave the azo bonds (Fig. 15), and intermediates, by-products, and degradation products of these dyes can become potential health hazards, owing to both their toxicity and carcinogenity [33, 34].

In Europe, there is a regulation [18] which bans the use of hazardous azo dyes and marketing of textile or leather articles whose elaboration was made from substances that can release the amines listed in Table 8.

Environmental concern about the discharge of wastewaters containing azo dyes has focused the research efforts on natural dyes [35], which have several advantages such as low toxicity, cheap and easy availability. Natural dyes are obtained from plants (e.g., Alizarin, Catechu, Indigo, and Longwood), from animals (e.g., cochineal, kermes, and tyrian purple), and from certain naturally occurring minerals (e.g., Ocher and Prussian blue). Presently natural dyes have reached considerable potential value in the global leather market in the context of environmentally benign leather processing.

4 Conclusions

Even though the tanning process of leather is known since long ago, it is still today an active industrial sector subjected to continuous evolution through the use of new techniques and chemicals.

The starting raw materials used on the leather industry are the hide or skin of animals, which are subsequently processed until the final product is obtained. The tannery operation thus consists of converting the raw skin, a highly perishable material, into leather, a stable material, which can be subsequently used in the manufacture of a wide range of goods. The whole process involves a sequence of complex chemical reactions and mechanical operations. Among these, tanning is the fundamental step, which gives to leather its stability and essential character. Preserving pelts by tanning and performing various steps of pretreatment and posttreatment generates a final product with specific properties: stability, appearance, water resistance, temperature resistance, elasticity and permeability for perspiration and air, etc. As described in this chapter, a great variety of additives is used both to facilitate the manufacturing process and to improve the final product properties.

During the last years, the leather industry has become an important commercial value for some developing and emerging countries like China, India, or Brazil, all being major producers of raw leather that is subsequently distributed worldwide for further manufacturing of final articles (i.e., shoes, clothing, or other goods). Therefore, leather production and trade have reached a global dimension, and their environmental outcomes should be considered at that scale.

Leather industry has been typically characterized by its environmental impact due to the high loads of chemicals released, which include among others heavy metals such as trivalent chromium, acids, and bases all of them used as major reagents throughout the leather production process. More strict environmental regulations have definitely contributed to reduce such environmental burden, though there are huge differences depending on the countries.

Furthermore, there is a rising concern about the environmental fate and effects of additives, since most of the additives still remain on the final product. For example, alkylphenols used as surfactants (i.e., NP) are known to cause endocrine disruption in some aquatic species, and have been included in the list of priority compounds in Europe under the provisions of the Water Framework Directive (Directive 2008/105/EC) (Table 9). Halogenated phenols (trichlorphenol, PCP) widely used as biocides to prevent growth of organisms on skin are currently banned in developed countries due to their toxicity and carcinogenic properties. Some azo dyes classically used for obtaining leathers with exotic and bright colors have been restricted (e.g., acid red 035, acid black 077, acid orange 031, direct blue 015, or

Table 9 List of restricted aromatic amines used in the synthesis of azoic colorants according to Directive 2002/61/EC

CAS number	Substances	CAS number	Substances
92-67-1	Biphenyl-4-ylamine 4-Aminobiphenyl xenylamine	119-93-7	3,3′-Dimethylbenzidine 4,4′-Bi-*o*-toluidine
92-87-5	Benzidine	838-88-0	4,4′-Methylenedi-*o*-toluidine
95-69-2	4-Chloro-*o*-toluidine	120-71-8	6-Methoxy-*m*-toluidine *p*-cresidine
91-59-8	2-Naphthylamine	101-14-4	4,4′-Methylene-bis-(2-chloroaniline) 2,2′-Dichloro-4,4′-methylene-dianiline
97-56-3	*o*-Aminoazotoluene 4-Amino-2′,3-dimethylazobenzene 4-*o*-Tolylazo-*o*-toluidine	101-80-4	4,4′-Oxydianiline
99-55-8	5-Nitro-*o*-toluidine	139-65-1	4,4′-Thiodianiline
106-47-8	4-Chloroaniline	95-53-4	*o*-Toluidine 2-Aminotoluene
615-05-4	4-Methoxy-*m*-phenylenediamine	95-80-7	4-Methyl-*m*-phenylenediamine
101-77-9	4,4′-Methylene-dianiline 4,4′-Diaminodiphenylmethane	137-17-7	2,4,5-Trimethylaniline
91-94-1	3,3′-Dichlorobenzidine 3,3′-Dichlorobiphenyl-4,4′-ylenediamine	90-04-0	*o*-Anisidine 2-methoxyaniline
119-90-4	3,3′-Dimethoxybenzidine *o*-dianisidine	60-09-3	4-Amino azobenzene

direct red 061) because these compounds may degrade and release aromatic amines with known carcinogenic effects.

Assessing the environmental risks associated with additives along the whole product's life cycle at global scale is thus an issue of paramount relevance, which requires the development of proper methodologies. This is the main objective to be achieved by the RISKCYCLE project, a Coordination Action funded under the 7th Framework Program of the European Community.

References

1. Ramasami TS, Sreeram KJ, Gayatri R (1999) Sources of pollution in tanneries, mitigation and case studies in unit 1: legal requirements. Manual on design, operation and maintenance of tannery effluent treatment plants. RePO-UNIDO, Chennai, India
2. Integrated Pollution Prevention and Control (IPPC) (2001) European commission, reference document on best available techniques in the tanning of hides and skins
3. FAO World statistical compendium for raw hides and skins, leather and leather footwear 1990–2009. http://www.fao.org. Accessed date: 2011
4. International Trade Center (ITC) Calculations based on COMTRADE statistics. http://www.trademap.org/. Accessed date: 2011

5. Raghava Rao J et al (2003) Recouping the wastewater: a way forward for cleaner leather processing. J Clean Prod 11(5):591–599
6. Olle L, Bacardit A, Morera JM, Bartoli E, Argelich J (2008) Cross-linked polymers for aqueous finishing. Binders crosslinked with polyaziridine. Part I: behaviour of polyurethane. J Soc Leather Technol Chemists 92(3):96–102
7. Bacardit A, Olle L, Morera JM, Bartoli E, Fernandez D (2008) Crosslinked polymers for aqueous finishing. Binders crosslinked with polyaziridine. Part II: acrylic binders. J Soc Leather Technol Chemists 92(4):157–161
8. Bouchard J, Greif M (1987) Cross-linked polymers to provide better properties in leather finishing. J Am Leather Chemists Assoc 82(10):323–329
9. Wenzel W (1993) Acrylates, butadienes, and polyurethanes - the right binder for the right purpose. J Am Leather Chemists Assoc 88(9):320–325
10. Ma JZ, Lu H (2008) Elasticity studies on leather retanned with various types of acrylic polymers. J Am Leather Chemists Assoc 103(11):363–369
11. Bajza Z, Vinkovic Vrcek I (2001) Fatliquoring agent and drying temperature effects on leather properties. J Mater Sci 36(21):5265–5270
12. Santos LM, Gutterres M (2007) Reusing of a hide waste for leather fatliquoring. J Clean Prod 15(1):12–16
13. Levy J, Bagg C (1995) Problems associated with finishing hydrophobic leather. J Am Leather Chemists Assoc 90(4):118–125
14. Luo ZY, Fan HJ, Lu Y, Shi B (2008) Fluorine-containing aqueous copolymer emulsion for waterproof leather. J Soc Leather Technol Chemists 92(3):107–113
15. Gonzalez S, Petrovic M, Barcelo D (2007) Removal of a broad range of surfactants from municipal wastewater - comparison between membrane bioreactor and conventional activated sludge treatment. Chemosphere 67(2):335–343
16. Garcia-Reyero N, Requena V, Petrovic M, Fischer B, Hansen PD, Diaz A, Ventura F, Barcelo D, Pina B (2004) Estrogenic potential of halogenated derivatives of nonylphenol ethoxylates and carboxylates. Environ Toxicol Chem 23(3):705–711
17. Soares A, Guieysse B, Jefferson B, Cartmell E, Lester JN (2008) Nonylphenol in the environment: a critical review on occurrence, fate, toxicity and treatment in wastewaters. Environ Int 34(7):1033–1049
18. Commission Regulation (EC) No 552/2009 of 22 June 2009 amending Regulation (EC) No 1907/2006 of the European Parliament and of the council on the registration, evaluation, authorisation and restriction of chemicals (REACH) as regards Annex XVII
19. Ganeshjeevan R, Chandrasekar R, Sugumar P, Kadigachalam P, Radhakrishnan G (2005) Focused microwave aqueous extraction of chlorophenols from solid matrices and their analysis by chromatographic techniques. J Chromatogr A 1069(2):275–280
20. Stockholm convention on persistent organic pollutants (POPs) http://chm.pops.int/default.aspx. Accessed date: 2011
21. Yapici BM, Karaboz I (1997) The effect of two anti-fungal compounds on the growth of molds that frequently appear on tanned leather. J Am Leather Chemists Assoc 92(2):38–45
22. Vanderen JM, Weiss EF (1978) Controlling fungal growth on leather - correlation of TCMTB uptake and duration of mold resistance. J Am Leather Chemists Assoc 73(11):498–507
23. Mohamed OA, Abdel-Mohdy FA (2006) Preparation of flame-retardant leather pretreated with pyrovatex CP. J Appl Polym Sci 99(5):2039–2043
24. Donmez K, Kallenberger WE (1992) Flame resistance of Leather. J Am Leather Chemists Assoc 87(1):1–19
25. Hale RC, Alaee M, Manchester-Neesvig JB, Stapleton HM, Ikonomou MG (2003) Polybrominated diphenyl ether flame retardants in the North American environment. Environ Int 29(6):771–779
26. North KD (2004) Tracking polybrominated diphenyl ether releases in a wastewater treatment plant effluent, Palo Alto, California. Environ Sci Technol 38(17):4484–4488

27. Eljarrat E, Guerra P, Martinez E, Farre M, Alvarez JG, Lopez-Teijon M, Barcelo D (2009) Hexabromocyclododecane in human breast milk: levels and enantiomeric patterns. Environ Sci Technol 43(6):1940–1946
28. Bureau EI (2009) Tanning of hides and skins. Draft reference document on best available techniques
29. Covington AD (2009) Tanning chemistry. RSC, Cambridge, p xxxi, 483
30. Cioni F, Bartolucci G, Pieraccini G, Meloni S, Moneti G (1999) Development of a solid phase microextraction method for detection of the use of banned azo dyes in coloured textiles and leather. Rapid Commun Mass Spectrom 13(18):1833–1837
31. Sparr Eskilsson C, Davidsson R, Mathiasson L (2002) Harmful azo colorants in leather: Determination based on their cleavage and extraction of corresponding carcinogenic aromatic amines using modern extraction techniques. Journal of Chromatography A 2002, 955(2): 215–227
32. Ellis MK, McPherson GJ, Ashcroft N (1997) Investigation of the reductive metabolism of the azo dye 'Coriacide Black SB' in rats. J Soc Leather Technol Chemists 81(2):52–56
33. Yilmaz Ozmen E, Erdemir S, Yilmaz M, Bahadir M (2007) Removal of carcinogenic direct azo dyes from aqueous solutions using calix[n]arene derivatives. Clean – Soil, Air, Water 35(6): 612–616
34. Pielesz A, Baranowska I, Rybak A, Wlochowicz A (2002) Detection and determination of aromatic amines as products of reductive splitting from selected Azo dyes. Ecotoxicol Environ Saf 53(1):42–47
35. Rao JR, Prakash A, Thangaraj E, Sreeram KJ, Saravanabhavan S, Nair BU (2008) Natural dyeing of leathers using natural materials. J Am Leather Chemists Assoc 103(2):68–75
36. Gill CJ (1991) Low VOC topcoats for the leather industry. J Am Leather Chemists Assoc 456, 86(2):42–48

Additives and Other Hazardous Compounds in Electronic Products and Their Waste

Jenny Westerdahl, Mohammed Belhaj, Tomas Rydberg, John Munthe, Rosa Mari Darbra, Alba Àgueda, Susanne Heise, and Lou Ziyang

Abstract The demand for electronic equipment in society is increasing not only as a result of higher living standards around the world but also due to fashion. Many electronic articles are today disposed of before the end of their technical lifetime since they have become outdated. Each year, electronic products are sold for a value of more than $1 trillion. In electronic equipment, there are various compounds that are hazardous to both the environment and human health, such as various metals and organic compounds. These compounds may be emitted from the products during its life cycle. The end-of-life phase has been identified as problematic with respect to emissions of these potentially hazardous additives. The risk caused by the end-of-life treatment of electronic and electric waste can be minimized if treated under controlled condition. If the treatment is under uncontrolled conditions, as in the informal e-waste system in Asia and Africa, there is a large risk that negative effects will occur with regard to human health and the environment.

Keywords Electronic products, WEEE, ROHS, Additives, Hazardous substances, TPP, PBDE, Pb, Risks

J. Westerdahl (✉), M. Belhaj, T. Rydberg and J. Munthe
IVL Swedish Environmental Research Institute, Box 5302, 400 14 Göteborg, Sweden
e-mail: jenny.westerdahl@ivl.se

R.M. Darbra and A. Àgueda
CERTEC, Department of Chemical Engineering, Universitat Politècnica de Catalunya, ETSEIB, Diagonal 647, pav. G, 08028, Barcelona, Spain

S. Heise
Faculty Life Sciences, Hamburg University of Applied Sciences, Lohbrügger Kirchstraße 65, 21033 Hamburg, Germany

L. Ziyang
School of Environmental Science and Engineering, Shanghai Jiao Tong University, 800 Dongchuan Road, Shanghai 200240, PR China

B. Bilitewski et al. (eds.), *Global Risk-Based Management of Chemical Additives I: Production, Usage and Environmental Occurrence*, Hdb Env Chem (2012) 18: 57–82, DOI 10.1007/698_2011_114,
Published online: 6 August 2011

Contents

Abbreviations

CRT	Cathode ray tube
IDG	International Data Group
ITU	International Telecommunication Union
LCD	Liquid crystal display
MCV	Maximum concentration value
MPPI	Mobile Phone Partnership Initiative
PBDE	Polybrominated diphenyl ether
TPP	Triphenyl phosphate
UNEP	United Nations Environment Programme

1 Introduction

Electronic equipment consists of a wide range of products such as mobile phones, computers, television sets, music players, and toys. Here, electronic equipment is defined as products composed of different components that are interconnected, which together fulfill certain functions. Common for all of these products is that they contain printed circuit boards (PC-boards) that transmit and/or process analog and digital data.

The demand for electronic equipment in society is currently increasing. As a result, there are large global flows of electronic equipment, both of the products itself and of the electronic waste.

These products contain many different materials and additives that are used to give the products the required characteristics. These materials and additives can leach from the product during different stages of the product's life cycle and thereby enter into the environment, where they can pose a risk to human health and the environment. Here, three different electronic products have been studied in more detail, namely televisions, computers, and mobile phones. These products were selected since they are sold in large quantities and thereby constitute a large fraction of the electronic equipment currently on the market [1].

In this chapter, available information regarding global flows of the three selected electronic products, their material and additive content, and how electronic waste is treated has been compiled in order to identify risks caused by emissions of hazardous materials and additives from electronic equipment.

2 Global Flows of Electronic Equipment

During the last years, the demand for electronic equipment has increased and is predicted to keep increasing [2]. This is partly a response to higher living standards around the world, but also due to fashion, which is especially true for mobile phones and computers. Technological advances during recent years have caused older products to quickly become outdated. Many of these products are therefore disposed of before the end of their technical lifetime. At the same time, the lifespan of electronics shortens.

It is assessed that electronic equipment is sold for an estimated value of more than $1 trillion each year, thereby contributing substantially to the world economy [2].

2.1 Production and Trade of Electronic Equipment

2.1.1 Televisions

In a study by the International Telecommunication Union [3], it was concluded that in 2009, approximately 1.4 billion households in the world had one or more TVs. That corresponds to 78% of all households in the world. Many households in the industrialized countries have more than one TV. In United Kingdom, there were 25.6 million households with a TV in 2008, but the total number of TVs that same year was 60 million [4]. That gives an average of more than two TVs in each household with TV access. In 2003, the total number of TV sets in the world was close to 2 billion. Since then, the number of households with a TV has increased by approximately 5%.

Of the two main types of TVs (cathode ray tube (CRT) and flat panel displays), flat panel displays dominate the sales today. However, there is still a large amount of CRT TVs in use around the world. In 2009, approximately 210 million TVs were sold globally [5]. Of these, approximately 75% were flat panel displays.

2.1.2 Computers

The current fraction of households that have access to Internet and the fraction of Internet users both were approximately 30% worldwide in 2009 [6]. The number of computers in the world was approximately 0.8 billion in 2005. Since then, the number of people with a computer has increased. In 2009, approximately 300 million computers were sold [7] of which approximately 170 million were laptop computers [8]. According to Cobbing [9], the worldwide sales of computers were approximately 230 million units in 2006. During the last decade, there has been a shift in technology, now laptop computers dominate the sales of new computers in the world.

2.1.3 Mobile Phones

According to studies by ITU [3], there were approximately 4.5 billion mobile phone subscriptions worldwide in 2009. As an estimate, it can be assumed that there is approximately one mobile phone to each subscription, which would equal 4.5 billion mobile phones in use worldwide. The number of mobile phone subscriptions is expected to increase in the future; even though the increase rate may decrease compared to the previous decade, since the worldwide mobile phone market is getting closer to saturation. The mobile phone penetration differs between different regions of the world. In Europe, there is more than one mobile phone subscription per inhabitant, whereas in Africa only a third of the inhabitants have a mobile phone subscription [3].

In the industrialized countries, the average expected life span of a mobile phone is very short, usually less than two years [10]. Due to the increasing demand and the short life span of mobile phones, a large quantity of mobile phones is produced each year. According to yearly sales statistics from 2009, approximately 1.1 to 1.4 billion mobile phones were shipped globally [11, 12]. Mobile phone sales in 2006 were approximately 1 billion units according to Cobbing [9].

2.2 *Electronic Waste*

When electronic equipment is disposed of by their users, it enters into the waste stream. A common name for waste of electronic equipment is e-waste. Another term that covers both electrical and electronic waste is waste of electric and electronic equipment (WEEE). It has been estimated that 20–50 million tons of WEEE are generated annually worldwide, and the flows of e-waste are continuously increasing [10]. In the future, it is expected that the WEEE will increase about 3–5% annually, making e-waste the fastest growing waste stream in society.

The content of electronics waste can be very diverse, partly not only due to the many products included in the term electronics, but also due to the development of electronics over time, since the electronics waste stream consists of both old and new products. In Fig. 1, the global composition (weight %) of WEEE with regard to different product categories in the year 2005 is shown. There it can be seen that large household appliances, such as refrigerators and washing machines, dominate the waste stream, which is mainly due to the high weight of these products. Two other large fractions of waste come from consumer equipment and information technology (IT) and telecommunications equipment. The waste from computers

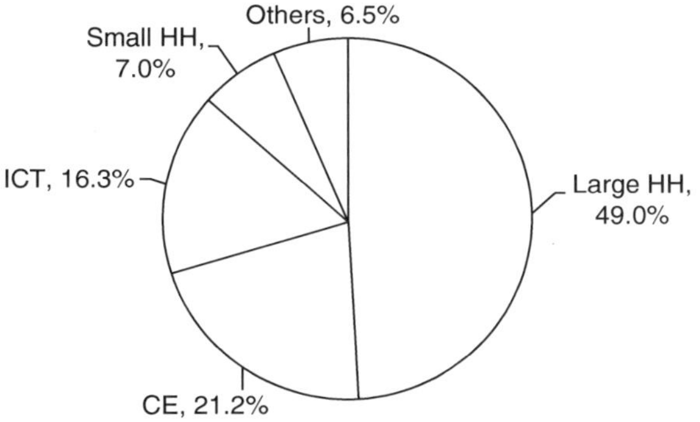

Fig. 1 Composition of WEEE in 2005. *Large HH* large household appliances, *Small HH* small household appliances, *CE* consumer equipment, *ICT* IT and telecommunications equipment. Source: [1]

and mobile phones is included in the category IT and telecommunications equipment while TVs belong to the category consumer equipment.

Of the estimated yearly amount of WEEE that is generated, i.e., 20–50 million tons [10], less than 10% is collected and sent to authorized recycling facilities, where hazardous and valuable materials are separated and treated under controlled conditions [2]. For the fate of the remaining part of the waste stream, no precise data is available [9]. It is believed that the WEEE is possibly stored, disposed of in other ways, for example together with municipal solid waste, or exported to be reused, recycled, or disposed of in other countries. In EU, the export of e-waste to developing countries is illegal. The Waste Shipment Regulation states that it is prohibited to export hazardous waste from OECD countries to non-OECD countries for recycling [9]. In the US, export of e-waste is still allowed, since they have not yet ratified the Basel Convention.

The fraction of e-waste that is disposed of together with municipal solid waste is either incinerated or landfilled. The fraction that is exported often ends up in Asian countries, such as India or China, or in Africa [9]. There, the waste is handled within the informal e-waste recycling industry. Many people earn their income only from the informal recycling industry, where the main goal is to recover valuable substances from the waste [13].

3 Material Composition of Electronic Equipment

Since there are a wide variety of products included in the concept electronics, the material composition of electronics as a group can be very complex. The main materials used in electrical and electronic equipment are metals, both ferrous and nonferrous, plastics, glass, and wood; see Fig. 2 [14]. A large number of different metals can be found in electronics. Many of these metal compounds are used in small amounts for doping materials to create semiconductors.

Additives are added to the components and materials used in electronic equipment to give them the required properties. The functions of the additives are diverse: Additives are, for example, used to make the materials able to withstand degradation, to become conductive, to make them flame retardant, or to make them flexible.

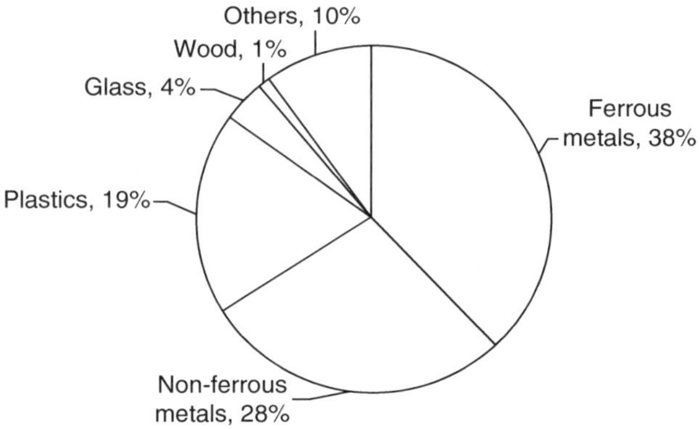

Fig. 2 Material composition of electric and electronic equipment (EEE). Source: Cui and Forssberg [14]

Some of the substances found in electronics, for example many of the metals, can be used both as additives and (more extensively) as materials in the different components.

In Sect. 3.1, the material composition of electronic equipment is described in detail. In Sect. 3.2, some of the additives that can be present in the materials constituting electronic equipment are explained.

3.1 Components and Materials in Electronic Equipment

In the following section, the material composition of three typical electronic products, televisions, computers, and mobile phones, is described on a component level. The data presented here are average data; there could be differences in the material composition of these electronic products depending on manufacturer and model of the product.

3.1.1 Televisions

There are two main types of TVs in use in today's society, CRTs and flat panel displays, such as liquid crystal displays (LCDs) and plasma TVs. Previously, CRTs dominated the market, but during the last decade, flat panel displays have increased in sales and are now sold in larger quantities than CRTs [5]. The material compositions of these two types of TVs differ to a great extent as a result of the different technologies used. The main components of a CRT TV are the cathode ray tube, which generates the picture, PC-boards, and wires as well as a plastic casing. In an LCD TV, the main components are the LCD module containing the liquid crystal display, PC-boards, light guides, etc., which together generate the picture. Other major components are frames, a stand, and a plastic casing. In Tables 1 and 2 below, the material compositions of a CRT and an LCD TV are described on a component level.

3.1.2 Computers

As for the TVs, there are two main types of computers, stationary desktop computers and laptop computers. Generally, laptop computers are smaller and

Table 1 Material composition of the components found in a CRT television (italic text represents information regarding subcomponents)

Component	Amount (w%)	Material
Case	18	Plastics (PC, ABS)
CRT explosion-protection unit	2	Fe
CRT unit	48	
Shadow mask	*4*	*Steel*
Panel glass	*28*	*Glass, Si, K, Na, Ba, Sr*
Funnel glass	*15*	*Glass, Si, Pb, K, Na, Al, Ca*
Electron gun	*1*	*Steel, glass, Cu, plastics*
Yoke	5	Cu, plastics, Fe
Metal parts	5	Fe
PC-boards	14	Flame retarded epoxy resin, ceramics, Cu, Fe, Sn, Ag
Wires	5	Plastics (PVC, thermoset plastics), Cu
Rubber parts	1	Rubber
Other plastic parts	2	Plastics

Sources: Lee et al. [15] and Méar et al. [16]

Table 2 Material composition of the components found in an LCD television (italic text represents information regarding subcomponents)

Component	Amount (w%)	Material
Case	18	Plastics (PC, ABS)
LCD module	42	
LCD panel	*19*	*Glass, LCD polymer, In, Sn*
Backlight films and light guide	*47*	*Plastics (PET, PE, PMMA, PC, polyester, polyacrylate)*
Inner frames	*2*	*Al, plastics (PC)*
Back plate	*31*	*Al, plastics (PC)*
Fluorescent lamp	*1*	*Glass, Hg*
Frames	11	Al, Fe, plastics (PC)
Stand	16	Al, Fe, plastics (PC, ABS)
Wires	4	Plastics (PVC, thermoset plastics), Cu
PC-boards	10	Flame retarded epoxy resin, ceramics, Cu, Sn, Ag, Zn

Sources: Felix and Letcher [17]

weigh less than the desktop computers. The main difference in material composition between these two types of computers is that laptops contain a larger fraction of PC-boards and other electronic components than a desktop. A desktop instead contains larger fractions of metals and plastics. Also, a desktop computer can use either a CRT display or a flat panel display, whereas all laptops use flat panel displays. In Table 3 below, the material composition of a laptop is shown.

3.1.3 Mobile Phones

The main components of a mobile phone are a handset consisting of a case, a display with a glass cover and a keypad, a printed circuit board with integrated chips, resistors, capacitors and wires, a battery, a microphone, and a speaker [19]. The major materials used in a mobile phone are plastics (40%), nonferrous metals (37%), glass and ceramics (12%), epoxy (6%), ferrous metal (3%), and other materials (2%) [19, 20]. Table 4 below presents a detailed description of the material composition of a mobile phone.

3.2 Additives in Electronic Equipment

Additives are added to the components within electrical and electronic applications in order to give the products certain desired properties. Their function is diverse; there are e.g., stabilizers, colorants, flame retardants, plasticizers, and conductive materials.

Table 3 Material composition of the components found in a laptop computer (italic text represents information regarding subcomponents)

Component	Amount (w%)	Material
Case	30	Plastics (PC, ABS)
LCD module	16	
LCD panel	*19*	*Glass, LCD polymer, In, Sn*
Backlight films and light guide	*47*	*Plastics (PET, PE, PMMA, PC, polyester, polyacrylate)*
Inner frames	*2*	*Al, plastics (PC)*
Back plate	*31*	*Al, plastics (PC)*
Fluorescent lamp	*1*	*Glass, Hg*
Frames	28	Al, Fe, plastics (PC)
Battery (NiCd, NiMh or Li-ion battery)	13	Ni, KOH, Cd, Co, Li, F
PC-boards	12	Flame retarded epoxy resin, ceramics, Cu, Sn, Ag, Zn
Wires	1	Plastics (PVC, thermoset plastics), Cu

Sources: Felix and Letcher [17] and Liang [18]

Table 4 Material composition of the components found in a mobile phone (italic text represents information regarding subcomponents)

Component	Amount (w%)	Material
Handset	50	
Case	*35*[a]	*Plastics (PC, PC/ABS), Ti, Cr*
Display	*50*[a]	*Glass, LCD polymers*
Keypad	*15*[a]	*Plastics, Ag*
Frames	5	Al, Fe, Cr, Ti, Mg, plastics (melamine)
PC-board	15	Flame retarded epoxy resin, ceramics, Plastics (PVC, thermoset plastics) Cu, Sn, Br, Mn, Ag, Ta, W, Zn
Battery (NiCd, NiMh or Li-ion battery)	20	Ni, KOH, Cd, Co, Li, F

[a]There can be large variations in the weight of the subcomponents in the handset due to differences between different manufacturers and models of mobile phones
Source: [19]

Additives are continuously improved – either in their composition, concentration, or way of application in order to fit the increasing need for consistent, well-dispersed additives in electrical and electronic applications, in which these additives should show low migration even under high temperature [21].

In the following sections, information regarding some different types of additives and their use in electronic equipment is presented.

3.2.1 Stabilizers

Polymers such as PVC react with atmospheric oxygen, forming free radicals in the process, which can degrade the polymer chains. Antioxidants are added to the PVC as stabilizers to prolong the product life by protecting the polymer from these radicals. To fulfill this function, the additive needs to be adapted to long-term use and should migrate very little under high temperature conditions [21]. The most common stabilizers used in PVC are lead, mixed metal, and organotin stabilizers. PVC is often used in electronic equipment for cables and wires. In most of these PVC wires and cables, lead-based stabilizers are now being phased out. More information about the phase out of lead can be found in Sect. 4.2.1.

3.2.2 Flame Retardants

Flame retardants are another type of additive that is widely used in plastics of electronic products. They are of increasing demand, due to the growing demand for electrical and electronic applications, current fire safety regulations, and miniaturization of electronics. In miniaturized and more powerful products, the individual circuits are more densely packed and thus lead to increased heat build-up, while at the same time integrated fans need to become smaller.

Additives used as flame retardants can be assigned to 4 different groups: (1) halogenated organic compounds such as hexabromocyclododecane (HBCDD) or polybrominated diphenyl ethers (PBDEs), (2) organic phosphorous compounds such as triphenyl phosphate (TPP) and phosphonate, (3) halogenated organic phosphorous compounds such as tri-(2,3-dibromopropyl)-phosphate, and (4) specific inorganic compounds such as aluminum trihydrate [22, 23].

The flame retarding effect of halogenated compounds results from halogen atoms, which are released in the fire due to the high temperature and which catch the radicals fuelling the fire. Other substances decompose with an endothermic reaction, e.g., by freeing water and thus reducing the temperature of the fire (e.g., alumina trihydrate). Substances such as TPP flame retard the plastic by forming char and thus create a barrier against heat and mass transfer [23–25].

Flame retardants are added to most plastics, and thus their production and consumption is high; in Germany, 100,000 tons of flame retardants were used in 1999, of which 30% were halogen-based [26]. Of these substances, 83% was used for electric devices [22], where components such as the casings and the PC-boards are most often flame retarded.

There are trends to restrict and limit the use of halogenated flame retardants due to environmental concerns, and some brominated compounds are forbidden in Europe. One way to reduce the amount of halogen flame retardants in the products is by using synergistic additives such as antimony oxide, alumina trihydrate, and magnesium hydroxide. However, alternative compounds, such as phosphate-based additives, are currently more expensive than halogenated flame retardants; they need to be added in larger volumes and are sometimes less efficient and partly not applicable with the polymers involved [21, 23].

3.2.3 Plasticizers

Plasticizers are used to increase the flexibility of a plastic polymer. One polymer in which plasticizers are commonly used is PVC, which is used for cables and wires in electrical and electronic equipment. Phthalates and phosphates have historically been two commonly used types of plasticizers, where bis(2-ethylhexyl)phthalate (DEHP) is the most commonly used substance. A functional disadvantage with DEHP is its high flammability, which often leads to an increased use of flame retardants in the product. Even though volatile plasticizers are more and more being replaced by larger molecules, DEHP may persist for economic reasons [22]. Some of the electronic producers have, however, banned the use of phthalates from their products due to environmental concerns [27, 28].

3.2.4 Conductive Additives

Additives are also used to provide different amounts of conductivity in electronics ranging from dissipating static charge to shielding from electromagnetic interference. Beryllium is an element, which has been used traditionally in motherboards in form of copper-beryllium alloys, since it has a good thermal and high electric conductivity [29]. With electronics becoming smaller in size, requirements for a lower static decay rate (which measures how quickly a charge is dissipated) and a minimum residual voltage become more pronounced. Carbon blacks have been used for about 20 years for electrostatic dissipation but some problems came up with regard to their impact on the mechanical properties of the product. New compounds of increasing use are carbon nanotubes. They have a minimal impact on the physical properties of polymers due to their size and the high efficiency [30]. However, concern regarding the toxicity of these carbon nanotubes has been raised [31].

3.2.5 Colorants

Other additives are pigments for colorization of polymers. Especially in electrical and electronic application, the persistence of these so-called colorants is crucial, as colors of wires and cables are used for identification and thus must not fade with time. In order to gain colors ranging from yellow to dark red, cadmium compounds were often used, partly mixed with other substances such as ZnS, $BaSO_4$, or HgS pigments that firmly bound within the plastics material and could not be separated [22]. While Dunning [32] estimated the world consumption of cadmium pigments to be approximately 2,500 tonnes in 2009 with 90% being used for polymers, European legislation has tried to reduce its application extensively (see Sect. 4.2.1).

4 Legislation with Regard to Hazardous Material in Electronics

4.1 Legislation

In February 2003, two intertwined legislations came into force in Europe, which had a big impact on the production of electronic and electrical appliances: the EU legislation restricting the use of hazardous substances in electrical and electronic equipment (Directive 2002/95/EC),[1] known as the "RoHS-Directive," and the "WEEE directive" on promoting the collection and recycling of such equipment (Directive 2002/96/EC).[2] The WEEE Directive aims at minimizing electrical and electronic waste by prevention, recycling, recovery, and re-use. The RoHS-Directive

[1] Directive 2002/95/EC of the European Parliament and of the Council on the Restriction of the Use of Certain Hazardous Substances in Electrical and Electronic Equipment.

[2] Directive 2002/96/EC of the European Parliament and of the Council on Waste Electrical and Electronic Equipment (WEEE).

Table 5 Maximum concentration values (MCVs) for homogenous materials in new Electrical and Electronic equipment as stated in the RoHS Directive

	Cd	Cr (VI)	Hg	Pb	PBB	PBDEs
MCV (weight %)	0.01	0.1	0.1	0.1	0.1	0.1

Source: Wäger et al. [33]

requires phasing out of specified hazardous substances from such products: heavy metals such as lead, mercury, cadmium, and hexavalent chromium and flame retardants such as polybrominated biphenyls (PBB) or polybrominated diphenyl ethers (PBDE) are to be substituted by safer alternatives. Member States are thus expected to make sure that from July 1, 2006 newly market EEE shall not contain any of these substances in concentrations above defined maximum concentration values (MCV) for homogenous material (see Table 5) unless specific exemptions apply. The flame retardant hexabromocyclododecane (HBCDD) and the plasticizers bis(2-ethylhexyl)phthalate (DEHP), butylbenzylphthalate (BBP) and dibutylphthalate (DBP) are discussed for future inclusion in the list of hazardous substances [34]. The German Federal Environment Agency also recommends the restriction of tetrabromobisphenol A (TBBPA) used as an additive [35].

Wäger et al. [33] compared the MCV values for the hazardous substances as stated in the RoHS with literature data on component concentration in plastics in European WEEE. With regard to heavy metals, chromium as well as mercury was mostly below, whereas cadmium and lead were in the range of the MCV [33]. With regard to brominated flame retardants (BFR), polybrominated biphenyls (PBB) were evaluated as being well below the MCV. The PBDEs, pentaBDE, octaBDE, and decaBDE were identified as being the brominated flame retardants that possibly occur in concentrations above the MCV in relevant plastic types.

4.2 *Phase Out of Hazardous Additives*

4.2.1 Heavy Metals

The application of lead and cadmium as additives has been mainly in PVC, as stabilizers and pigments, respectively [34]. Under the Vinyl 2010 Commitment, the European PVC industry has promised to phase out lead stabilizers completely by 2015, even though they are currently still the most widely used type globally [36]. The lead stabilizers are being replaced by e.g., "mixed metal" stabilizers such as calcium/zinc stabilizers for use in rigid PVC products and liquid "mixed-metal" stabilizers such as barium/zinc stabilizers for use in flexible PVC.

Use of cadmium has been widely restricted by the REACH regulation (Regulation No. 1907/2006 for Registration, Evaluation, Authorization and Restriction of Chemicals, as amended by Commission Regulation (EC) No. 552/2009). Thus, for example, its application as a stabilizer in PVC articles for packaging and as a pigment in a number of plastics and paints has been prohibited. With the new

Table 6 Classification of additives based on health and environmental effects

Class	Substance
Potentially unproblematic	Bisphenol A diphenyl phosphate, Resorcinol bis(diphenyl phosphate), Aluminum trihydroxide, Magnesium hydroxide
Potentially problematic	Triphenyl phosphate, Tricresyl phosphate, Diphenyl cresyl phosphate, Tetrakis(hydroxymethyl) phosphonium chloride, Antimony trioxide, Boron compounds (borates other than zinc borate)
Insufficient data	Diethylphosphinic acid (aluminum salt), Melamine, Red phosphorus, Ammonium polyphosphates
Not recommended	Polytetrafluoroethylene, Zinc borate

Source: Illinois Environmental Protection Agency [38]

Commission Regulation (EU) No. 494/2011, effective January 10, 2012, the ban will be extended to all plastic products under REACH. Reacting to the pressure on phasing out hazardous substances from products, some companies have reformulated their resins to achieve high-chroma red, orange, and yellow colors without heavy metals [37].

4.2.2 Brominated Flame Retardants

There are a number of ways to reduce the risk from fire for products that do not require chemicals and for which toxicity is not a concern, such as redesigning products to be less fire-prone, use of inherently fire-resistant fibers and light-weight metals. In a report to the Governor and the General Assembly, the Illinois Environmental Protection Agency reviewed and evaluated potential alternatives to flame retardants, especially decaBDE. With regard to chemical alternatives to BFR, the substances were assigned to four different classes on the basis of identified health and environmental effects, see Table 6 [38].

The report concludes that effective alternatives exist for most of the plastics uses of decaBDE, but states that chemical alternatives tend to be more costly, even though equally effective. It also states that toxicity data gaps exist for many of the main potential alternatives and further research on environmental consequences and human health effects will be necessary. These conclusions are supported by studies of the Danish Environmental Agency [39] and the German Federal Environment Agency [40].

5 Waste Treatment of Electronic Equipment

Due to the many hazardous substances contained in e-waste, it is essential that e-waste is treated in a way that allows for hazardous substances to be removed before final disposal. By recycling e-waste, the hazardous substances can be separated from the waste stream and treated in an appropriate manner, thereby minimizing the risk of adverse effects to the environment and to human health. During the recycling process, valuable components and substances can also be collected, which can reduce the need for extracting new raw materials for electronics production [13].

There is both controlled treatment of e-waste in authorized recycling facilities in the developed countries as well as uncontrolled treatment within the informal recycling industry in Asia and Africa. The main difference between these two systems is that within the controlled treatment, emissions to the environment and exposure of hazardous substances in the working environment are limited, whereas no such measures are taken within the uncontrolled treatment. In the following sections, the recycling processes within these two systems are described in more detail.

5.1 Controlled Treatment

There are four general types of processes involved in the controlled recycling of e-waste. Those are manual dismantling and sorting, mechanical shredding and separation, metallurgical processes, and recycling of plastics.

The first step of the controlled recycling process is the manual dismantling and sorting. In this process, whole components that are reusable, recyclable, or valuable are recovered from the waste stream. Components containing hazardous substances are also recovered from the waste for further special treatment. Examples of such components are batteries, mercury-containing components, CRT-glass, and LCDs [14]. The components and materials aimed for refinement are then shredded to reduce the size of the components/materials. After shredding, the materials are sorted in different fractions based on weight, size, shape, density, and electrical and magnetic characteristics [13]. The final output streams from these two processes are often a magnetic fraction, an aluminum fraction, a copper fraction, various plastic fractions, and whole components for reuse or further treatment. Residues from the recycling process are often incinerated or landfilled.

To further upgrade and refine the material fractions obtained from the two previous processes, metallurgical processes can be used [13]. Here, two main types of processes can be applied, pyrometallurgical processes where the metals are melted and/or hydrometallurgical processes in which the metals are dissolved. Often, the hydrometallurgical process is used for the final refining of the material, following the pyrometallurgical processes. The metallurgical processes can be performed by the metallurgical industry, which normally extract metals from ores. There, the e-waste can be used as feedstock for the processes. The plastic fractions can also be further refined. In this process, the fractions are further separated, using, for example, sieves and density separation. Paints and coatings are also removed using further grinding, solvent stripping, high temperature washing, etc. The final output fractions from the recycling can then be sold and the materials can be turned into new products. However, due to the unknown additive composition of inhomogeneous recycled plastics, high quality use of the recycled plastics is often hindered. Today, a large share of the plastics is instead incinerated [41].

5.2 Uncontrolled Treatment

In the informal recycling industry, the processes are not as standardized. Still, it is possible to give a general description of the processes.

As in the controlled recycling process, the first step is manual disassembly. The components that are retrieved for further recycling are mainly materials containing copper, steel, plastics, aluminum, printer toners, and PC-boards [42]. These components are then further treated to recover the valuable substances in them. CRTs are often broken to remove the copper yokes, which are then collected for further treatment and copper recovery. Solder and individual components on PC-boards are recovered by heating the boards and thereby melting the solder. After the melting of the solder, the components can be removed and the melted solder can be collected.

To recover metals, primitive metallurgical processes are used, where PC-boards and other components are dissolved in strong acid solutions that are heated over small fires [43]. To dissolve the metals, the acid solution is swirled for hours by the workers. To recover metals that are encased in plastics, such as wires, the e-waste is often burned in open fires. Plastic materials are also recycled. As for the controlled recycling, the plastics are shredded into smaller pieces. After that, the fragments are separated manually, for example, by color, followed by further grinding. The plastic can then be melted and extruded to be used in new products. The fraction of the e-waste that cannot be recycled is often dumped directly into the environment.

The people working in the informal recycling industry generally do not have any respiratory protection when working and the ventilation is often insufficient.

6 Risks from Treatment of Electronic Waste with Special Consideration of Additives

6.1 Emission and Exposure

During the recycling and waste treatment of electronic equipment, many of the additives in the equipment can potentially be emitted to the environment and pose a risk to both the environment and human health. Generally, these risks are lower when the e-waste is treated under controlled conditions than under uncontrolled conditions. In the following sections, emissions and exposure associated with treatment of electronic waste are described.

6.1.1 Traditional Waste Treatment

When electronic equipment is disposed of together with municipal solid waste, there are no specific measures taken to limit the emissions of hazardous substances found in electronic equipment specifically.

When electronic equipment is landfilled, there is a risk of hazardous substances leaching or evaporating from the landfill [13]. Additives such as brominated flame retardants, phthalate plasticizers, Pb and Cd, as well as other metals such as Cu, Ni, Sb, and Zn are likely to be found in leachate from landfills. Volatile substances such as metallic or methylated mercury species are instead evaporated and can be found in landfill gas [44]. When the e-waste is incinerated in a municipal solid waste incinerator, various metals (e.g., Cu, Pb, and Sb) and organic compounds (e.g., dioxins and PAHs) are emitted via the exhaust gases [13]. Dioxins are created when substances such as halogenated flame retardants are incinerated. The incineration of PVC plastic, which is often used in electronics, also contributes to the formation of dioxins. However, emissions of dioxins and other pollutants can be minimized in modern waste incinerators by the use of post combustion steps and flue gas treatment systems.

6.1.2 Controlled Recycling

When dismantling the electronic equipment, the workers can be exposed not only to dust containing Pb and Ba-oxide from broken CRTs but also to dust gathered in the electronic equipment, which can contain various compounds [13]. There are also risks of cuts from glass in the waste as well as emissions of volatile compounds such as mercury from broken components. During shredding, pyrometallurgical processes and plastics recycling, dust and fumes are generated. They may contain various metals and substances such as brominated flame retardants and phthalates. During the hydrometallurgical processing, emissions of acid fumes may pose a risk to the workers. These emissions can cause both occupational and environmental risks.

6.1.3 Uncontrolled Recycling

During the uncontrolled recycling of e-waste, many of the substances emitted are the same as for the recycling under controlled conditions. The difference is that during the uncontrolled recycling, there are no measures taken to reduce the emissions, no appropriate treatment of hazardous substances that are disposed of and no protective equipment for the workers. As in the controlled recycling, dismantling can expose the workers not only to dust containing Pb, Ba-oxide, but also to mercury from broken components. Shredding and plastic recycling generates dust and fumes that contain various metals (e.g., Pb, Zn, Cu, Sn, Sb, Cd, Ni, and Hg) and organic compounds [13]. These substances are emitted directly to the surrounding environment. When recycling the PC-boards, workers are directly exposed to fumes from the solder and other components such as flame retardants in the PC-boards. Since this process is often performed over an open fire, the dioxins that are generated during the incineration are emitted to the environment. When recovering metals by acid extraction, the workers are exposed to acid fumes during the process [45]. Burning or dumping of residual materials may also cause serious environmental impacts. As for the burning of PC-boards, burning of residual materials generate emissions of metals and organic compounds, especially dioxins, that are emitted to the air and soil. Waste that is dumped directly on the ground contributes to leaching of metals and organic substances to the surrounding soil, groundwater, and surface water. All of these emissions can cause serious risks both to the environment and to the health of humans working at or living close to the facilities. Some of these emissions even contribute to the global contamination of hazardous substances.

6.1.4 Release of Additives to the Environment During Recycling

E-waste can contain heavy metals and PBDE, which can be released by leaching when disposed of in landfills, especially in those landfills, which do not have a protective liner. In that case – depending on pH, organic matter content, temperature, adsorption – desorption processes, complexation, uptake by biota, degradation processes, and the intrinsic chemical properties of a substance, these hazardous substances can be transported by rainwater to other aquatic systems or end up in soil. For a review on environmental fate of hazardous substances from EEE during recycling, see Sepúlveda et al. [46] and Tsydenova & Bengtsson [47]. Whereas the selected additives discussed in this paper in principal have a low solubility, lead ions and lower brominated congeners of flame retardants are mobile to some extent and will be more bioavailable in the dissolved phase (or in the case of methylated mercury – in the gas phase) whereby incineration may increase the mobility of heavy metals, and particularly of Pb [48]. When these additives are transported in a less soluble form, e.g., bound to particles, they may accumulate in soil and or sediment and thus enter the food chain or are metabolized. This is especially a problem with mercury, which can be methylated by bacteria to methyl-Hg, which is much more bioavailable than the elemental form.

Dismantling activities release dust particles that are loaded with heavy metals, organophosphates [49], and flame retardants [50] into the atmosphere. They can undergo wet or dry deposition and settle within a certain distance from the recycling location depending on their size.

Thermal processes have additional hazardous effects, as they can lead to the formation of polybrominated and polychlorinated dioxins and furans from brominated flame retardants. Dioxins are among the most toxic substances that humankind produces, see Watson et al. [51] for review.

6.2 *Effects and Impacts*

In this section, human health and environmental effects caused by three different additives used in electronic equipment (lead, polybrominated diphenyl ethers, and triphenyl phosphate) are presented. It is important to highlight that lead can be used as an additive or as a component in electronic and electrical devices. In any case, in this section the effects of this substance on human health and environment are discussed. Lead and polybrominated diphenyl ethers (PBDE) were chosen for further study as they most likely persist to exceed the maximum concentration values in those electronic products that are currently part of the global trade. In addition, triphenyl phosphate has been addressed as another flame retardant with a potential to replace halogenated compounds.

6.2.1 Human Health Effects

Lead

Several studies on lead (Pb) levels in human populations have been carried out during these last years [52–54]. A high percentage of these recent studies have been developed based on statistics from the Chinese population, including two reviews of blood levels in Chinese children and suggesting that lead health problems constitute a rising problem in this country [55, 56].

The adverse effects associated with elevated Pb concentrations have been widely studied and documented. For example, exposure to lead can induce anemia, irreversible nephrotoxic effects, and depression of thyroid activity. A comprehensive report about the toxicological effects of Pb on humans is available in Tukker et al. [57]. The US EPA has established a reference dose for Pb of 0.1 mg kg^{-1} day^{-1} (Korre et al. 2002).

PBDE

PBDE levels in humans have been studied for the last years. Only several PBDE congeners tend to be found in this type of measurements. These are BDEs 47, 99, 100, 153, 154, being BDE-47 the dominant. On the contrary, the congener BDE-209 seems to be dominant in measurements of house dust and indoor air [58].

Although no studies have been done on the effects of PBDEs on humans, laboratory studies suggest that PBDEs may affect human health. The toxic endpoints likely to be the most sensitive for the PBDEs are thyroid hormone disruption and neurobehavioral toxicity. According to several reports of the US EPA [59–62], there is some lack of reliable information about the potential carcinogenicity and the chronic toxicity of BDEs 47, 99, and 153. However, there is a suggestive evidence of carcinogenic potential for BDE-209 and endocrine disrupting effects have been observed for BDE-209 as well. The US EPA has established reference doses for BDEs 47, 99, and 153 (0.1 μg kg^{-1} per^{-1} day) and for BDE-209 (7 μg kg^{-1} day^{-1} -for noncancer/oral exposures-), but the confidence level associated with these values is low.

TPP

Humans may be exposed to TPP through the environment, from the use of consumer products and at workplace. Several important issues about the potential health effects of TPP are pointed out in the report of the OECD [63], available information being based on data obtained from laboratory animals, such as rats and hens. For example, according to this report, there are no long-term carcinogenicity bioassays available and a Non-Observed Effect Level (NOEL) value of 161 mg kg^{-1} day^{-1} has been deduced from rat data.

Moreover, according to the information contained in the IUCLID Dataset of the European Commission (ECB, 2000), the minimum LD_{50} value reported for TPP is 1,300 mg/kg. For comparison purposes, this value is 260 times higher than the LD_{50} value reported for mice from mercuric species (5 mg/kg) [64].

6.2.2 Environmental Effects

Lead

Lead can complex with electron-rich ligands in many organic compounds, including amino acids in proteins. Low concentrations of Pb^{2+} inhibit enzymes. Lead is chemically similar to nutrient minerals with a 2+ charge, including iron, calcium, and zinc [65]. There is strong evidence for fish that lead leads to ionoregulatory distress Pb seems to be highly capable of outcompeting Na^{+} and Ca^{2+} for ion transport sites at the freshwater fish gill [66]. It may also interfere with normal physiological functions by displacing elements from their normally occupied metabolic sites, and it interferes with synaptic transmission in the peripheral nervous system [65].

Currently, there is no evidence that lead is biomagnified in higher trophic levels of the aquatic food chain. Biomagnification seems to be limited to the phytoplankton or to filtering organisms such as mussels and oysters.

Predicted no effect concentrations (PNECs), below which no risk to life forms is expected, have been calculated as 14.5 μg/l for freshwater and 8.3 μg/l for marine waters. For sediments, a value of 53.4 mg Pb/kg sediment dry wt has been proposed [65].

PBDE

Accumulated amounts in electric products still in use in 1994 in the Nordic countries were estimated to be 250 metric tons of pentaBDE and 5,500 metric tons of octaBDE and decaBDE combined [67]. PBDEs tend to be stable and persistent in nature. They are often associated with soils and sediments due to their high hydrophobicity and relatively low volatility. However, air and water particulate phases constitute important transport media for the dispersion of these contaminants on local, regional, and global scales. According to Watanabe and Sakai [68], the lower brominated PBDEs are more volatile, water soluble and bioaccumulate more than higher brominated PBDEs. Many congeners have been found to accumulate in living organisms and biomagnify in food chains.

Reviews on brominated flame retardants in the environment have been given by de Wit [69] and Yogui and Sericano [70]. PNEC values for environmental matrices have so far not been derived.

TPP

Taking into account a vapor pressure of 8.35×10^{-4} Pa, a water solubility of 1.9 mg/l and a log K_{ow} of 4.6 for triphenyl phosphate (TPP), the following environmental distribution has been calculated (Mackay Level 1 model): 43.9% to soil, 41.0% to sediment, 14.3% to water, 0.7% to air, 0.07% to suspended matter, and 0.03% to biota. Thus, TPP is mainly distributed to soil and sediment [71].

On the basis of the experimental results included the IUCLID Dataset of the European Commission (ECB, 2000), a moderate bioaccumulation potential for TPP in fish can be assumed. The maximum bioconcentration factor (BCF) reported there is 271, which is 37 times lower than the lowest BCF of organic mercury reported for aquatic species (range: 10,000–85,700) by the US Agency for Toxic Substances and Disease Registry [72].

Risk limits for TPP for the compartments water, soil, and sediment on the basis of ecotoxicological data for the aquatic environment have been established by the Dutch National Institute for Public Health and the Environment [73] in form of Maximum Permissible Concentrations (MPCs). The MPCs in surface water and in soil/sediments are, 0.17 μg/l and 95 $\mu g/kg_{dw}$, respectively.

7 Conclusions

Electronic equipment consists of a wide range of products for which the demand is increasing. So are the global flows and stocks of electric and electronic products. To give the electric and electronic equipment the desired properties, additives are added to the materials. Some of these substances are hazardous to both human health and the environment.

These additives can be emitted from the products during the life cycle of the equipment, i.e., during the production, use phase and/or end-of-life treatment. End-of-life treatment has been identified as a life cycle phase where high emissions of additives can occur. The amount of emitted additives from the end-of-life treatment is to a large extent affected by how the e-waste is treated. If a controlled treatment is applied, the emissions can be minimized and so will the risks to human health and the environment. When the e-waste is treated in the uncontrolled system, for example in Asia or Africa, there is a large risk that it will be dismantled manually and burned in open fires by people with insufficient protection and appropriate equipment.

Growing awareness of risks involved in trading waste globally has led to the Basel Convention on the Control of Transboundary Movements of Hazardous Wastes and Their Disposal, usually known as the Basel Convention, which entered into force in 1992. This international treaty aims at preventing the shipping of hazardous waste, including e-waste, from developed to less developed countries, which have ratified the Basel Convention and the Basel Ban Amendment, adopted in 1995. Also, initiatives such as "Solving the E-waste problem" (StEP) founded by various UN organizations address the problem of exporting of e-waste to countries where informal recycling techniques pose a risk to humans and the environment. Nevertheless, large volumes of e-waste are still shipped to poorer countries by being falsely declared as "reusable."[3]

Since many of the additives used in electronic products can be hazardous to human health and the environment, uncontrolled treatment of e-waste, which often is the case in those countries, may cause negative effects for the people working with the waste treatment as well as the surrounding environment. For many of the additives, information regarding emission, fate, exposure, and toxicity is scarce. To be able to better assess the risks connected to emissions of additives from electronic equipment and to suggest appropriate management actions to reduce the risks, more data on releases and toxic effects caused by these substances are needed.

References

1. United Nations University (2007) 2008 review of directive 2002/96 on Waste Electrical and Electronic Equipment (WEEE), Final Report. United Nations University, Bonn
2. LaDou J, Lovegrove S (2008) Export of electronics equipment waste. Int J Occup Environ Health 14:1–10
3. ITU (2010) World Telecommunication/ICT Development Report 2010 - Monitoring the WSIS targets – a mid-term review
4. ITU (2010) World Telecommunication/ICT Indicators Database 2010, 14th Edition
5. DisplaySearch (2010) Global LCD TV shipments reached 146M units in 2009, faster growth than 2008 (press release). http://www.displaysearch.com/cps/rde/xchg/displaysearch/hs.xsl/100222_gobal_lcd_tv_shipments_reached_146m_units_in_2009_faster_growth_than_2008.asp. Accessed 14 Dec 2010
6. ITU (2010) Market information and statistics. http://www.itu.int/ITU-D/ict/statistics/. Accessed 17 Dec 2010
7. Gartner (2010) Gartner says worldwide PC shipments in fourth quarter of 2009 posted strongest growth rate in seven years (press release). http://www.gartner.com/it/page.jsp?id=1279215. Accessed 14 Dec 2010
8. DisplaySearch (2010) Notebook, netbook and slate PC demand expected to propel portable PC market to 26% Y/Y growth in 2010 (press release). http://www.displaysearch.com/cps/rde/xchg/displaysearch/hs.xsl/100405_notebook_netbook_slate_pc_demand_expected_to_propel_portable_pc_market_growth_in_2010.asp. Accessed 14 Dec 2010

[3] European Parliament - 2–3 February 2011, Brussels plenary session "MEPs demand better waste management".

9. Cobbing M (2008) Toxic tech: not in our backyard, uncovering the hidden flows of e-waste. Greenpeace International Report
10. UNEP (2005) E-waste, the hidden side of IT equipment's manufacturing and use. Environment Alert Bulletin, UNEP
11. DisplaySearch (2010) Strong growth in mobile phone display shipments in Q4 pushes 2009 into positive growth (press release). http://www.displaysearch.com/cps/rde/xchg/displaysearch/hs.xsl/100506_strong_growth_in_mobile_phone_display_shipments_in_q4_pushes_2009_into_positive_growth.asp. Accessed 14 Dec 2010
12. IDG (2010) IDG press release – mobile phone shipments rebound to double-digit growth in fourth quarter, According to IDC (press release). http://www.idg.com/www/pr.nsf/0/220E6A3FF9F1E835852576BA0055D2E6. Accessed 14 Dec 2010
13. Lundstedt S (2011) Recycling and disposal of electronic waste – health hazards and environmental impacts. Swedish EPA, Report 6417. ISBN 978-91-620-6417-4
14. Cui J, Forssberg E (2003) Mechanical recycling of waste electric and electronic equipment: a review. J Hazard Mater B99:243–263
15. Lee CH, Chang CT, Fan KS, Chang TC (2004) An overview of recycling and treatment of scrap computers. J Hazard Mater 114:93–100
16. Méar F, Yot P, Cambon M, Ribes M (2006) The characterization of waste cathode-ray tube glass. Waste Manag 26:1468–1476
17. Felix J, Letcher B (2009) Prestudy – recycling of flat panel displays, Vinnova project no. 2008–04230
18. Liang J (2009) WEEE waste LCD liquid crystal burning pyrolysis. Master dissertation, Southwest Jiaotong University (in Chinese)
19. MPPI (2006) Guidance document on the environmentally sound management of used and end-of-life mobile phones Shredder Residue from an Electronic Waste Recycling Facility and in Soils from a Chemical Industrial Complex in Eastern China. Environ Sci Technol 43:7350–7356
20. Wu BY, Chan YC, Middendorf A, Gu X, Zhong HW (2008) Assessment of toxicity potential of metallic elements in discarded electronics: a case study of mobile phones in China. J Environ Sci 20:1403–1408
21. Markarian J (2002) Additives in the North American electrical and electronics market. Plastics, Additives and Compounding 4(9):12–15
22. Harrant M (2002) Umweltrelevante Inhaltsstoffe in Elektro- und Elektronikgeräten. Bayerisches Landesamt für Umweltschutz. Augsburg, April 2002. 62 pp
23. Zweifel H, Maier RD, Schiller M (2009) Plastics additives handbook, 6th edn. Carl Hanser Verlag, Munich
24. Brossas J (1989) Fire retardance in polymers: an introductory lecture. Polym Degrad Stab 23(4):313–325
25. Camino G, Costa L (1988) Performance and mechanisms of fire retardants in polymers–a review. Polym Degrad Stab 20(3–4):271–294
26. Leisewitz A, Kruse H, Schramm E (2001) Erarbeitung von Bewertungsgrundlagen zur Substitution umweltrelevanter Flammschutzmittel. Band I - Ergebnisse und zusammenfassende Übersicht. Umweltbundesamt, Berlin, 209 pp
27. Sony Ericson (2010) The Sony Ericsson lists of banned and restricted substances. http://www.sonyericsson.com/cws/download/1/904/292/1292579180/SEBannedandRestrictedSubstancelist.pdf. Accessed 4 Mar 2011
28. Nokia (2011) Nokia substance list. http://www.nokia.com/environment/strategy-and-reports/substance-and-material-management. Accessed 4 Mar 2011
29. Horne RE, Gertsakis J (2006) A literature review on the environmental and health impacts of waste electrical and electroni equipment. RMIT University, prepared for the Ministry for the Environment, Government of New Zealand. Melbourne, 24 pp
30. Markarian J (2005) Increased demands in electronics drive additive developments in conductivity. Plastics, Additives and Compounding 7(1):26–30

31. Donaldson K, Aitken R, Tran L, Stone V, Duffin R, Forrest G, Alexander A (2006) Review – carbon nanotubes: a review of their properties in relation to pulmonary toxicology and workplace safety. Toxicol Sci 92(1):5–22
32. Dunning P (2009) Cadmium pigments. In: Faulkner EB, Schwartz RJ (eds) High performance pigments. Wiley-VCH verlag GmbH & Co KG, Weinheim
33. Wäger P, Schluep M, Müller E (2010) RoHS substances in mixed plastics from Waste Electrical and Electronic Equipment. Final Report, 17.9.2010. EMPA, St. Gallen, 99 pp
34. Premier Farnell (2010) RoHS Directive Technical Guide. Farnell Guide to RoHS Compliance. 2 June 2010. 17 pp
35. German Federal Environment Agency (2009) Stellungnahme des Umweltbundesamtes zum Vorschlag der EU-Kommission vom 3. Dezember 2008 zur Revision der Richtlinie 2002/95/ EG zur „Beschränkung der Verwendung bestimmter gefährlicher Stoffe in Elektro- und Elektronikgeräten“ (RoHS-RL). Umweltbundesamt (UBA). Dessau-Roßlau, 14 pp
36. Markarian J (2007) PVC additives – what lies ahead? Plastics Additives and Compounding 9(6):22–25
37. Murner C (2006) Plastics, electronics the environment. Plastics Technology Magazine no information given (reprint) - www.geplastics.com: 56–61
38. Illinois Environmental Protection Agency (2007) Report on alternatives to the flame retardant decaBDE: evaluation of toxicity, availability, affordability, and fire safety issues. Illinois Environmental Protection Agency. A report to the Governor and the General Assembly. 86 pp
39. Lassen C, Havelund S, Leisewitz A, Maxson P (2006) Deca-BDE and Alternatives in Electrical and Electronic Equipment. Danish Ministry of the Environment. Environmental Project No. 1141 2006, Miljøprojekt 93 pp
40. German Federal Environment Agency (2007) Bromierte Flammschutzmittel in Elektro- und Elektronikgeräten: Das Flammschutzmittel Decabromdiphenylether (DecaBDE) ist durch umweltverträglichere Alternativen ersetzbar. Umweltbundesamt (UBA). Fachpapier. 11 pp
41. van Oers L, van der Voet E, Grundmann V (2011) Additives in the plastics industry. In: Bilitewski B, Darbra RM, Barceló D (eds) The handbook of environmental chemistry – global risk-based management of chemical additives I: production, usage and environmental occurrence. Springer, Heidelberg
42. Pucketh J, Smith T (eds.) (2002) Exporting harm: the high-tech trashing of Asia. Basel Action Network and Silicon Valley Toxics Coalition
43. Brigden K, Labunska I, Santillo D, Allsopp M (2005) Recycling of electronics waste in China and India: Worklplace and environmental contamination. Greenpeace International Report
44. Lindberg SE, Wallschläger D, Prestbo EM, Bloom NS, Price JL, Reinhart D (2001) Methylated mercury species in municipal waste landfill gas sampled in Florida, USA. Atm Environ 35:4011–4015
45. Tsydenova O, Bengtsson M (2009) Environmental and human health risks associated with the end-of-life treatment of electrical and electronic equipment. Institute for Global Environmental Strategies. (IGES)
46. Sepúlveda A, Schluep M, Renaud FG, Streicher M, Kuehr R, Hagelüken C, Gerecke AC (2010) A review of the environmental fate and effects of hazardous substances released from electrical and electronic equipments during recycling: examples from China and India. Environ Impact Assessment Rev 30(1):28–41
47. Tsydenova O, Bengtsson M (2011) Chemical hazards associated with treatment of waste electrical and electronic equipment. Waste Manag 31(1):45–58
48. Gullett B, Linak W, Touati A, Wasson S, Gatica S, King C (2007) Characterization of air emissions and residual ash from open burning of electronic wastes during simulated rudimentary recycling operations. J Mater Cycl Waste Manag 9:69–79
49. Bi X, Simoneit BRT, Wang Z, Wang X, Sheng G, Fu J (2010) The major components of particles emitted during recycling of waste printed circuit boards in a typical e-waste workshop of South China. Atmos Environ 44(35):4440–4445

50. Ma J, Addink R, Yun S, Cheng J, Wang W, Kannan K (2009) Polybrominated dibenzo-p-dioxins/dibenzofurans and polybrominated diphenyl ethers in soil, vegetation, workshop-floor dust, and electronic shredder residue from an electronic waste recycling facility and in soils from a chemical industrial complex in eastern china. Environ Sci Technol. 43(A): 7350–7356
51. Watson A, Bridgen K, Shinn M, Cobbing M (2010) Toxic transformers; - a review of the hazards of brominated & chlorinated substances in electrical and electronic equipment. Greenpeace. Greenpeace Research Laboratories Technical Note 01/2010 48 pp
52. Olivero-Verbel J, Duarte D, Echenique M, Guette J, Johnson-Restrepo B, Parsons P (2007) Blood lead levels in children aged 5–9 years living in Cartagena, Colombia. Sci Total Environ 372:707–716
53. Costa de Almeida G, Umbelino de Freitas C, Barbosa F, Tanus-Santos J, Gerlach R (2009) Lead in saliva from lead-exposed and unexposed children. Sci Total Environ 407:1547–1550
54. Wang H, Han M, Yang S, Chen Y, Lui Q, Ke S (2011) Urinary heavy metal levels and relevant factors among people exposed to e-waste dismantling. Environ Int 37:80–85
55. Wang S, Zhang J (2006) Blood lead levels in children, China. Environ Res 101:412–418
56. He K, Wang S, Zhang J (2009) Blood lead levels of children and its trend in China. Sci Total Environ 407:3986–3993
57. Tukker A, Buijst H, van Oers L, van der Voet E (2005) Risk to health and the environment related to the use of lead in products. TNO Report. STB-01-39 (Final) (2005)
58. US EPA (2006) Polybrominated diphenyl ethers (PBDEs). Project Plan, 2006
59. US EPA (2008) Toxicological review of 2,2',4,4'-tetrabromodiphenyl ether (BDE-47). EPA/635/R-07/005F, 2008
60. US EPA (2008) Toxicological review of 2,2',4,4',5-pentabromodiphenyl ether (BDE-99). EPA/635/R-07/006F, 2008
61. US EPA (2008) Toxicological review of 2,2',4,4',5,5'-hexabromodiphenyl ether (BDE-153). EPA/635/R-07/007F, 2008
62. US EPA (2008) Toxicological review of decabromodiphenyl ether (BDE-209). EPA/635/R-07/008F, 2008
63. OECD SIDS (Organisation for Economic Co-operation and Development - Screening Information Dataset) (2002) Triphenyl phosphate CAS Nº: 115-86-6. SIDS Initial assessment report for SIAM 15. UNEP Publications, 2002
64. US NLM (US National Library of Medicine) (2009) Hazardous Substances Data Bank (HSDB). US, 2009. http://toxnet.nlm.nih.gov. Accessed 31 Mar 2010
65. Anonymous (2005) Substance Data Sheet: Lead and its compounds. Common Implementation Strategy for the Water Framework Directive. Environmental Quality Standards (EQS), Brussels, 31 July 2005. 40 p
66. Rogers JT, Richards JG, Wood CM (2003) Ionoregulatory disruption as the acute toxic mechanism for lead in the rainbow trout (Oncorhynchus mykiss). Aquat Toxicol 64:215–234
67. Hedelmalm P, Carlsson P, Palm V (1995) A survey of the contents of material and hazardous substances in electric and electronic products. TemaNord 1995:554, Nordic Council of Ministers, Copenhagen, Denmark, cited in de Wit (2002)
68. Watanabe I, Sakai S-I (2003) Environmental release and behavior of brominated flame retardants. Environ Int 29(6):665–682
69. de Wit CA (2002) An overview of brominated flame retardants in the environment. Chemosphere 46(5):583–624
70. Yogui GT, Sericano JL (2009) Polybrominated diphenyl ether flame retardants in the U.S. marine environment: a review. Environ Int 35(3):655–666
71. Bayer AG (2002) Calculation of the environmental distribution of triphenyl phosphate acc. to Mackay Fugacity Model Level I, v2.11, 2001 Calculation of the indirect photolysis of triphenyl phosphate in air according to AOPWIN, v1.89, 2001 Calculation of the BCF for fish by SCRBCFWIN v2.14, 2000

72. ATSDR (Agency for Toxic Substances and Disease Registry) (1999). Toxicological profile for mercury. US Department of Health and Human Sciences
73. Verbruggen EMJ, Rila JP, Traas TP, Posthuma-Doodeman CJAM (2005) Posthumus, R. Environmental Risk Limits for several phosphate esters, with possible application as flame retardant. RIVM, Report No. 601501024/2005, 2005
74. ECB (European Chemicals Bureau) (2000) IUCLID (International Uniform ChemicaL Information Database) Dataset. Triphenyl phosphate. European Commission, 2000. (Last visited: 31/03/2010) Available at: http://ecb.jrc.ec.europa.eu/iuclid-datasheet/101020.pdf
74. Korre A, Durucan S, Koutroumani A (2002) Quantitative-spatial assessment of the risks associated with high pb loads in soils around Lavno, Greece. Applied Geochemistry 17:1029–1045

Additives in the Textile Industry

R.M. Darbra, J.R. González Dan, J. Casal, A. Àgueda, E. Capri, G. Fait, M. Schuhmacher, M. Nadal, J. Rovira, V. Grundmann, D. Barceló, A. Ginebreda, and D. Guillén

Abstract Many different additives are used in the textile industry and each of them has its specific function: protection from ultraviolet light, oil, water repellency, etc. Clothing and textile products are produced, consumed, and incinerated or disposed in landfills over the world. This means that the fate of their additives can be widely spread, and the potential impact of these chemicals on the environment and human health can, therefore, be present on a global scale. Perfluorocarbon compounds, triclosan, and brominated compounds have been selected as representative

R.M. Darbra (✉), J.R. González Dan, J. Casal, and A. Àgueda
CERTEC, Department of Chemical Engineering, Universitat Politècnica de Catalunya, ETSEIB, Diagonal 647, 08028 Barcelona, Spain
e-mail: rm.darbra@upc.edu

E. Capri, and G. Fait
Institute of Agricultural and Environmental Chemistry, Università Cattolica del Sacro Cuore, Via Emilia Parmense, 84, 29100 Piacenza, Italy

M. Schuhmacher, and J. Rovira
School of Chemical Engineering, Universitat Rovira i Virgili, Av. Països Catalans 26, 43007 Tarragona, Spain
and
Laboratory of Toxicology and Environmental Health, School of Medicine, Universitat Rovira i Virgili, Sant Llorenç 21, 43201 Reus, Spain

M. Nadal
Laboratory of Toxicology and Environmental Health, School of Medicine, Universitat Rovira i Virgili, Sant Llorenç 21, 43201 Reus, Spain

V. Grundmann
Technische Universität Dresden, Pratzchwitzer Str. 15, 01796 Pirna, Germany

D. Barceló, A. Ginebreda, and D. Guillén
Department of Environmental Chemistry, IDAEA-CSIC, Jordi Girona 18–26, 08034 Barcelona, Spain

B. Bilitewski et al. (eds.), *Global Risk-Based Management of Chemical Additives I: Production, Usage and Environmental Occurrence*, Hdb Env Chem (2012) 18: 83–108,
DOI 10.1007/698_2011_101,
Published online: 21 June 2011

additives of this sector. A review of their main environmental consequences and human health damage has been undertaken. The information gathered shows that these compounds can be toxic, very persistent in the environment, and bioaccumulable.

Keywords Environmental and human health concerns, Hexabromocyclododecane, Perfluorocarbon compounds, Textile sector, Triclosan

Contents

1 Introduction

Additives used in the production and manufacture of textiles may pose a potential harm to the environment and human health. The impacts caused by these chemicals can be related to the production process as well as to the end product itself.

In this chapter, a global perspective of the textile sector is given. Data concerning the final fate of the used textiles are provided. Then, a review of the major finishing chemicals (additives) used in the textile industry is presented. They are divided according to their function in the final textile product (softeners, flame retardants, dyers, etc.).

Chemicals applied during the finishing process remain in the end product and may pose a potential risk to the environment and human health during the use, disposal, and/or recycling of the products. Therefore, the potential harmful action of these chemicals is highlighted in Sect. 4.

Three of the main textile additives have been studied in more detail in this chapter. The additives selected for evaluation were: repellent finishers (perfluorocarbon compounds), antimicrobial finishers [triclosan (TCS)], and flame retardants (brominated compounds).These additives were selected after taking into consideration the information gathered on the finishing chemicals. Criteria such as their usage, public interest, potential adverse effects to the environment, and human health were taken into account when selecting them. For these substances, information related to their potential risk to the environment and human health has been collected.

2 The Textile Sector

The textile sector prepares natural and man-made materials for use in clothing, carpets, furniture manufacture, interior decoration, etc. Natural and man-made fibers are spun into yarns and threads, woven or knitted into fabrics, and finished and sewed into final products [1].

In the textile sector, most of the chemicals are applied during the finishing process. This process enhances the appearance, durability, and serviceability of fabrics and it involves different steps: fabric pretreatment (desizing, washing, scouring, bleaching, etc.), coloring (dyeing and/or printing), and functional finishing (e.g., applying finishes for flame retardation or water proofness) [2].

In accordance with the World Trade Organization [3], manufactured textile products can be divided into two basic subsectors: textiles and clothing. The textiles subsector includes carpets, furnishing and upholstery fabrics, household linen, and technical textiles, such as textiles for construction, agriculture, vehicles, and packing. The clothing subsector includes the woven fabrics and knitted fabrics for sportswear, rainwear, fashion articles, work wear, etc. In agreement with this division, textile and clothing exports of selected regions and economies according to their destination are shown in Fig. 1 for data of 2009. From this figure, it can be stated that during 2009 the main exporter was the Asian region, followed by Europe. Furthermore, most of the economic movements associated with European countries were exports within the countries of the region (around 80%), whereas in the Asian region and in North American countries the portion of exports to other countries was significant (respectively, around 26% and 24% to Europe and North America from Asian countries, and around 11% and 15% to the Asian region and South and Central America from North American countries).

Since RISKCYCLE project is focused on the end-of-life of the products, it is very important to know more information about the final fate of the textiles. Information on this aspect has been found in available data on annual municipal solid waste generation in the USA. In 2008, the amount of textiles in municipal waste generated in the USA was about 12.37 million tons, which comprised about 5% of municipal waste. From this amount, only a 15% was recovered and the remaining 85% was discarded (in landfills or incinerated) [4]. These proportions of recovered and discarded textiles have been studied in the USA from the beginning of the twenty-first century. As it can be seen in Fig. 2, the recovered textile percentage was lower between 1960s and 1990s (3–11%) than in the recent years.

In the UK, according to the data from Allow et al. [5], similar figures were estimated for the percentage of textile materials recovered and discarded in 2004. More specifically, about 13% of the textile waste went to material recovery and 87% was discarded (13% incinerated and 74% landfilled). From the total flow for material recovery, two thirds were estimated to be exported for recycling and reuse abroad.

All these information lead to the conclusion that most of the textile products end up in landfills and only a very low percentage is recovered. The additives used in

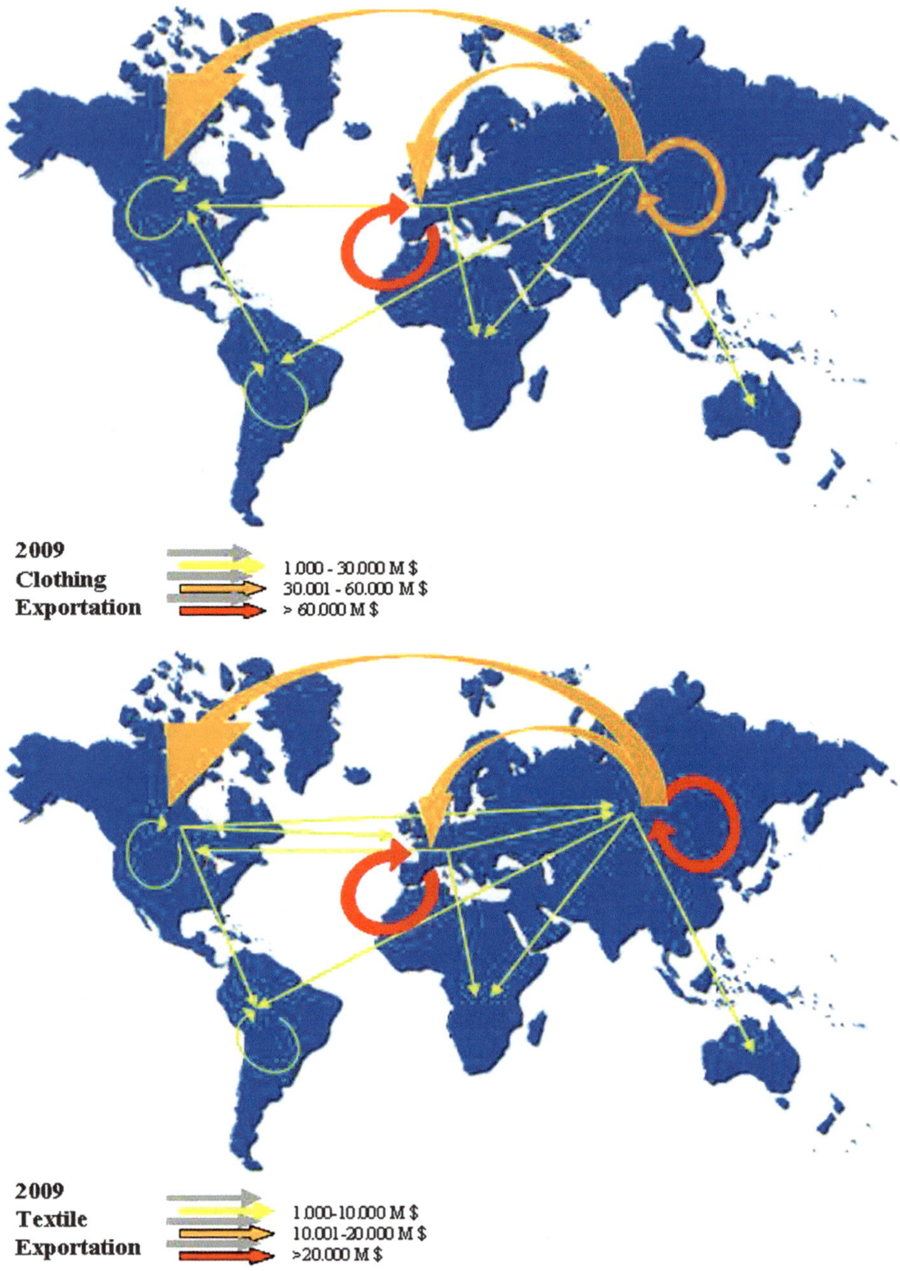

Fig. 1 Diagram of world exportations of manufactured textile products: (**a**) clothing; (**b**) textiles (in million US $). Regions considered: Africa, Asian region (Asia, Commonwealth of Independent States, Middle East), Europe, Australia with New Zealand, North America, and South and Central America. Source: [3]

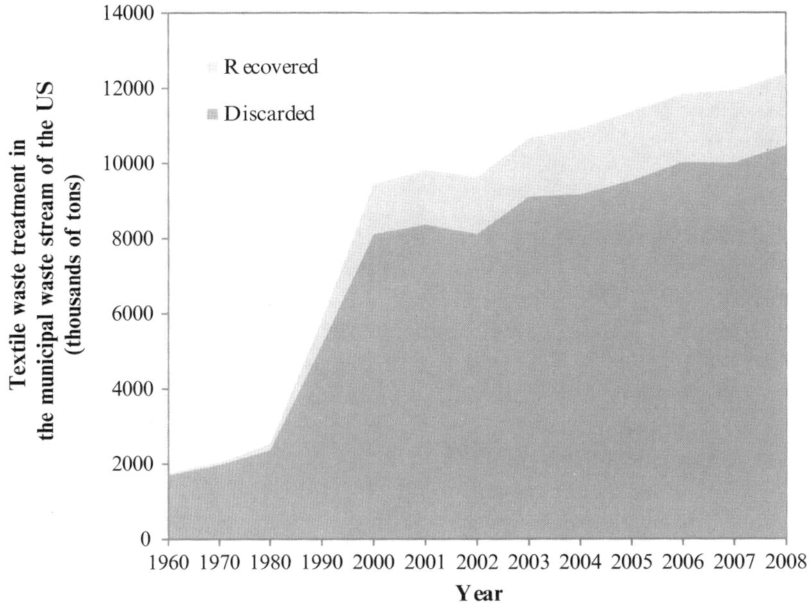

Fig. 2 Accumulation plot of textile waste treatment (recovery and discard) in the municipal waste stream of the USA (discards include combustion with energy recovery). Source: [4]

their production can have a potential effect on the environment and human health when released into the soil (from landfills), air (from incineration), or groundwater (from landfills). Therefore, it is important to know what types of additives are used and which potential effects to the population and the nature they can have. In the following section, a brief summary of the main additives used in the textile sector is presented.

3 Additives in the Textile Industry

As aforementioned, chemical additives in the textile sector are mainly introduced in the finishing process. This process involves the fabric treatment, coloring, and the functional finishing [2]. In this section, chemicals used for coloring textiles are presented, and then the main additives used in the functional finishing are also introduced.

3.1 Chemicals Used for Coloring the Textiles

Dyeing is a method for coloring textile material in which a dye is applied to the substrate in a uniform manner to obtain an even shade [6].

World production of colorants is ca. one million tons per year, of which about 50% are textile dyes [7]. Textiles may be dyed in fiber form (before spinning), as spun yarns, after the finished material has been woven or knitted into textile fabric or, in the case of apparel, in garment form after cutting and sewing [8].

Textile dyes may be classified according to their solubility, chemistry, or based on the fibers for which they have affinity. When classification is based on solubility, a distinction is made between the two groups: dyes (which are deemed to be soluble in the application medium), and not readily or insoluble pigments (which are normally applied only to specific areas, instead of coloring the whole fabric). When classification is by coloring groups (chromophores), i.e., from the chemical angle, a distinction is made between azo dyes, anthraquinone dyes, metal complex dyes, and others [9]. Table 1 gives a classification of the most important dyes according to the application, i.e., information is given about to which fibers they are normally applied and about their fiber affinity (information extracted and summarized from US EPA Manual [8]).

Textiles are commonly exposed to light, friction, and washing. These impacts may change any of their color characteristics, transfer their colorants to adjacent materials, or both. For this reason, all colored fabrics and textile articles are finished to improve color fastness. Several finishers are used in the textile industry to improve color fastness: aromatic compounds (e.g., benzophenone, benzotriazole, and phenyl triazine compounds), hindered phenol and amine stabilizers, polyammonium-based compounds (e.g., polydiallyldimethyl ammonium chloride – DADMAC), peroxidases, formaldehyde condensation products from urea and melamine, syntan, reactive fixation compounds (e.g., epoxy derivatives with cationic structures), etc. [10].

3.2 Functional Finishes

Textile finishes can be divided according to their function in esthetic and functional finishes. The first ones modify their appearance, whereas the later improve their performance under specific use conditions. Textiles finishes can also be divided according to the process from which they are obtained in: chemical and mechanical finishes [11]. The mechanical finishes can be achieved without the addition of any compounds, for example, calendaring involving heat, moisture, and pressure. Chemical finishing involves the application of compounds as finishing agents, which may or may not include chemical reactions [11]. These kinds of finishers are normally applied in the form of aqueous solution or emulsion and may be applied via a variety of techniques, the main one being the pad mangle [12]. Another way in which textile finishers can be classified is according to the degree of performance in temporary, semipermanent, and permanent finishers [13].

This subsection will focus on the functional finishers that give the final properties to the textile product. More than ten types of functional finishers have been

Table 1 Main dye classes and characteristics (information extracted from [8])

Dye class	Description	
Acid	Fibers	Nylon, wool, silk, some modified acrylic textiles
	Dye-fiber affinity	Ionic bounds between the sulfonic acid part of the dye and the basic amino groups in wool, silk, and nylon fibers
Naphtol (azoic dyes developed on the fiber)	Fibers	Cotton, rayon, cellulose acetate, linen, jute, hemp, sometimes polyester
	Dye-fiber affinity	Azoic dyes are made up of two chemically reactive compounds. The reaction of the two compounds in the fiber produce the colored azochromophore
Basic (cationic)	Fibers	Silk and wool (using a mordant), synthetic fibers (acrylic, modacrylic, modified nylons and polyesters)
	Dye-fiber affinity	Ionic bonds are formed between the cation in the dye and the anionic site on the fiber
Direct	Fibers	Cotton, rayon, linen, jute, hemp, silk, nylon fibers, mixtures of fibers
	Dye-fiber affinity	Dyes are absorbed into hydrophilic fibers as the fibers expand in the water solution. The dye molecules are held in place through Van der Waals forces and hydrogen bonds
Disperse	Fibers	Polyester and other synthetics (oleophylic fibers), cellulose acetate nylon (i.e., regenerated cellulose fibers), nylon, acrylic fibers
	Dye-fiber affinity	They transfer into the synthetic fiber polymer because of their high solubility in the substrate
Fiber reactive	Fibers	Cellulosic fibers (cotton, rayon), wool, silk, nylon
	Dye-fiber affinity	They form covalent chemical bonds with the fiber and become part of the fiber
Mordant	Fibers	Wool, silk, nylon
	Dye-fiber affinity	The fiber is first treated with a metal and the metal exchanged fiber (mordanted fiber) is contacted with a solution containing a complexing agent, typically a chelating agent, called mordant dye. The reaction between the ligand and the exchanged metal forms a metal-chelate complex on the surface of the fiber, which is immobilized
Pigment	Fibers	All fibers
	Dye-fiber affinity	Pigments are usually mixed with a binder (aqueous dispersion of cross-linkable mixed polymers) that hardens upon drying, forming an opaque coating

(*continued*)

Table 1 (continued)

Dye class	Description	
Sulfur	Fibers	Cotton, rayon, blends of cellulosic and synthetic fibers (nylon, polyester)
	Dye-fiber affinity	Sulfur dyes are reduced to a water-soluble form before application to the fiber and then they color by absorption. The reducing agent in the dyeing of textile material with sulfur dyes is sodium hydrosulfite ($Na_2S_2O_4$), which is used in combination with glucose, although in some cases sulfur dyes are still being reduced using sulfides (Na_2S, NaHS) and polysulfides (Na_2S_8)
Vat	Fibers	Cotton, cellulosic fibers
	Dye-fiber affinity	They are either supplied in water-soluble reduced leuco form or reduced with a reducing agent (e.g., sodium hydrosulfite). Then they are allowed to migrate into the fiber. When this migration is complete, the substrate is rinsed to remove surface dye, then the dye is oxidized back to its water insoluble form within the fiber

distinguished by Schindler and Hauser [10]. In Table 2, the most commonly used finishers are presented following the excellent classification done by these authors. In this table, the information given on the function, classification, and chemicals involved in the textile sector has been summarized from [10].

From Table 2, it can be seen that there exist many different substances that are involved in the finishing process of textiles. The chemicals present in this table add new proprieties to the final textile product. For example, softeners proportionate comfort in the daily use of the textiles. Besides the chemicals presented in Table 2, silicone-based compounds, also used as elastomers and water repellent finishers, have also been used to formulate textile softening chemicals since the 1960s [14]. An example is polydimethylsiloxane.

Biocides are substances that aim at destroying, deterring, preventing the action, or otherwise controlling the effect of any harmful organisms by chemical or biological means [15]. When talking about textiles, their function is to control bacteria, fungi, mildew, etc. The typical biocides used in the textile industry are [2]: antimicrobial finishes (e.g., isothiazolinonederivates, TCS), moth proofing (e.g., pyrethoids, sulcofuronderivates), and preservation agents for the improvement of the storage stability of the textile auxiliaries (e.g., chlorinated and non-chlorinated isothiazolinone derivates, chlorinated benzene).

Table 2 Functional finishes, description, classification, and chemicals involved (information extracted from [10])

Finishes and description	Classification	Chemicals
Softeners Used to prevent static cling and give textiles a soft hand, smoothness, more flexibility, and better drape and pliability	Cationic	*N*,*N*-distearyl-*N*,*N*-dimethyl ammonium chloride (DSDMAC)
		Quaternary ammonium salts with one or two alkyl chains
		Amine salts
		Imidazolines
		Di-(stearylcarboxylethyl)-hydroxyethylmethylammoniummethylsulfate
		A triethanolamine ester quaternary
	Anionic	Alkylsulfate salt
		Alkylsulfonate salt
	Amphoteric	Alkyldimethylamine oxide softeners
		Betaine softeners
	Nonionic	Polyethylene
		Ethoxylated fatty compounds
Biocides Finishing textiles with biocides refers to applying antimicrobial or insect resist treatments to protect the textile itself from damage caused by mold, mildew, or rot-producing microorganisms, to protect the textile user against pathogenic or odor-causing microorganisms, and to protect wool and other animal fibers from attack by larvae of certain moths and beetles	Antimicrobials	Tributyltin oxide (TBTO) (deleted in many countries)
		Dichlorophene
		3-Iodopropynylbutyl carbamate
		Benzimidazol derivatives
		Triclosan
		Salicylanilides
		Alkylolamide salts of undecylenic acid
		Organo-silver compounds and silver zeolites
		Octadecylaminodemethyltrimethoxysilylpropylammonium chloride
		Polyhexamethylenebiguanide (PHMB)
		Chloramines into the fiber
		Chitosan
	Insect resist	Chlorinated triphenylmethanes
		Chlorphenylids
		Sulcofenurons
		Flucofenurons
		Dieldrin (banned in most countries)
		Permethrin
		Hexahydropyrimidine

(*continued*)

Table 2 (continued)

Finishes and description	Classification	Chemicals
Hand-building finishes The main effects of hand-building finishes are fullness, which is the feeling of increased bulk or weight, and stiffness or resistance to bending	Nondurable	Starch derivatives Polyvinyl alcohol structures
	Durable	Polyvinyl acetate Ethylene-vinyl acetate copolymer Polyvinylmethyl ether Acrylic copolymers (polyacrylates and polymethacrylates) Formaldehyde-containing thermosetting polymers (precondensates of urea or melamine with formaldehyde)
Flame-retardant finishes Flame retardants are chemicals applied to fabrics to inhibit or suppress the combustion process	Phosphorus-based	Diammonium phosphate Ammonium polyphosphate Tetrakis(hydroxymethyl)phosphoniumchloride (THPC) and urea *N*-methyloldimethylphosphonopropionamide with trimethylolmelamine and phosphoric acid Cyclic organophosphonate Alkyl dioxaphosphorinane disulfide
	Sulfur-based	Ammonium sulfamate Alkyl dioxaphosphorinane disulfide
	Bromine-based	Ammonium bromide Tetrabromophthalic anhydride (TBPA) Hexabromocyclododecane (HBCD) Antimony trioxide combined with decabromodiphenyl ether (DecaBDE) and a binder
Repellent finishes Finishes dedicated to repel water, oil, and dry dirt can be modified to have a wide range of properties to fit the different demands of the users and intended purpose	Paraffin-based	Emulsions containing aluminum or zirconium salts of fatty acids (usually stearic acid)
	Stearic and melamine-based	Compounds formed by reacting stearic acid and formaldehyde with melamine (e.g., Phobotex FTC)
	Silicone-based	Polydimethylsiloxane products Silanol and silane with tin octoate as catalyst
	Fluorocarbon (FC)-based	Perfluoroalkyl groups incorporated into acrylic or urethane monomers Fluorocarbon polymers (e.g., hyperbranched polymers containing short fluorocarbon chains)

Antistatic finishes Antistatic finishes are chemicals that are applied to the surface of synthetic fabrics to control the tendency of these fabrics to accumulate static charge	Nondurable finishes	Esters of phosphoric acid
		Ditallow dimethyl ammonium chloride
		Dihydrogenated tallow dimethyl ammonium chloride
		Ethoxylated fatty esters, alcohols, and alkylamines
		Mixtures of cationic and nonionic surfactants
	Durable finishes	Cross-linked polyamines and polyglycols
		Polyhydroxypolyamines
		Polyalkylene and polyacrylic copolymers
Easy-care and durable-press finishes of cellulosics Implemented to reduce swelling and shrinkage, and to improve wet and dry wrinkle recovery, smoothness of appearance after drying, and retention of intentional creases and pleats	Formaldehyde-containing products	Dimethylether of DMU (dimethylol urea) (U/F)
		Trimethylol melamine (TMM) (M/F)
		Hexamethylol melamine (HMM) (M/F)
		1,3-Dimethylol-4,5-dihydroxyethylene urea (DMDHEU) (U/glyoxal/F)
	Nonformaldehyde-containing products	Glyoxal/glycol mixtures
		N,N'-dimethyl-4,5-dihydroxyethylene urea (DMeDHEU) (U/glyoxal)
		Polycarboxilic acids (e.g., maleic acid, citric acid, butane tetra carboxylic acid)
Other finishes	Elastomeric finishes	Silanol and methyl hydrogen silane
	UV protection finishes	Benzophenone and benzotriazole compounds, and phenyl salicylates

Hand-building finishes can be classified according to whether they are durable or nondurable. This depends on the type of chemicals that compose them (e.g., polyvinyl alcohol structures – nondurable – and polyvinyl acetate – durable).

Flame-retardant finishes are used to control the ignition of the textiles in case of fire. In places of high risk where upholstered furniture and textiles are present such as hospitals, prisons, and hotels, it is very important to comply with the fire safety regulations [16]. Therefore, the use of these additives is highly required.

Repellent finishers are introduced in the textiles to avoid water and oil contact. Fluorocarbons polymers are excellent additives for oil repellence [10]. They have optimal chemical and thermal stability and can be applied at a lower add-on than any other repellent finishes [17]. Water repellence is achieved using different chemical compounds such as silicone-based products and paraffin-based compounds.

The antistatic agents can function by reducing the generation of electrostatic charges, by increasing the conductivity of the materials to which they are applied, or by both mechanisms [18].

Since the late 1980s, the market place has shown a steady increase in the demand for easy-care and wrinkle-resistant 100% cotton apparel [19]. There exists an increasing fear due to the potential carcinogenicity of the formaldehyde release from the *N*-methylol agents used as this type of finishes [10]. For this reason, other additives such as glyoxal/glycol mixtures [20], glyoxal urea-based derivatives such as that based on *N,N′*-dimethyl-4,5-dihydroxyethylene urea (DMeDHEU), and polycarboxylic acids forming ester links with cellulose have been developed as formaldehyde-free finishing agents [14].

As it can be seen, each one of these finishes has its own function and they are necessary for the formation of the final textile products. The commented additives are the main ones used in the textile industry. It is obvious that there exist other textile finishes (e.g. UV protection, elastomeric); however, it is not the objective of this chapter to go into more detail in this general classification of the textile additives.

In the next section, a general overview of the environmental and human health effects of the main additives is presented.

4 Environmental and Human Health Concerns of Textile Additives

Several compounds mentioned in the former subsections are prone to originate environmental and human health impacts. Some of the associated concerns are briefly discussed in Table 3, in which the main consequences of the diverse additives are presented, according to the information available in [8, 10].

Table 3 Human health and environmental consequences of several additives used in the textile sector (information extracted from [8] and [10])

Consequence	Type of additives	Description of the consequence
Carcinogenic effects on humans	Dyestuffs	Due to the possible presence of carcinogenic naphthylamines in effluents, the use of naphtol dyes has declined over the years
		Reactive fixation compounds for color fastness (epoxy derivatives) may cause cancer or mutagenicity if they come into contact with workers
Systematic effects on humans	Biocides	Some consumers may develop dermatitis from prolonged skin contact with antimicrobial finishes
	Flame retardant finishes	DecaBDE belongs to the category of the polybrominated diphenyl ethers (PBDEs) and they have the potential to disrupt thyroid hormone balance and contribute to a variety of developmental deficits
Water pollution and toxicity for aquatic organisms	Dyestuffs	Basic (cationic) dyes exhibit high aquatic toxicity
		The wastewater of vat and sulfur dyes may contain excess of sodium hydrosulfite ($Na_2S_2O_4$), which is classified in some countries among the most hazardous class of waters
		Benzophenones and benzotriazole derivatives may pose mutagenic threats and endocrine disruption to aquatic organisms
		Cationic products used to improve color fastness (e.g., DADMAC) can cause fish toxicity
	Softeners	Many conventional quaternary ammonium compounds pose high toxicity to fishes
	Hand-building finishes	Starch derivatives increase the biological oxygen demand from wastewaters and often produce high levels of total suspended solids, which are difficult to remove or settle
	Flame-retardant finishes	Phosphorous, antimony, and zirconium compounds, and halogenated organic flame retardants, especially the aromatic ones, in wastewater may impose threat to the environment
	Repellent finishes	The residual baths from silicon-based repellents finishing application processes are toxic to fish

(continued)

Table 3 (continued)

Consequence	Type of additives	Description of the consequence
	UV protection finishes	Benzophenones and benzotriazole derivatives may pose mutagenic threats and endocrine disruption to aquatic organisms
	Biocides	Ecological problems can appear if the residual finish is improperly disposed of at the finishing plant and the desirable microbes in the waste treatment facility are destroyed
	Softeners	Silicone softeners can contain variable amounts of volatile siloxane oligomers, which, together with volatile emulsifiers, may cause atmospheric pollution problems in the waste air
Air pollution	Flame-retardant finishes	TBPA is suspected to generate polybrominated dioxins under burning conditions
	Repellent finishes	The exhaust air of the drying and curing processes of fluorocarbon finishes often contains high amounts of volatile organic compounds (VOC)
	Softeners	Silicone softeners can contain variable amounts of volatile siloxane oligomers, which, together with volatile emulsifiers, may cause atmospheric pollution problems in the waste air
Persistence and bioaccumulation in the environment	Flame-retardant finishes	DecaBDE persist in the environment and bioaccumulate in organisms
	Softeners	Poor biodegradability of quaternary ammonium compounds
	Dyestuffs	Cationic products are almost nonbiodegradable
	Repellent finishes	Possible residual content of PFOS and PFOA in fluorocarbon polymers which can have a moderate toxic effect in mammals, is bioaccumulative, and very persistent to abiotic and biotic degradation

5 Effects and Exposure Assessment of PFOA/PFOS, TCS, and HBCD

Three textile additives have been reviewed in this section to depict the type of effects they may pose to the environment and human health, and to give some references about the monitoring levels reported in the environment and in human samples. These additives were selected taking into consideration the information gathered on the finishing chemicals. Criteria such as their usage, public interest, potential adverse effects to the environment, and human health were taken into account when selecting them. The selected substances have been: perfluorinated compounds (PFCs) used in the manufacture of oil and water repellent polymers [perfluorooctanoic acid (PFOA) and perfluorooctane sulfonate (PFOS)], triclosan (TCS) and hexabromocyclododecane (HBCD), a flame retardant.

5.1 PFOA and PFOS

Fluorocarbon-based polymers are used to confer oil and water repellence to textiles. Perfluorinated substances are characterized to contain a total fluorinated carbon chain, where hydrogen atoms have been substituted by fluorine atoms (see Fig. 3). This substitution contributes to the PFCs stability, due to the electronegativity of fluorine (large energy of the C–F bound).

Two of the main PFC are the PFOA and the PFOS. The first one belongs to the perfluorocarboxylic acids (PFCAs) that are fully fluorinated carboxylic acids. The PFOA is one of the most studied of this group due to its similarity with the PFOS.

Possible residual content of PFOA and PFOS has moved toward the center stage of public interest. These two compounds may also be potential products from thermal or biodegradation of fluorotelomer-based substances.

As an example of the interest of these chemicals, PFOS is one of the new candidate substances under revision [21] to be included in the list of priority substances within the European Water Framework [22]. In May 2009, PFOS was added to Annex B of the Stockholm Convention on Persistent Organic Pollutants. PFOS is extremely persistent, bioaccumulative, and toxic [23].

PFOA, also called "C8," is a synthetic chemical and fluorosurfactant. Companies use PFOA to make fluoropolymers, substances with special properties that have thousands of important manufacturing and industrial applications, such as

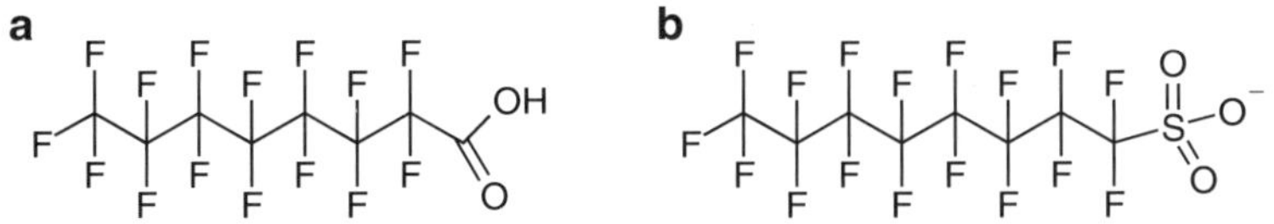

Fig. 3 (**a**) Perfluorooctanoic acid (PFOA) and (**b**) perfluorooctane sulfonate (PFOS)

aerospace, automotive, building/construction, chemical processing, electronics, semiconductors, and textile industries. Fluoropolymers confer valuable properties, including fire resistance and oil and water repellence. They are used to provide nonstick surfaces in waterproof and breathable membranes kitchenware, for clothing.

The main production process of PFOS/PFOA-related substances includes the yielding of PFCAs. PFCAs have been manufactured as salts by two main synthesis routes: electrochemical fluorination (ECF) and fluorotelomer iodide oxidation, as shown in Fig. 4 [24]. From 1947 to 2002, ECF was used to manufacture the large part (80–90%) of perfluorooctanoate ammonium salt (APFO). The main production sites were the USA and Belgium, followed by Italy and, in a smaller scale, Japan. The remaining 10–20% of APFO was produced from 1975 to the current days by oxidation of perfluorooctyl iodide in Germany and Japan.

Several environmental and toxicological characteristics of PFOS/PFOA are described in the following list:

- Unlike other persistent organic pollutants (POPs), such as polychlorinated biphenyls (PCBs), polychlorinated dibenzo-*p*-dioxins (PCDDs), and polychlorinated dibenzofurans (PCDFs), PFOA has a great potential to dissolve in water.
- The anticipated formation of an emulsified layer between the octanol and water surface interface has made, until now, the determination of log K_{ow} impossible.
- PFOS is extremely persistent, bioaccumulative, and toxic (Stockholm Convention on POPs [24]).
- The pharmacokinetic properties of PFOA have been studied in some detail. Animal studies of PFOA exposure have shown that it is well absorbed orally, but poorly eliminated and it is not metabolized.
- PFOA tends to accumulate in blood, liver, and kidneys [25–27]. Levels of PFOA and other PFCs have also been reported in umbilical cord blood [28] and in human breast milk [29].
- The half-lives in human blood serum have been estimated to be 3.8 and 5.4 years for PFOA and PFOS, respectively [30].
- The persistence of PFCs in the environment and their slow elimination rates in humans means they are likely to accumulate in people.

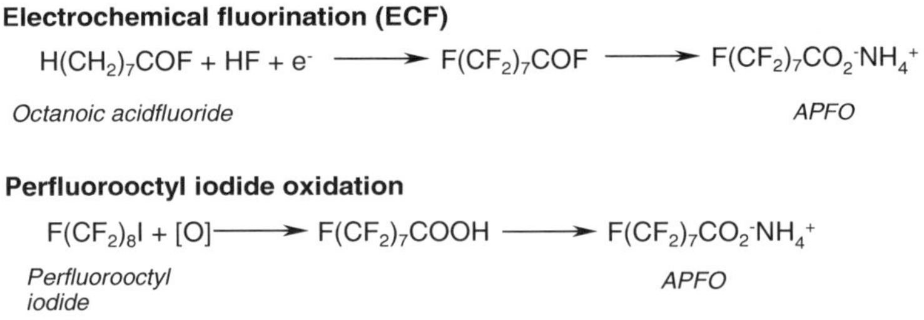

Fig. 4 Perfluorooctanoate ammonium salt (APFO) synthesis routes [24]

- Epidemiologically, no clear correlations have been observed between PFC serum concentrations in human populations and adverse health effects or illnesses [31, 32]. However, some correlations have been reported with fecundity, as measured by time to pregnancy (TTP) [33], and between PFOA/PFOS and cholesterol levels [34].
- Other epidemiological studies have demonstrated inhibitory effects on human neonatal growth at the background exposure levels of PFOS and/or PFOA in general population [35, 36].
- For mammalian species, PFOA and its salts have been found to cause cancer in rats and adverse effects on the immune system in mice. In addition, PFOA and its salts have displayed reproductive or developmental toxicity in rodents at moderate levels of exposure [37, 38].
- PFOA is considered a probable human carcinogen by the US Environmental Protection Agency (EPA) [39].

With regard to monitored levels of exposure to PFOA and PFOS, it has been observed that textile consumer products made with fluoropolymers and fluorinated telomers may contain trace amounts of PFOA and other related perfluorinated chemicals as impurities. In the 2009 USEPA study of 116 products – purchased between March 2007 and May 2008 – PFOA concentrations ranged from nondetectable to 462 ng/g (Table 4).

Yamada et al. [41] observed that polyester/cellulose fabrics treated with a fluorotelomer-based acrylic polymer were destroyed under typical municipal incineration conditions, and that an undetectable amount of PFOA was formed. Therefore, textiles treated with such a fluorotelomer-based acrylic polymer disposed in municipal waste and incinerated were expected to not be a significant source of PFOA in the environment.

The highest PFOS and PFOA concentrations in human blood samples have been measured in workers used in fluorine production plants [42]. In most cases, workers occupationally exposed presented serum levels of both PFOA and PFOS approximately one order of magnitude higher than those reported in the general population [37]. These results point out the occupational exposure as the main exposure pathway of both PFOS and PFOA.

Some discrepancies exist between the importance of the drinking water and the dietary exposure routes. Ericson et al. [43] estimated that approximately the dietary intake of PFOS in Catalonia was one-half of the PFOS ingestion through drinking

Table 4 PFOA concentrations in textile consumer products [40]

Product	Range (ng PFOA/g product)
Pretreated carpeting	ND (<1.5) – 462
Treated apparel	5.4–161
Treated upholstery	0.6–293
Treated home textiles	3.8–438
Membranes for apparel	0.1–2.5 ng/cm^2

water. On the contrary, Fromme et al. [44] stated the diet as the main exposure pathway, followed by drinking water and house dust.

Trudel et al. [45] calculated the contribution to exposure of several indoor exposure pathways, and hand-to-mouth transfer from treated carpets seemed to be an important exposure route to PFOA, especially in infants, toddlers, and children from North America and Europe.

5.2 *Triclosan*

Biocides are chemical substances that are necessary for the control of organisms that cause damage to natural or manufactured products. These organisms can pose risks to humans, animals, and environment in a variety of ways. Therefore, the European Directive 98/8/EC [15] aims at regulating the placing of biocidal products on the market. The scope of this Directive is very wide, covering 23 different product types.

In the textile industry, biocides are used to control bacteria, fungi, mildew, algae, and the problems of deterioration, staining, odors, and health concerns that they cause. Textile fabrics are made of two or three type of fibers according to their origins: natural fibers (e.g., cotton, wool), natural man-made fibers (e.g., acetate, rayon), and synthetic man-made fibers (e.g., polyester, poliamide). In general, textiles made from natural fibers are more susceptible to biodeterioration than man-made fibers [46].

One of the most used biocides is TCS (5-chloro-2-(2,4-dichlorophenoxy)phenol) (Fig. 5). TCS is a chlorinated aromatic compound that has been used extensively for more than 30 years [47]. TCS is a nonionic compound, used as synthetic broad-spectrum antimicrobial agent in the form of a white powder, possessing mostly antibacterial along with some antifungal and antiviral properties [48].

TCS is widely used in many consumer and professional healthcare products, such as soaps, detergents, toothpastes, disinfectants, cosmetics, and pharmaceuticals [49], and it has also been incorporated into fabrics and plastics [50].

Due to its widespread use, TCS featured among the seven most frequently detected compounds in a US survey of 139 streams across the states [51]. In Europe,

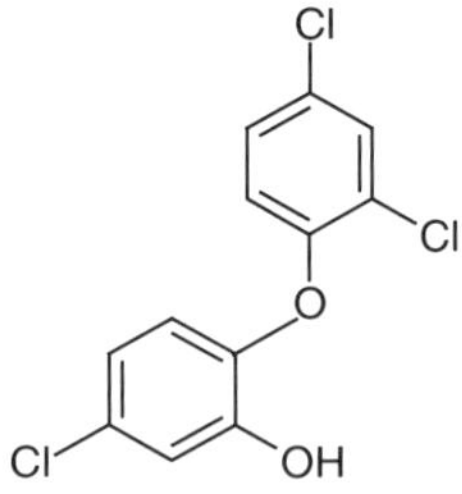

Fig. 5 Triclosan (TCS)

350 tons of TCS is used per year, and approximately 1,500 tons is produced annually worldwide [52].

TCS is regulated by the EPA, the Food and Drug Administration (FDA), and the Consumer Product Safety Commission (CPSC). Europe recently banned TCS in all items expected to come into contact with food and has set limits on how much TCS can be in cosmetics [53].

Several environmental and toxicological characteristics of TCS are described according to the information contained in US EPA [54]:

- It has low solubility in water (12 mg/L) and it is relatively nonvolatile (vapor pressure 7×10^{-4} Pa).
- It is stable to hydrolysis at pH 4–9.
- In aquatic environments, it is expected to adsorb to suspended solids (being immobile) and sediments.
- It may bioaccumulate, posing a concern to aquatic organisms.
- TCS is highly toxic to aquatic organisms (48 h median $EC_{50} = 0.39$ mg/L for *Daphnia magna*) and to bacteria as well [55, 56].
- The photodegradation of TCS into dissociated forms maybe a possible elimination process for this compound in surface water.
- Aerobic degradation is one of the most efficient biodegradation pathways.
- TCS has a low acute toxicity by the oral and dermal routes of administration (moderate irritation to eyes and skin).
- Repeated exposure by inhalation route may result in inflammation of the respiratory tract.
- TCS was negative for mutagenicity in vitro and in vivo.
- Liver toxicity to rats was noted after repeated oral administration.
- No evidence of carcinogenic effect was observed in rats/hamsters, although liver tumor has been observed in mice.
- No adverse effects have been observed in two-generation reproductive toxicity tests in rats.
- Tests with rats and rabbits showed no evidence of pre- or postnatal developmental toxicity.

These characteristics indicate that TCS is not likely to contaminate surface or ground waters due to its immobility in soils, its susceptibility to photodegradation, and its potential biodegradation in soil and water. The majority of published studies on the occurrence of TCS in wastewater treatment plants (WWTPs), treatment plant efficiency, and open water measurements of TCS suggest that aerobic biodegradation is one of the major and most efficient biodegradation pathways [57]. However, Ricart et al. [58] have demonstrated the persistence of TCS through WWTPs and its potential effects on both bacterial and algal communities.

US EPA has concluded that chronic aquatic risks are unlikely originating from consumer uses of TCS-treated textile items due to the low probability of TCS being released into household wastewater and surface waters. Therefore, the agency has assumed that the antimicrobial uses of TCS are unlikely to introduce significant quantities of TCS into household wastewater and eventually in surface water.

5.3 *Hexabromocyclododecane*

Flame retardants are compounds added or applied to a material to increase the fire resistance of that product [59]. Brominated flame retardants (BFRs) are organobromide compounds which are used to inhibit the inflammability of a material. They are widely used in commercial products such as textiles [60].

The HBCD is one of the most extended flame retardants. It is a brominated alicyclic hydrocarbon. Three different diastereomers (α-, β-, γ-HBCDD), which exist as pairs of enantiomers (Fig. 6), are mainly found in technical HBCD. The final distribution of the diastereomers in the technical product varies with a range of about 70–95% γ-HBCDD and 5–30% α- and β-HBCDD.

HBCD is industrially used as flame retardant in different life cycle steps: production, formulation, and industrial use with the aim to increase the flame resistance of different end products. HBCD was used for the first time by the BASF company in their production of flame-retarded polystyrene foam in the late 1980s; however, this additive has been on the market since the 1960s [61]. World production of HBCD in 2001 was of 16,700 tons, and the industrial use of HBCD has increased during the last decade, making it the third most used BFR worldwide at this time [62]. HBCD can be used on its own or in combination with other flame retardants, e.g., antimony trioxide and decabromodiphenyl ether (decaBDE).

Today the main downstream uses of HBCD are in the polymer and textile industries; the four main product types in which it is used are: expandable polystyrene (EPS), extruded polystyrene (XPS), high impact polysterene (HIPS), and polymer dispersion for textiles. According to industry information [61], practically 90% of the HBCD is used in the poly styrene (PS) production, whereas 10% is used in polymer dispersions on cotton or cotton/synthetic blends. Textile end products containing HBCD are used both professionally and by consumers. They have a relatively long service life and are disposed of by different means (incinerated, recycled, put on landfill, or left in the environment).

Several environmental and toxicological characteristics of HBCD are described below, according to the information gathered in European Chemicals Bureau (ECB) [61]:

- HBCD is not readily biodegradable.
- It has a very high potential to adsorb onto soils and sediment, and a low potential to leach through soil.

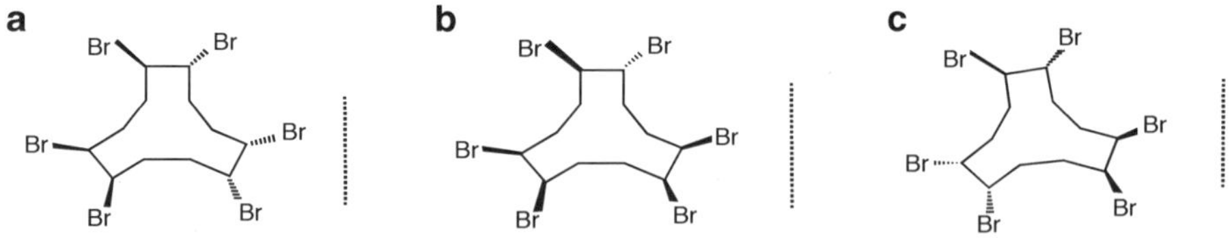

Fig. 6 Main diastereomers (pairs of enantiomers) in technical product of HBCDD: (**a**) α-HBCDD; (**b**) β-HBCDD; (**c**) γ-HBCDD. The *lines* indicate a mirror plane [61]

- Evaporation seems to be a less important route of dispersion.
- Sewage treatment plants eliminate 80% of HBCD.
- Studies confirm the bioconcentration and bioaccumulation of HBCD in the trophic chain.
- HBCD has a very low acute toxicity by the oral and dermal routes.
- It is mildly irritating to the eye but neither irritating nor sensitizing to skin.
- No repeated dose studies with inhalation are available.
- HBCD lacks significant genotoxic potential in vitro and in vivo.
- It is not possible to assess the carcinogenic potential of HBCD.
- A recent study has indicated that neonatal HBCD exposure may cause developmental neurotoxic effects.

Data about contamination levels of HBCD in diet [63] and in several environmental compartments [64–66] already exist. Those most relevant have been compiled in the risk assessment report of the ECB [61]. However, a quantitative analysis of HBCD in textile consumer products has only been conducted by Kajiwara et al. [67] recently, and data about the concentration of HBCD in samples from disposal operations (landfill, incineration) or recovery operations (recycling) are scarce or nonexistent, according to the information reported by the ECB [61]. HBCD concentrations analyzed by Kajiwara et al. [67] in flame-retarded upholstery textile samples ranged from 22,000 to 43,000 mg/kg (i.e., 2.2–4.3%).

6 Conclusions

The textile industry is spread over the world and implies the movement of products and additives at the same global scale. As a consequence, a product that is used in Spain can be produced in China and recycled in India. This shows the complexity of the chemical flows around the world.

As presented in this chapter, additives used in the textile sector are mainly applied during the finishing process and have many different uses, such as improving the antistatic properties of the textile and protecting it from microbial agents or fire. In any case, they have a specific function inside the final product.

Most of these additives may pose an environmental and health damage during the textile production, the product's use or at the end of its life. Taking into consideration the three additives studied in more detail in this chapter, examples on their potential damage are easy to find. For instance, PFCs are extremely persistent, bioaccumulative, and toxic compounds. TCS is highly toxic to aquatic organisms, and HBCD, one of the most extended flame retardants, is not readily biodegradable since it has a very high potential to adsorb to soils and sediments.

The assessment of the damage of these additives has been estimated in some cases during the production and use. However, difficulties arise when trying to characterize their risk at the end of the product's life. Data about the exposure levels to these substances during their last lifecycle stages (disposal or recovery) are scarce.

Therefore, it would be very interesting to be able to carry out research at this stage. Since most of the textile products end in landfills, all the additives present in clothing or textiles are going to end up mainly on the soil or groundwater, originating the associated pollution and being able to affect the environment and the population.

Efforts to select an appropriate methodology to assess the risk of additives at the end of the product's life are being made in the EU-funded RISKCYCLE project. The results of this research will be communicated to the competent authorities expecting from them to take the required actions on the use of the most dangerous additives.

References

1. Graedel T, Howard-Grenville J (2005) Textiles and leathers. In: Graedel T, Howard-Grenville J (eds) Greening the industrial facility: perspectives, approaches and tools. Springer, New York
2. Organisation for Economic Co-operation and Development (OECD) (2004) Emission scenario document on textile finishing industry. OECD Environmental Health and Safety Publications. http://www.oecd.org/officialdocuments/displaydocumentpdf/?cote=env/jm/mono%282004%2912&doclanguage=en. Accessed 22 Dec 2010
3. World Trade Organization (WTO) (2010) International trade statistics 2010. WTO. http://www.wto.org/english/res_e/statis_e/its2010_e/its10_toc_e.htm. Accessed 22 Dec 2010
4. US Environmental Protection Agency (EPA) (2011) Municipal Solid Waste (MSW) in the United States: facts and figures. http://www.epa.gov/osw/nonhaz/municipal/msw99.htm. Accessed 23 Mar 2011
5. Allow JM, Laursen SE, Malvido de Rodriguez CM, Bocken NMP (2006) Well dressed? The present and future sustainability of clothing and textiles in the United Kingdom. Institute for Manufacturing, University of Cambridge. http://www.ifm.eng.cam.ac.uk/sustainability/projects/mass/uk_textiles.pdf. Accessed 24 Mar 2011
6. Lacasse K, Baumann W (2004) Textile chemicals: environmental data and facts. Springer Verlag, New York
7. Nousiainen P (1997) Modern textile dyeing takes note of the environment. Kem Kemi 24(5):376–380
8. US EPA (1996) Manual. Best management practices for pollution prevention in the textile industry. EPA/625/R-96/004
9. BundesinstitutfürRisikobewertung (BfR) (2007) Introduction to the problems surrounding garment textiles. BfR. http://www.bfr.bund.de/cm/230/introduction_to_the_problems_surrounding_garment_textiles.pdf. Accessed 23 Dec 2010
10. Schindler WD, Hauser PJ (2004) Chemical finishing of textiles. Woodhead, Philadelphia
11. Tímár-Balázsy Á, Eastop D (1998) Chemical principles of textile conservation. Butterworth-Heinemann, Oxford
12. Miraftab M (2000) Technical fibres. Handbook of Technical Textiles. Wood head publishing Ltd; Cambridge, UK. pp. 29–30
13. Senthil Kumar R (2006) Textile finishing. KCT Coimbatore. http://es.scribd.com/doc/32352341/Textile-Finishing-Introduction Accessed 6 April 2011
14. Bajaj P (2002) Finishing of textile materials. doi:10.1002/app.2262
15. European Parliament and European Council (EPEC) (1998) Directive 98/8/EC of the European Parliament and of the Council of 16 February 1998 concerning the placing of biocidal products on the market. Official J Eur Commun. http://ec.europa.eu/environment/biocides/pdf/dir_98_8_biocides.pdf. Accessed 23 Dec 2010

16. European Flame Retardants Association (EFRA) (2007) Making textiles safer against fire. http://www.cefic-efra.com/Objects/2/Files/Making%20textiles%20safer%20against%20fire.pdf. Accessed 23 Mar 2011
17. Ceria A, Hauser PJ (2010) Atmospheric plasma treatment to improve durability of a water and oil repellent finishing for acrylic fabrics. doi:10.1016/j.surfcoat.2009.09.077
18. Abdel-Halim ES, Abdel-Mohdy FA, Al-Deyab SS, El-Newehy MH (2010) Chitosan and monochlorotriazinyl-β-cyclodextrin finishes improve antistatic properties of cotton/polyester blend and polyester fabrics. doi:10.1016/j.carbpol.2010.04.077
19. Hashem M, Ibrahim NA, El-Shafei A, Refaie R, Hauser P (2009) An eco-friendly – novel approach for attaining wrinkle-free/soft-hand cotton fabric. doi:10.1016/j.carbpol.2009.06.004
20. Lee ES, Kim SI (2005) Effect of additives on durable-press cotton fabrics treated with a glyoxal/glycol mixture. doi:10.1002/app.20751
21. Directive 2008/105/EC of the European Parliament and of the Council (Directive 2008/105/EC). http://eur-lex.europa.eu/LexUriServ/LexUriServ.do?uri=OJ:L:2008:348:0084:0097:EN:PDF
22. Directive 2000/60/EC of the European Parliament and of the Council (Directive 2000/60/EC). http://eur-lex.europa.eu/LexUriServ/LexUriServ.do?uri=OJ:L:2000:327:0001:0072:EN:PDF
23. United Nations (UN) (2009) Text and annexes as amended in 2009. Secretariat of the Stockholm Convention on Persistent Organic Pollutants. http://chm.pops.int/Convention/tabid/54/language/es-CO/Default.aspx. Accessed 23 Dec 2010
24. Prevedouros K, Cousins I, Buck R, Korzeniowski S (2006) Sources, fate and transport of perfluorocarboxylates. doi:10.1021/es0512475
25. Ericson I, Gómez M, Nadal M, Van Bavel B, Lindström G, Domingo JL (2007) Perfluorinated chemicals in blood of residents in Catalonia (Spain) in relation to age and gender: a pilot study. doi:10.1016/j.envint.2007.01.003
26. Seacat AM, Thomford PJ, Hansen KJ, Olsen GW, Case MT, Butenhoff JL (2002) Subchronic toxicity studies on perfluorooctanesulfonate potassium salt in cynomolgus monkeys. doi:10.1093/toxsci/68.1.249
27. Seacat AM, Thomford PJ, Hansen KJ, Clemen LA, Eldridge SR, Elcombe CR, Butenhoff JL (2003). Sub-chronic dietary toxicity of potassium perfluorooctanesulfonate in rats. doi:10.1016/S0300-483X(02)00511-5
28. Inoue K, Okada F, Ito R, Kato S, Sasaki S, Nakajima S, Uno A, Saijo Y, Sata F, Yoshimura Y, Kishi R, Nakazawa H (2004) Perfluorooctanesulfonate (PFOS) and related perfluorinated compounds in human maternal and cord blood samples: assessment of PFOS exposure in a susceptible population during pregnancy. doi:10.1289/ehp.6864
29. Llorca M, Farré M, Picó Y, López M, Álvarez JG, Barceló D (2010) Infant exposure of perfluorinated compounds: levels in breast milk and commercial baby food. doi:10.1016/j.envint.2010.04.016
30. Olsen GW, Burris JM, Ehresman DJ, Froelich JW, Seacat AM, Butenhoff JL, Zobel LR (2007) Half-life of serum elimination of perfluorooctanesulfonate, perfluorohexanesulfonate and perfluorooctanoate in retired fluorochemicals production workers. doi:10.1289/ehp.10009
31. Alexander BH, Olsen GW (2007) Bladder cancer in perfluorooctanesulfonyl fluoride manufacturing workers. Annals of Epidemiology 17(6):471–478
32. Emmett EA, Zhang H, Shofer FS, Freeman D, Rodway NV, Desai C, Shaw LM (2006) Community exposure to perfluorooctanoate: relationships between serum levels and certain health parameters. doi:10.1097/01.jom.0000233380.13087.37
33. Olsen GW, Butenhoff JL, Zobel LR (2009) Perfluoroalkyl chemicals and human fetal development: an epidemiologic review with clinical and toxicological perspectives. doi:10.1016/j.reprotox.2009.02.001
34. Steenland K, Tinker S, Frisbee S, Ducatman A, Vaccarino V (2009) Association of perfluorooctanoic acid and perfluorooctanesulfonate with serum lipids among adults living near a chemical plant. doi:10.1093/aje/kwp279
35. Apelberg BJ, Witter FR, Herbstman JB, Calafat AM, Halden RU, Needham LL (2007) Cord serum concentrations of perfluorooctanesulfonate (PFOS) and perfluorooctanoate (PFOA) in relation to weight and size at birth. doi:10.1289/ehp.10334

36. Fei CY, McLaughlin JK, Tarone RE, Olsen J (2007) Perfluorinated chemicals and fetal growth: a study within the Danish National Birth Cohort. doi:10.1289/ehp.10506
37. Lau C, Anitole K, Hodes C, Lai D, Pfahles-Hutchens A, Seed J (2007) Perfluoroalkyl acids: a review of monitoring and toxicological findings. doi:10.1093/toxsci/kfm128
38. OECD (2006) Report of an OECD workshop on perfluorocarboxylic acids (PFCAs) and precursors. OECD Environment, Health and Safety Publications. OECD Environmental Health and Safety Publications. http://www.oecd.org/officialdocuments/displaydocumentpdf/?cote=env/jm/mono%282007%2911&doclanguage=en. Accessed 22 Dec 2010
39. California EPA (2010) Groundwater information sheet. Perfluorooctanoic acid (PFOA) & related compounds. State Water Resources Control Board, Division of Water Quality, GAMA Program. http://www.swrcb.ca.gov/water_issues/programs/gama/docs/pfoa.pdf. Accessed 18 Jan 2011
40. Gou Z, Liu X, Krebs KA (2009) Perfluorocarboxylic acid content in 116 articles of commerce. US Environmental Protection Agency. http://www.epa.gov/nrmrl/pubs/600r09033/600r09033.pdf. Accessed 22 Dec 2010
41. Yamada T, Taylor PH, Buck RC, Kaiser MA, Giraud RJ (2005) Thermal degradation of fluorotelomer treated articles and related materials. doi:10.1016/j.chemosphere.2005.03.025
42. Olsen GW, Logan PW, Hansen KJ, Simpson CA, Burris JM, Burlew MM, Vorarath PP, Venkateswarlu P,Schumpert JC, Mandel JH (2003) An occupational exposure assessment of a perfluorooctanesulfonyl fluoride production site: biomonitoring. doi:10.1080/15428110308984859
43. Ericson I, Martí-Cid R, Nadal M, Van Bavel B, Lindström G, Domingo JL (2008) Human exposure to perfluorinated chemicals through the diet: intake of perfluorinated compounds in foods from the Catalan (Spain) market. doi:10.1021/jf0732408
44. Fromme H, Tittlemier SA, Völkel W, Wilhelm M, Twardella D (2009) Perfluorinated compounds – exposure assessment for the general population in western countries. doi:10.1016/j.ijheh.2008.04.007
45. Trudel D, Horowitz L, Wormuth M, Scheringer M, Cousins IT, Hungerbuhler K (2008) Estimating consumer exposure to PFOS and PFOA. doi:10.1111/j.1539-6924.2008.01017.x
46. Tissier C, Chesnais M, Migné V (2001) Supplement to the methodology for risk evaluations of biocides. Emission scenario document for biocides used as preservatives in the textile processing industry. Product type 9&18. INERIS-DRC-01-25582-ECOT-CTi/VMi-n°01DR0176. 21 pp
47. Oliveira R, Domingues I, Grisolia CK, Soares A (2009) Effects of triclosan on zebrafish early-life stages and adults. doi:10.1007/s11356-009-0119-3
48. Orhan M, Kut D, Gunesoglu C (2006) Use of triclosan as antibacterial agent in textiles. Indian J Fibre Text Res 32:114–118
49. Ying G, Kookana RS (2007)Triclosan in wastewaters and biosolids from Australian wastewater treatment plants. doi:10.1016/j.envint.2006.09.008
50. Orhan M, Kut D, Gunesoglu C (2007) Use of triclosan as antibacterial agent in textiles. Indian J Fibre Tex Res 32:114–118
51. Kolpin DW, Furlong ET, Meyer MT, Thurman EM, Zaugg SD, Barber LB, Buxton HT (2002) Pharmaceuticals, hormones and other organic wastewater contaminants in U.S. streams, 1999–2000: a national reconnaissance. Environ Sci Technol 36:1202–1211
52. Singer H, Muller S, Tixier C, Pillonel L (2002) Triclosan: occurrence and fate of a widely used biocide in the aquatic environment: field measurements in wastewater treatment plants, surface waters, and lakes sediments. Environ Sci Technol 36:4998–5004
53. European Commission (2010) Commission decision on the non-inclusion of 2,4,2'-trichloro-2'-hydroxyydiphenyl ether in the Union list of additives which may be used in the manufacture of plastic materials and articles intended to come into contact with foodstuffs Directive 2002/72/EC. http://eur-lex.europa.eu/LexUriServ/LexUriServ.do?uri=OJ:L:2010:075:0025:0026:EN:PDF Accessed 15 February 2011

54. US EPA (2008) 5-Chloro-2-(2,4-dichlorophenoxy)phenol (Triclosan): risk assessment for the Reregistration Eligibility Decision (RED) Document. Case No 2340
55. Delorenzo ME, Keller JM, Arthur CD, Finnegan MC, Harper HE, Winder VL, Zdankiewicz DL (2008) Toxicity of the antimicrobial compound triclosan and formation of the metabolite methyl-triclosan in estuarine systems. doi:10.1002/tox.20327
56. Farré M, Asperger D, Kantiani L, González S, Petrovic M, Barceló D (2007) Assessment of acute toxicity of triclosan and methyl triclosan in wastewater basedon the bioluminescence inhibition of *Vibrio fischeri*. doi:10.1007/s00216-007-1779-9
57. US EPA (2010) Triclosan facts. http://www.epa.gov/oppsrrd1/REDs/factsheets/triclosan_fs.htm. Accessed 23 Dec 2010
58. Ricart M, Guasch H, Alberch M, Barceló D, Bonnineau C, Geiszinger A, Farré M, Ferrer J, Ricciardi F, Romaní AM, Morin S, Proia L, Sala L, Sureda D, Sabater S (2010) Triclosan persistence through wastewater treatment plants and its potential toxic effects on river biofilms. doi:10.1016/j.aquatox.2010.08.010
59. Environmental Health Criteria Monographs (EHC 192) (1997) International Programme on Chemical Safety (IPCS), INCHEM. Flame retardants. http://www.inchem.org/documents/ehc/ehc/ehc192.htm. Accessed Oct 2010
60. Agencia Española de Seguridad Alimentaria y Nutrición (AESAN) (2010) BFR (Brominated-FlameRetardants). http://www.aesan.msc.es/AESAN/web/cadena_alimentaria/subdetalle/org_bfr.shtml. Accessed Oct 2010
61. European Chemicals Bureau (ECB) (2008) Risk assessment. Hexabromocyclododecane. Final report. Office for Official Publications of the European Communities. http://ecb.jrc.ec.europa.eu/documents/Existing-Chemicals/RISK_ASSESSMENT/REPORT/hbcddreport044.pdf. Accessed 22 Dec 2010
62. Lower N, Moore A (2007) The impact of a brominated flame retardant on smoltification and olfactory function in Atlantic Salmon (*Salmo salar* L.) smolts. Mar Freshw Behav Physiol 40(4):267–284
63. EljarratE, De la Cal A, Raldua D, Duran C, Barceló D (2005) Brominated flame retardants in *Alburnusalburnus* from Cinca River Basin (Spain). doi:10.1016/j.envpol.2004.06.017
64. D'Hollander W, Roosens L, Covaci A, Cornelis C, Reynders H, Van Campenhout K, De Voogt P, Bervoets L (2010) Brominated flame retardants and perfluorinated compounds in indoor dust from homes and offices in Flanders, Belgium. doi:10.1016/j.chemosphere.2010.07.043
65. Kunisue T, Takayanagi N, Isobe T, Takahashi S, Nakatsu S, Tsubota T, Okumoto K, Bushisue S, Shindo K, Tanabe S (2008) Regional trend and tissue distribution of brominated flame retardants and persistent organochlorines in raccoon dogs (*Nyctereutesprocyonoides*) from Japan. doi:10.1021/es071565z
66. Yu Z, Peng P, Sheng G, Fu J (2008) Determination of hexabromocyclododecanediastermers in air and soil by liquid chromatography–electrospray tandem mass spectrometry. doi:10.1016/j.chroma.2008.02.082
67. Kajiwara N, Sueoka M, Ohiwa T, Takigami H (2009) Determination of flame-retardant hexabromocyclododecanediastereomers in textiles. doi:10.1016/j.chemosphere.2008.11.046

Lubricants and Additives: A Point of View

Diego Baderna, Elena Boriani, Fabio Dalla Giovanna, and Emilio Benfenati

Abstract Lubricants are used to reduce the friction between two surfaces, improving efficiency and reducing wear. The most important application of lubricants is in the automotive sector where lubes are used to avoid damages to the engines. Additives are essential components of lubricants: they are added to lube to improve its physical and chemical properties. Additives also play a key role in the reduction of environmental burden by saving important resources, avoiding leakage losses and reducing exhaust gas emissions. Some concerns have been expressed regarding the reuse and disposal of used lubricating oils due to their possible toxicity for human and ecosystems. Most ecological effects are induced by the mineral base oils because additives are normally insoluble in water. Focusing on used oils, it is possible to say that the toxicity of these products derives from their improper disposal. Heavy metals, PCBs, dioxins, and PAHs can be released into the environment as a result of lube combustion. Modern lubricants do not contain halogenated additives; hence, very small amount of halogens may be present as contaminants from chemical processing. Moreover, modern re-refining techniques, like thermal deasphaltation, reduce the environmental impact and the toxicity of the used oils that otherwise could be disposed using incorrect ways with higher environmental load.

Keywords Additives, Lubricants, Metals, Oil re-refining, PAHs

D. Baderna (✉), E. Boriani, and E. Benfenati
Laboratory of Environmental Chemistry and Toxicology, Mario Negri Institute, Via Giuseppe La Masa, 19, 20156 Milan, Italy
e-mail: diego.baderna@marionegri.it

F. Dalla Giovanna
VISCOLUBE SPA, Via Tavernelle 19, 26854 Pieve Fissiraga (Lodi), Italy

B. Bilitewski et al. (eds.), *Global Risk-Based Management of Chemical Additives I: Production, Usage and Environmental Occurrence*, Hdb Env Chem (2012) 18: 109–132,
DOI 10.1007/698_2011_99, © Springer-Verlag Berlin Heidelberg 2011,
Published online: 16 March 2011

Contents

1 Introduction

A lubricant (or lube) is a substance introduced between two moving surfaces to reduce the friction between them, improve efficiency, and reduce wear.

The most important application for lubricants is in the sector of motor oil where lubes are used mainly to protect the internal combustion engines in motor vehicles and powered equipment. In addition to automotive applications, lubricants are used for many other purposes such as cooking, biomedical applications (e.g., lubricants for artificial joints), ultrasound examination, internal examinations, and the use of personal lubricants for sexual purposes.

A simple classification is based on the physical status of the lubes: liquid, solid, and gas.

Liquid lubricants are the most common and used type. Some examples are *water*, *lanolin*, *mineral oil*, *vegetable oils*, and *synthetic oils*.

Solid lubes, also known as dry lubricants, are the second most common type. They are used for lubrication under extreme conditions where the bearing surfaces in tribological contact must be effectively separated. They can be classified in structural lubricants and mechanical lubes. Some examples of structural lubes are *graphite*, *hexagonal boron nitride*, *molybdenum disulfide*, *tungsten disulfide*, *cobalt chloride*, *cerium fluoride*, *talc*, *borax*, and *mica*, while *polytetrafluoroethylene* (PTFE, also known as Teflon), *perfluoroalkoxyl*, *polyethylene*, *polypropylene*, *polyurethane*, and *polyamide* are examples of the mechanical type.

Finally, gas lubes like technical gases, steam or liquid–metal vapors, and air are used for specific technical engines due to their lower viscosity and higher compressibility if compared to liquid lubricants. In this chapter, we will focus mainly on liquid lubes.

Environmental issues and applications that require lubricants to operate under severe conditions will cause an increase in the use of synthetics. Owing to performance and maintenance reasons, many applications that have historically relied on petroleum-derived lubricants are shifting to synthetic lubricant-based products.

Legislation is driving changes to fuel composition and lubricant components, and therefore future lubricant developments will be constrained compared to what has been done in the past. Registration, Evaluation, Authorisation and Restriction of Chemicals (REACH) in the European Union (EU) is placing constraints on the incentive to develop new molecules that will serve as additives. The cost of introduction of new proprietary materials will be the burden of the company that develops the new material. For many common additives that are produced by several manufacturers, they will share costs to generate any needed data on the toxicology or biodegradability of the materials.

Regarding chemical composition, lubes are made of base oils and additives. In terms of volume, base oils are the most important components: they can account for 70–99% of lubricant formulation [1].

The chemical characterization of base oils has been subjected to drastic changes from the 1950s to nowadays according to the development of new engineering techniques and accounting of environmental, health, and safety criteria. Base oils might be synthetic or derived from petroleum crude oil (also known as mineral oils). It has been estimated that about 10% of base oils will be synthetic products in 2015.

Mineral oils can be classified into three categories based on the determination of carbon chemical bond: aromatic, naphthalenic, and paraffinic. A further refinement in the characterization of base oils is the identification of component molecular families in the saturated and aromatic fractions by chemical analyses. Saturated fractions include six classes (alkanes, 1-ring, 2-ring, 3-ring, 4-ring, and 5-ring naphthenes), while aromatic fractions are divided into seven groups (monoaromatics, diaromatics, triaromatics, tetraaromatics, pentaaromatics, thiopeno aromatics, and unidentified aromatics) [1].

According to the UNEP guideline, a used oil is an oil from industrial and nonindustrial sources which has been used for lubricating and has become unsuitable for its original purpose due to the loss of original properties and to the presence of contaminants or impurities [2]. Used mineral-based crankcase oil is similar to unused oil except that it contains additional chemicals that are produced or that build up in the oil when it is used as an engine lubricant. Used lubes have many of the characteristics of unused oil: it smells like unused oil and contains the chemicals found in unused oil.

Contaminants and impurities appear as a result of physical and chemical reactions occurring during the use due to high temperature or high mechanical strains. Main occurring reactions are nitration, cracking of polymers, oxidation, and decomposition of organometallic compounds. The most important compounds that can be detected as impurities include fuel, water, metals, metal oxides, and other combustion products that accumulate in the oil [3].

An important difference between new and used motor oil is the heavy metal content, originating from the fuel and from motor wear. This difference is extremely important because many of the metals are harmful to human health and living organisms.

1.1 From Crude Oil to Base Oil: The Refining Process

The process of converting crude oil into a finished base oil is named as refining. Refining requires different steps: distillation, deasphalting, solvent extraction, dewaxing, and finishing (Fig. 1) [4].

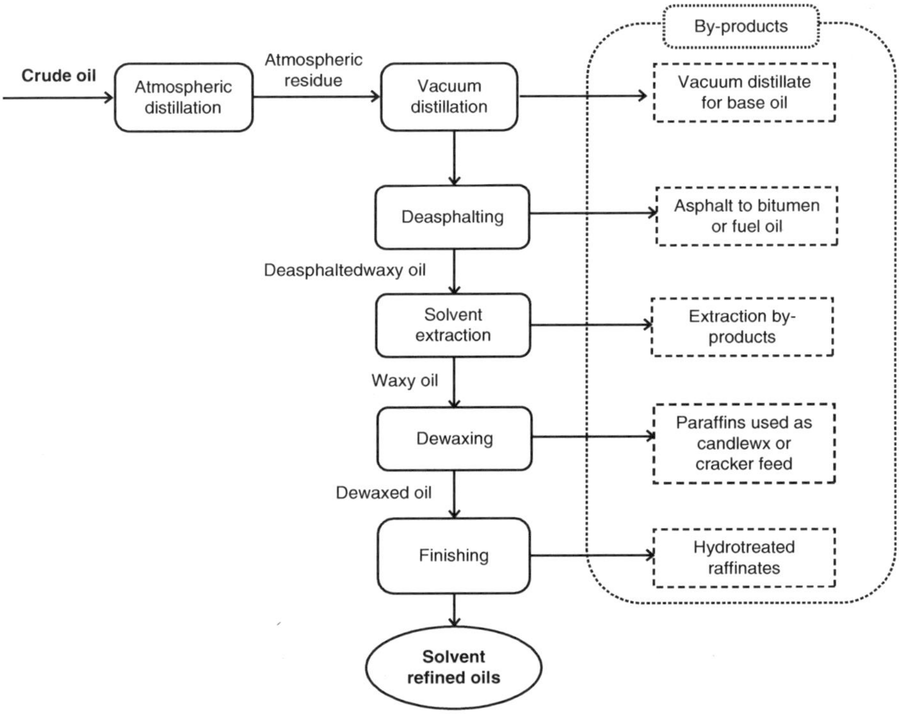

Fig. 1 Lubes processes

Distillation is the primary method to separate the useful fractions for lubes production from crude oil. Distillation includes atmospheric and vacuum distillation. Atmospheric distillation separates the lighter components from the crude oil, resulting in the presence of the lubricant components in the atmospheric residue. The atmospheric residue is injected in a furnace before entering the lower part of the vacuum column and then subjected to vacuum distillation to remove the components required for lubricants. Vacuum distillation cuts mainly determine the viscosity and flashpoint of later base oils.

Distillation cannot separate highly viscous hydrocarbons (also known as bright stocks), which can be used as lube base oil components, from the asphalt. The residue from vacuum distillation is a very viscous material containing large amounts of asphaltic and resinous components. Low molecular weight hydrocarbons are used as solvent to dissolve the useful compounds for lubes while leaving the asphaltic material as a separate phase. Liquid propane is the most used solvent.

Deasphalting process separates hydrocarbons basing on their solvent solubility: the extracted product, commonly known as deasphalted oil, is made of paraffinic components and compounds with low molecular weight, while the heavy fraction and the aromatic compounds remain in the asphalted matrix which can be used to produce bitumen or fuel oil. Solvent extraction replaces acid treatment as method for improving oxidative stability and viscosity characteristics of base oils. The solvent selectively dissolves the undesired aromatic components preserving saturated useful components such as alkanes. The process creates both base oils and an aromatic-rich extract. The selectivity of the extraction media is very important because of the cancerogenicity of the polycyclic aromatics.

Solvent extraction is then followed by solvent dewaxing to remove long-chain and high melting point paraffins. These compounds, in fact, can negatively affect the oil properties. Dewaxing is made by crystallization of paraffins at low temperature and separation by filtration. The separated paraffins can be sold as candle wax or used as cracker feed in fuel catalytic crackers.

Finally, a finishing stage follows the two above-mentioned refining stages to remove trace impurities. The hydrofinishing process is a mild hydrotreating used to improve color, odor, and ultraviolet stability of the oil removing in particular organonitrogen molecules.

1.2 Synthetic Base Oils

Synthetic base oils are available on an industrial scale since 1931, but their use on a large technical scale has increased only slowly because of their relative higher costs: if we consider as 1 the cost of mineral oils, the relative cost of synthetic oils varies from 2 to 800 [5].

Synthetic lubricants include *hydrocarbons*, *esters*, *polyalkylene glycols*, *polysiloxanes*, and *polyethers*. These lubes are prepared by reaction of a few well-defined chemicals in contrast to the mineral oil-based oils that contain different hydrocarbons and their derivates.

Most of the synthetic hydrocarbons are synthesized from *ethylene*. This group includes *polyalphaolefins* (PAO), *polyinternalolefins* (PIO), *polybutenes*, *alkylated aromatics*, and *chlorofluorocarbons*.

Synthetic esters were developed to satisfy the demand of lubes for aircraft engine oils. This family of compounds is represented by *dicarboxylic acid esters* (such as *di(2-ethylhexyl)adipate)*, *polyol esters*, and *phosphate esters* (*triaryl*, *trialkil*, and *alkylaril phosphates*).

Polyalkene glycols are derived from the reaction between epoxides and compounds containing active hydrogen in the presence of a basic catalyst. They are used as additives in lubes for metal-cutting fluids.

Silicone oils or *polysiloxanes* can be liquids or solids and they are straight-chain polymers of the *dimethylsiloxane* or *phenylmethylsiloxane series*. These products are mainly used as base oils for lubricating greases and for high-performance applications.

Finally, the group of *polyethers* includes *perfluorinated polyethers*, used as spacecraft lubes, and *polyphenyl ethers*, derived from the reaction of phenols with halogenated aromatic chemicals.

1.3 Lubricant Sales and Market

The market for lubricant additives is expected to grow [6]. New emerging countries such as China and India represent highly populated markets that are expected to see growth in infrastructure, and therefore a growth in industrial equipment and number of vehicles.

Data from 2004 indicate that about 40 million tons of lubricants were consumed worldwide. Analyzing the classification of these products, it is possible to assume that 53% of the total lubricants was used for automotive, 32% for industrial applications, 5% as marine oils, and 10% as process oils [1]. North America and Europe are the most important users and are responsible for about 50% of world lubricant consumption.

Different data are included in the report made by Kline on the global used oil in 2009 [7]. The global lubricant demand for 2009 was estimated at around 33 million tons with America and Pacific Asia as most demanding users. About 22.5 million tons of used oil could be potential generated at the global scale with a loss of about 10.5 million tons of lubes due to evaporation, leakages, spills, and other factors. Only 73% of the total generated oil was collected.

Collected oils can be used for re-refining fuel or other functions: about 16% (2.6 million tons) of the collected oils was sent to re-refining plants, while 78% (about 13 million tons) was used as fuel.

Focusing on re-refined products, just about 1.6 million tons of base stock were produced (61% of the total). Other main products were naphtha, diesel, and asphalt.

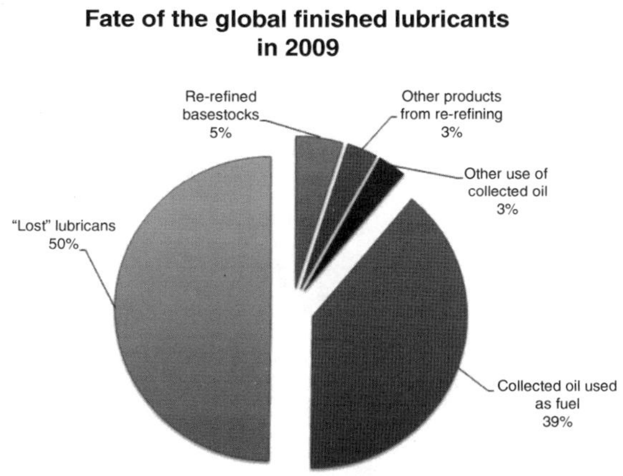

Fig. 2 Global production and fate of lubricants in 2009

Summarizing the data from Kline, from 33 millions of finished lubes, only 1.6 million tons are produced (Fig. 2).

1.4 *Oil Recycling in Europe*

The lubricants demand in EU27 is estimated to be about 5.7 million tons. According to the data kindly provided by GEIR (Groupement Européen de l'Industrie de la Régénération), in Europe there are 35 different plants working on used oil recycling which can treat more than 40% of the global re-refined base stock supply. Five of the most important re-refiners are operating in the Western Europe and account for more than 40% of the total supply: Safety-Kleen (UK), Avista oil (Germany), Puralube (Germany), Viscolube (Italy), and Eco Huile (France).

Focusing on the European landscape, EU27 collects on the average 35% of total consumed lubricants. Germany, UK, France, Italy, and Spain are the most lube-consuming countries, but, at the same time, they are the countries that collected the highest amount of used oils.

Data about EU27 waste oil utilization in 2006 are reported in Fig. 3. Starting from a total collected waste oil of approximately two million tons, most of the oil used is for burning purpose replacing coal and heavy fuel oil, while about one-third is re-refined to produce base oils.

To develop and maximize the used oil collection is the interest of re-refining in Europe for two linked reasons:

- The "lube-to-lube" re-refining is considered the best option to save a precious raw material.
- More than 70% of European re-refining capacity can produce high-quality grade products that can substitute virgin mineral-based oils.

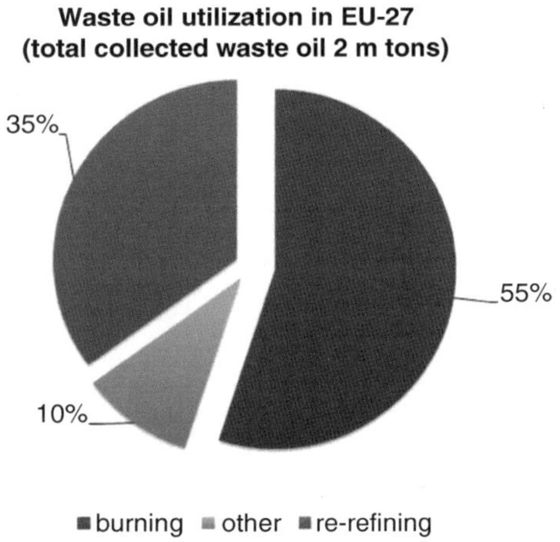

Fig. 3 Percentage distribution of waste utilization in Europe

2 Recycling of Used Oil

Used lubes are created when the product performances are compromised, especially when additives are subjected to aging processes and byproducts are present in the oil. They can represent an important problem for the environment if not correctly disposed.

There are different options to recycle or reuse used oils: first, it is possible to conserve the original properties of the oil and consequently directly reuse; the second option is to recover its heating value.

Re-refining could be considered the most environment-friendly way for oils disposal. Moreover, these techniques for disposal of used oils contribute to the reduction of the consumption of virgin oils: the process, in fact, is not able to create new complete lubes from used one but it can restore the value of base oil which can be added with additives to produce new lubes.

Re-refining is an energy-efficient process: less energy is required to produce a gallon of re-refined base stock than to produce a base stock from crude oil [8]. Energy cost of re-refining is lower than those of virgin base oils because the latter need different processes while the first only a purification and a finishing step (Fig. 4).

Base oils can be recycled *n*-times without any problem, and about 2.5 L of re-refined lubricants can be produced from 4 L of used oil, contributing to save fossil resources.

The EC Directive 87/101, concerning the disposal of waste oils, suggests that regeneration is generally the most rational way of reusing waste oils considering the energy savings which can be obtained. The legislation also mentions that regeneration must be preferred to combustion where technical, economical, and organizational

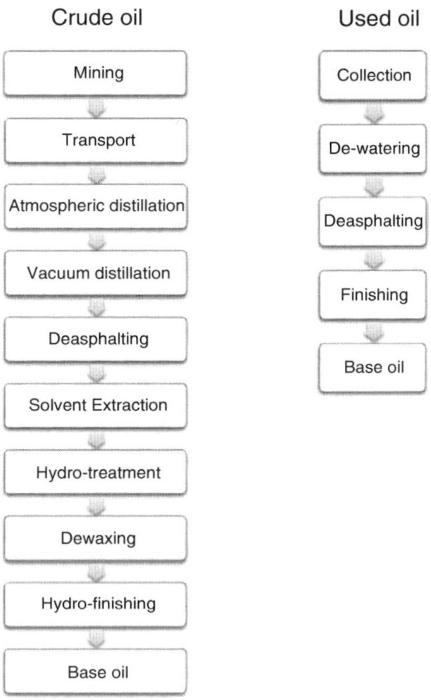

Fig. 4 Refining processes to obtain base oils from crude (*left*) and used oil (*right*)

constraints allow it. Regeneration is described as "any process whereby base oils can be produced by refining waste oils removing contaminants, oxidation products, and additives contained in such oils." Moreover, the directive states in the article 2 that EC members have to take the necessary measures to ensure that waste oils are collected and disposed without causing any avoidable harm to the humans and the environment. It has been demonstrated that in the countries where re-refining exists, the rate of waste oil collection is improved with great benefit for the environment.

If waste oils are not regenerated, members have to verify that any combustion of used oils is carried out under environmentally acceptable conditions (Art. 3) accounting of air pollution which can exceed the level prescribed by existing provisions.

The legislation also allows to stock waste oils that cannot be regenerated or burned and forbids the discharge of the used lubes into inland surface water, groundwater, territorial sea, drainage systems, and soil.

Focusing on regenerated oils (Art. 7), the directive states that members have to take the necessary measure to ensure that:

- The regeneration process plant will not cause damage to the environment.
- The process-derived base oils do not constitute a toxic and dangerous waste.

3 Additives

Additives are synthetic chemicals used to improve different lube parameters: they can boost existing properties, eliminate adverse characteristics, or introduce new properties in the base oil. They can be added to base oils at trace levels up to 20–30% of total weight.

Main functions of additives are: antioxidants, viscosity modifiers, pourpoint depressants, detergents, dispersants, antifoam agents, antiwears, friction modifiers, and antirust [9]. A brief description of the main additives is reported below.

3.1 Antioxidants

Antioxidant compounds are used to reduce/delay oil-aging processes, which can influence significantly the lifetime of lube base stocks. They can be classified into two main categories on the basis of their antioxidant mechanism: radical scavengers (or primary antioxidants) and peroxide decomposers (or secondary antioxidants).

Radical scavengers react with the radical oxidation products forming stabilized and nonreactive entity which block the propagation of radical oxidants. As a result, the autoxidation reaction is stopped. Instead, peroxides decomposers are able to prevent the propagation of oxidative reaction converting hydroperoxides into non-radical products.

There are different chemical categories which can be used as antioxidants: phenolic compounds, aromatic amines, and chemicals containing sulfur and phosphorus.

Phenolic antioxidants are typically phenols substituted with tertiary alkyl groups in the positions 2 and 6. Most common examples of this category are *2,6-di-tert-butylphenol (2,6-DTB), 2,6,-di-tert-butyl-4-methylphenol (BHT), and polynuclear phenols (such as 4,4′-methylenebis(2,6-di-tert-butylphenol))* (Fig. 5).

Aromatic amines can be an example of radical scavengers used in greases and polar lubricants. This group includes different oil-soluble secondary aromatic amines such as the *alkylated diphenylamines* (Fig. 6) and the polymeric *2,2,4-trimethyldihydroquinoline*.

Zinc dithiophosphates are the most spread compounds containing phosphorus used as radical scavengers. These antioxidants are a very important additive because

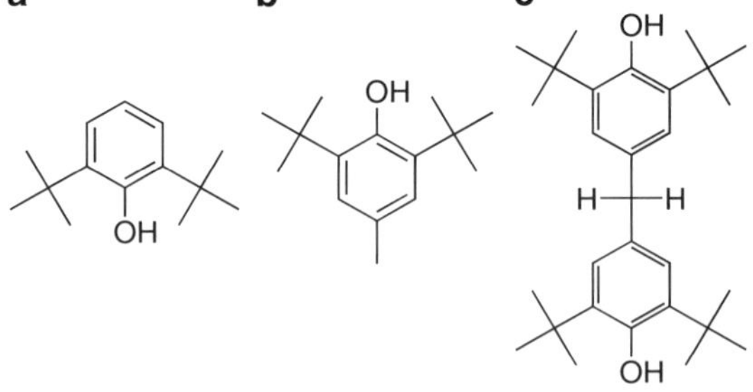

Fig. 5 (**a**) 2,6-DTB, (**b**) BHT, (**c**) 4,4′-methylenebis(2,6-di-tert-butylphenol)

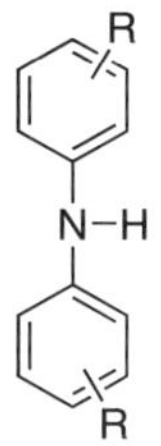

Fig. 6 Generic diphenylamines

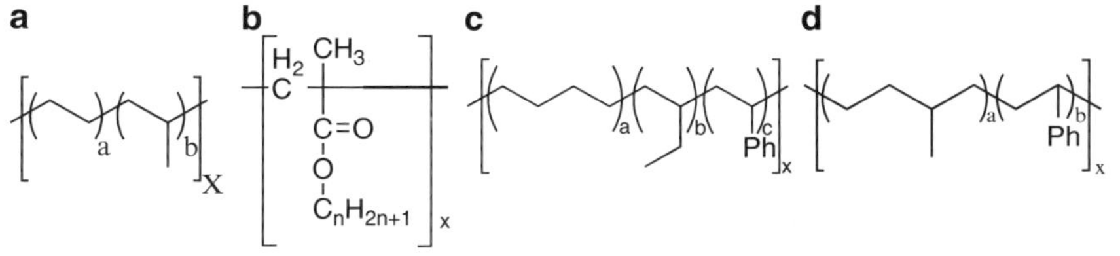

Fig. 7 Viscosity modifiers: (**a**) olefin copolymers, (**b**) polyalkylmethacrylate, (**c**) copolymers of styrene-isoprene, (**d**) hydrogenated styrene-butadiene

of their multifunctional properties such as antiwear, anticorrosion, and additive for extreme pressure.

Organosulfur compounds can be considered as typical examples of peroxide decomposers. This additive category includes *dialkyl-, diaryl- and poly- sulfides, modified thiols, dialkyldimercaptothiadiazoles, dialkyltdithiocarbamates*, and others.

3.2 *Pourpoint Depressants*

Focusing on lubes, the pour point can be defined as the minimum temperature of a liquid after which, on decreasing the temperature, the liquid ceases to flow.

Polyalkyl methylacrylates, polyacrylates, and *ethylene vinyl acetate copolymers* are commonly used as pourpoint depressants.

They are polymers closely linked to chemicals used as viscosity modifiers but they differs for the selected monomer building block. The mechanism of depression is based on their capability to interact with the crystals of paraffinic components resulting in an alteration of the morphology of that crystals.

3.3 *Viscosity Modifiers*

Viscosity is an important property of lubes, also used to classify them. The role of viscosity improvers is to give a desired viscosity index to lubricants.

These compounds are polymers classified on the basis of the chemical nature of their monomers (Fig. 7): *olefin copolymers, polyalkylmethacrylates, polyisobutylene, hydrogenated copolymers of styrene-isoprene*, and *hydrogenated styrene-butadiene*.

3.4 Detergents and Dispersants

Detergents and dispersants constitute 2–15% of the additives [3]. These compounds keep oil-insoluble combustion byproducts in suspension and prevent the agglomeration of the oxidation products into solid particles. Commonly, these chemicals are metal-containing compounds with a large oleophilic hydrocarbon portion and a polar hydrophilic region. Polar group can link lube contaminants, while the tail section act as a solubilizer in the base oil.

Most important classes of detergent are *phenates* (Fig. 8), *salicylates* (Fig. 9), *sulfonates*, and *thiophosphonates* (Fig. 10).

Phenates are synthesized by the reaction of alkylated phenols with sulfur chloride and metal oxides. The most widely use types are the *calcium phenates*. Regarding sulfonates, calcium compounds are relatively cheap additives with good general performances, while magnesium compounds are also excellent anticorrosion additives but they are sensible to thermal decomposition.

Ashless dispersants are free of metal compounds derived from hydrocarbon polymers. The most known examples of this family are *polybutenes* such as *polyisobutene succinimide* (PIBSI).

Fig. 8 Calcium phenates

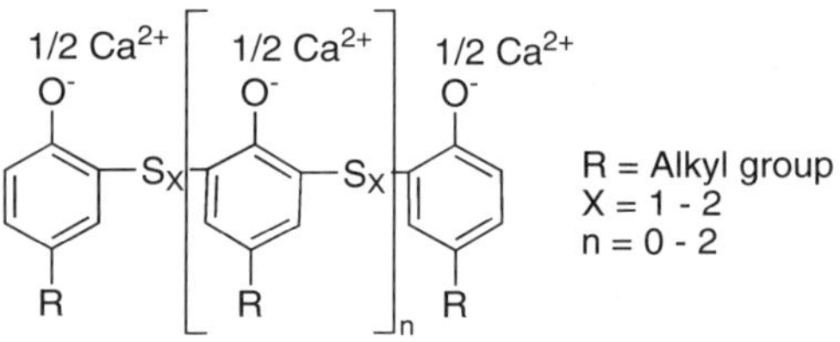

Fig. 9 General structure of salicylates

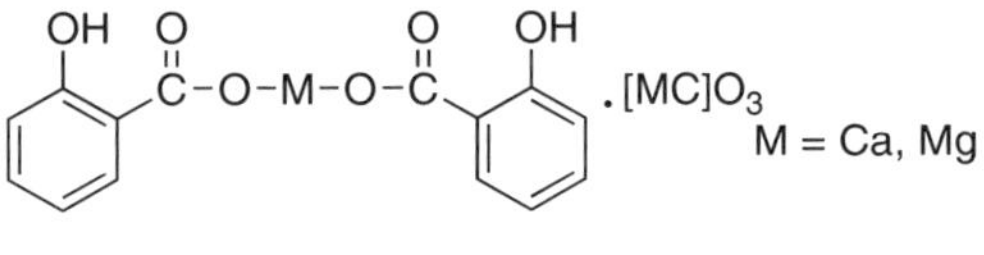

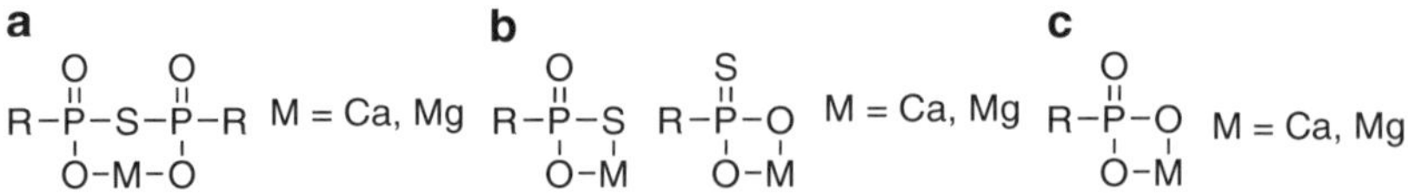

Fig. 10 Thiophosponates: (**a**) thiopyrophosphonate, (**b**) thiophosphonates, (**c**) phosphonate

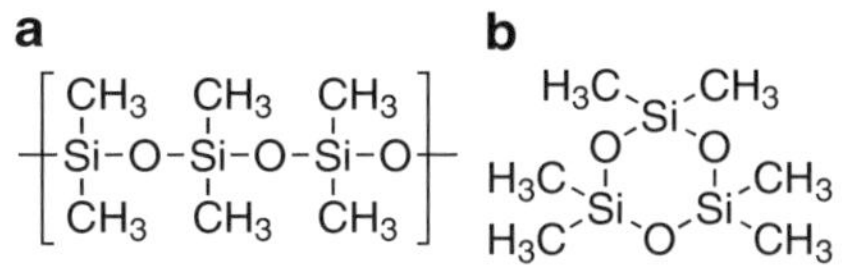

Fig. 11 Silicon defoamers: (**a**) linear, (**b**) cyclic

3.5 Antifoam Agents

Antifoam additives are used to prevent the foaming of lubes that can cause enhanced oxidation and consequently lube aging. Antifoam agents can be classified into two main categories: the silicon defoamers (Fig. 11) and the silicone-free agents. Liquid silicones are efficient antifoam agents at very low concentration (max. 100 mg/kg), while silicone-free agents, such as *polyethylene glycols* and *polymethacrylates*, antifoam compounds used in metalworking processes.

3.6 Antiwear and Extreme Pressure Additives

Antiwear agents and extreme pressure products are added to lube base oil to prevent welding of the moving and contacting parts of a machinery. The function of these additives is to form a covering layer on the metal surface of engine and gears. Antiwear compounds are designed to reduce wear in a moderate stress-exposed running systems, while extreme pressure additives are used to increase wear. Many of these chemicals cannot be easily classified into a precise "role" of anti- or prowear.

A possible classification based on chemical components is possible:

- Phosphorus compounds such as *trialkyl- and triaryl- phosphates*, *etoxylated mono- and di- alkylphosphoric acids*, *phosphonates*, and *phosphines*
- Sulfur and phosphorus compounds such as *zinc-*, *molybdenum-*, *ammonium-*, and *lead dialkyldithiophosphates*
- Sulfur- and nitrogen-containing chemicals, e.g., *dialkyl-2,5-dimercapto-1,3,4-thiadiazole (DMTD)*

3.7 Friction Modifiers

Friction modifiers are chemicals used to prevent stick–slip oscillations and noises acting directly on the frictional forces. These compounds can be classified into different categories based on their function: mechanically working (such as *molybdenum disulfide*, *graphite*, *PTFE*), adsorption layers forming (*fatty acid esters*, *amines*, *imides*, *long-chain carboxylic acids*), and friction polymers forming (including *methacrylates*, *unsaturated fatty acids*, and *sulfurized olefins*).

3.8 Antirust Compounds

Petroleum sulfonates, synthetic alkylbenzene sulfonates, carboxylic acid derivates such as oxidized paraffins, alkylated succinic acids, lanolin, and 4-nonylphenoxyacetic acid (Fig. 12) are the most commonly used antirust additives.

3.9 Nanomaterials

Some nanomaterials might be used as additives. Carbon black, white carbon black (superfine silicic acid particles), fullerenes, and 40-nm diameter microspheres of SiO_2 could be used as dispersant, antiwear, and extreme pressure additives. Different toxicological studies performed by IARC, RTECS, EPA, and WHO on carbon black have demonstrated the toxic effects and its possible carcinogenic potential (IARC classification 2B) [10–13]. Other nanomaterials are still under investigation to clarify their toxic potential and cancerogenicity in humans with particular attention on skin, respiratory systems, and eyes.

3.10 Zinc Dithiophosphates

A special attention must be reserved to zinc dithiophosphates (ZDTPs) that are used in the lubricant industry as low-cost and multifunctional additives for engine oils, transmission and hydraulic fluids, and greases.

ZDTP is an organometallic compound having four sulfur atoms coordinated to the zinc atom (Fig. 13) [14]. Pure ZDTPs with alkyl groups of four or less carbons are solid at ambient temperature and tend to have limited or no solubility in petroleum base stocks, while ZDTPs with more than five aryl or alkyl groups are liquid at ambient temperature.

ZDTPs are excellent antiwear agents, antioxidants, and corrosion inhibitors but also mild extreme pressure additives. The compounds can act simultaneously in these functions. The antioxidant functionality of ZDTP can be linked to its affinity for the peroxy radicals and hydroperoxides.

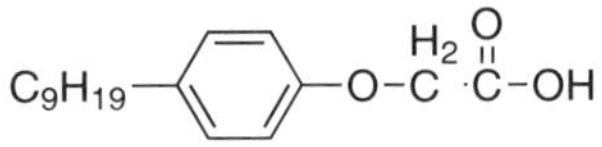

Fig. 12 4-Nonylphenoxyacetic acid

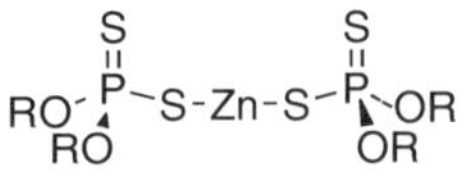

Fig. 13 Zinc dithiophosphate (ZDTP)

These compounds have many applications: they are used alone or in synergic combination with other additives in engine oils, hydraulic fluids, and extreme pressure applications as antiwear and antioxidant agents and in metalworking fluids, gear oils, and greases as extreme pressure additives.

Thermal and oxidative stabilities of ZDT depend on the type of alcohol used in the compound synthesis. Most reactive compounds are derived from secondary alcohols with low molecular weight, and the most stable ones derive from aryl and primary alkyl alcohols.

4 Environmental Impact of Lubricants

Lubricants and functional fluids are omnipresent due to their widespread use. Generally, they pollute the environment in small, widespread amounts and rarely in large, localized quantities [15]. Unfortunately, every year, about 40–50% of the approximately five million tons of used lubricants in Europe is released into the environment as a result of leaks, emissions, spillages, or other problems.

Considering lubricants in themselves, lubes are not particularly problematic from the environmental point of view if compared to other chemical products: in fact, the environmental damage is caused by the incorrect disposal of mineral oil-based lubes.

Used mineral-based crankcase oil may enter the water or soil when it is not properly handled, treated, or disposed of, for example, pouring into sewers or directly onto the ground. It may also enter the environment as a result of disposal in landfills, engine oil leaks, automobile or truck exhaust gas emissions, and application onto rural roads for dust control [16].

The uncontrolled release of used oils into environment can induce important damages for the ecosystem: for example, if oils reach natural waters, like rivers or lakes, each liter of lube can cover about 1,000 m^2 of water surface with a thin film that blocks the replenishment of dissolved oxygen at the water–air interface and resulting into an ecological damage for aquatic species [17]. Moreover, the hydrocarbon components of the oil that enter surface water can settle to the bottom contaminating sediments for many years.

The discharge of lubes on soil causes huge and long-term pollution due to the presence of lead, chlorinated solvents, and PCBs [18].

Releases of used mineral-based crankcase oil in urban stormwater runoff have been reported. Used mineral-based crankcase oil from engine leaks accumulates on parking lots, in garages, and in the central area of roadway lanes. Stormwater runoff from heavily traveled roadways and bridges is often contaminated with used mineral-based crankcase oil, with aliphatic hydrocarbon concentrations ranging from 0.20 to 24 mg/L (200–24,000 μg/L) [19–22]. Mineral oil products can be relatively rapidly biodegraded by the environmental microorganism, but, generally, these natural degrading systems are overwhelmed by the volume of the losses.

The movement in the environment of the chemicals found in used mineral-based crankcase oil depends on their individual properties. The hydrocarbon components of the oil generally stick to the soil surface and do not move through the soil. In this matrix, hydrocarbons may stay in the soil for a long time because they do not dissolve in water and do not generally break down. If oil is spilled, some hydrocarbons can evaporate into the air with different kinetics.

Regarding chemical and biological processes, such as microbial degradation and oxidation, that normally degrade hydrocarbons are less effective on used oil because the high molecular weight hydrocarbons and less-soluble aromatic chemicals are less susceptible to these environmental processes. Microbial degradation is dependent on different parameters like the hydrocarbon fractions present in the oil or other site-specific conditions such as the number of microorganisms present in the soil, sediment or water, and the environmental conditions, in particular temperature.

As a result of microbial degradation, metals can be released into the aqueous phase which can consequently increase not only the toxicity of the metals but also their environmental fate influencing the persistence and the mobility.

After degradation, heavy metals may then accumulate in plants, soil, sediments, surface water, and groundwater.

The combustion of used oils is a possible way of waste removal due to the chance of energetic recovery. The combustion products of used mineral-based crankcase oil include lead (4 g/kg of exhaust oil), cadmium zinc, chromium, aluminum, nickel, copper, other metal particles, sulfur, nitro-compounds, sulfur dioxide, phosphorus, calcium, hydrochloric acid, and nitrogen oxides [23]. Chlorinated compounds and PCBs can be also released into the air as a result of oil combustion. All these products may be released into the atmosphere when the oil is used as primary or supplemental fuel in industrial steam boilers, domestic oil burners, utility steam boilers, rotary cement kilns, or waste disposal incinerators.

The impact on waters of correctly disposed oils can lead to the hydrocarbon fractions obtained from physicochemical treatments. The process byproducts can be classified into two main categories: the acid slimes, derived from the treatment with sulfuric acid, and the exhaust active earths used in the filtration and bleaching processes.

Acid slimes are 20–25% of the weight of treated oils, and their composition is variable containing sulfuric acid, hydrocarbons, heavy metals, sulfur, phosphorus, and nitrogen. Slimes are usually stock into landfills, but they can be also used as unconventional fuels at cement plants due to their calorific value (about 4,000 kcal/kg of slimes).

The exhaust bleach earths adsorb about 30–60% of oil organic fractions. They are commonly disposed in landfills or used as unconventional fuels. Their environmental impact is limited if compared to the acid slimes because of the produced quantity and because earths are mainly composed by inert materials like silicates.

A possible way to estimate the environmental impact of lubes can be based on biodegradation estimation and ecotoxicity testing. Focusing in particular on ecotoxicology, different parameters and criteria must be considered: bacteria toxicity using *Pseudomonas* type, the algae toxicity, fish bioassays using *Leuciscus idus*, the

Daphnia test as models for the aquatic toxicity, and wheat, cress, and rape seeds test to evaluate the terrestrial soil and plants toxicity. All these bioassays must be performed according to OECD or DIN guidelines to compare data. Additional studies can be done to improve the toxicological profile including mammal and human toxicity and occupational safety investigations.

A recent life cycle assessment (LCA) study [24] provides an updated vision of the ecological and energetic aspects of re-refining of used oil analyzing five advanced techniques adopted by four companies operating in Europe and one in the USA. The studied techniques were respectively based on hydrotreatment technology (Cyclon Hellas SA), vacuum distillation and hydrofinishing (Evergreen), hydrogenation of base oils and catalytic hydrotreatment (PURALUBE), modern solvent extraction (Mineralöl-Raffinerie Dollbergen), and, finally, thermal deasphalting and hydrofinishing (Viscolube). The assessment was done considering the environmental impact and the environmental benefit due to the substitution of primary products by the above-mentioned techniques. Fossil energy resource depletion, global warming, terrestrial nitrification, acidification, and toxicity (with respect to carcinogenic pollutants and fine particulates) were the analyzed environmental impact categories. The assessment of the overall impact or benefit of re-refining in relation to the substitution of primary products demonstrates that both systems are directly comparable, and the overall balance is in favor of re-refining because the recycling process causes less environmental impact compared to the process used to obtain base oil from crude oil.

The second goal of the LCA study is to clarify whether re-refining might be more beneficial to the environment than combustion. In contrast to most of the LCA studies performed in the past and concluded with a no difference evaluation between regeneration and combustion, results indicate that clear advantages concerning fossil resources, acidification, and toxic air pollutants are in favor of regeneration, while the discrepancies related to nitrification between the two systems are low.

Concluding, it is possible to assume that lubricants (new, used, and re-refined ones) can alter environmental quality due to their composition characterized by heavy metals, polycyclic aromatics hydrocarbons (PAHs), fuel and solvent residues, and other nonmetallic impurities. Three major aspects can be addressed to limit environmental damages:

- Control to minimize losses
- Minimizing the impact of the above-mentioned losses
- Efficient collection and treatment of waste materials

5 Impacts of Lube on Human Health

Used mineral-based crankcase oil is a complex mixture of low and high (C15–C50) molecular weight aliphatic (paraffins, olefins, and acetylene hydrocarbons and their derivatives) and aromatic hydrocarbons, lubrication additives, metals, and various organic and inorganic compounds [23]. Toxicity of used mineral-based crankcase oil has not been fully investigated, but it is possible to assume that the toxicity of used oil is higher if compared with unused oils. The toxicity is mainly attributable to some contaminants such as PAHs, lead, molybdenum, chromium, zinc, cadmium, copper, and silicon. These pollutants can originate as a consequence of the use or as degradation products of additives. The degree of chemical change and accumulation of contaminants in the oil increases with use and varies depending on the type of fuel used and the mechanical properties of the engine.

The chemicals found in used mineral-based crankcase oil vary depending on the type of engine oil used, if gasoline or diesel fuel was used, the mechanical condition of the engine that the oil came from, the various sources of used mineral-based crankcase oil (e.g., automobiles, airplanes, trains, ships, tractors, lawn mowers), and the number of miles driven between oil changes [23].

5.1 In vivo Studies

Acute exposure to used oils can result in important adverse effects on pulmonary efficiency, mainly due to the heavy metals content. For example, cadmium can lead to pulmonary edema and, in some case, chemical pneumonitis [23]. Some in vivo studies were performed since 1950 to evaluate the overall toxicity due to the inhalation of used oils, resulting in nose and throat irritation and chest tightness. Also animal acute tests (1 h) were done to evaluate possible adverse effects, but no effects on pulmonary functions were obtained using exposure concentration of about 200 mg/m^3.

Studies on hematotoxicity and vascular systems conducted in 1970 have evidenced a significant correlation between the lead level and delta-aminolevulinic acid dehydratase (δ-ALAD) activity in the blood of both normal subjects and autoworkers. In the blood of autoworkers, the δ-ALAD activity was depressed because of the higher blood lead levels that was significantly higher than that of control subjects. δ-ALAD is one of the most sensitive and one of the first enzymes in the pathway for heme synthesis, and it is an early biological indicator of subclinical lead poisoning in humans and birds [23]. Moreover, lower than normal hematocrit and mean corpuscular hemoglobin were observed in several mechanics and apprentice mechanics in an epidemiologic study examining health effects among auto shop workers.

Regarding other systemic effects, skin and eye irritations have been reported probably due to direct contact of aerosols or volatile components with these tissues. No studies were located regarding cancer in animals after inhalation exposure to used mineral-based crankcase oil.

Focusing on oral exposure, no deaths were observed in rats for up to 14 days after ingestion of 22,500 mg/kg of used mineral-based crankcase oil. However, increased mortality was observed among cattle believed to have accidentally ingested discarded, used crankcase oil mainly due to lead and molybdenum poisoning resulting in a central nervous system (CNS) hypoxia [23]. Possible adverse long-term effects on CNS due to oral exposure to used oils are shown to be induced in cattle. CNS disorders were attributed to lead poisoning (blindness, muscle twitching, hyperirritability, depression, and convulsions) and molybdenum-induced copper deficiency (muscle tremors, weakness, and hypochromic microcytic anemia).

A possible no observed adverse effects level (NOAEL) of 8 ml/kg can be obtained from 2-week studies conducted on rabbit due to the initial onset of hepatic symptoms following dermal exposure.

5.2 Toxicokinetics

Only a very limited number of animal studies were performed and no human studies were done on the toxicokinetics of used mineral-based crankcase oils.

Data derived from cattle poisoning indicate that metals from used oil may be absorbed and distributed to various tissues following oral exposure, and feces are the most significant pathway for excretion. Concerning single toxic, results shown that lead is stored in the skeletal and soft tissue pool, cadmium is accumulated in the kidneys with an half-life of about 30 years, while molybdenum is stored in the liver and rapidly excreted in the urine and in the bile.

PAH-DNA adducts induced in the skin and lungs of male mice dermal exposed to used crankcase oil indicate that absorbed PAHs are distributed to various tissues. As PAHs are lipophilic compounds, they are mainly stored in adipose tissue and secreted in milk.

5.3 Carcinogenic Potential

Analyzing carcinogenic potential of used oils, PAHs can be considered as a possible matter of concern. PAHs, in fact, are well-known high reactive procarcinogens which have to be bioactivated either directly or indirectly to ultimate carcinogens. Bioactivation takes place mainly in liver but also lungs and skin participate in the process: PAHs is metabolized by the CYP450 system into epoxides that can bind to DNA to form DNA adducts, resulting in necrosis and/or cancer [25].

A huge amount of animal studies were conducted on cancer induction by used crankcase oils to obtain information for human health protection. Dermal application of these products to the shaved backs of mice produced DNA adducts in the skin and the lungs, indicating that used mineral-based crankcase oil was genotoxic. In some cases, a good correlation within total PAH contents and adducts levels were found, while focusing on single compounds the most significant correlations was obtained by Carmichael et al. in 1990 [26] with the reactive metabolites such as the chrysene, benzo(*a*)pyrene, benzo(g,h,i)perylene, benzo(*c*)phenanthrene, and benzo (g,h,i)fluoranthene.

Others evidences were produced in 1989 [27] by testing oil from cars driven the longest distances before removing the oil. The greatest tumor incidence was observed in mice exposed to oil from cars driven the longest distances before removing the oil, while no tumors were observed in mice exposed to unused motor oil, indicating that carcinogens accumulated in the oil during its use. The increase in carcinogenicity was attributed to accumulation of PAHs in the oils because the tumor incidence correlated with PAH content of the oil. Moreover, PAHs can affect the toxicity of other chemicals and promote carcinogenesis in the liver, induce immunosuppression, and negatively affect reproduction system [25].

Also some chronic studies (80–104 weeks) has been done in mouse showing serious effects after dermal exposure to 60–1,667 mg/kg of used crankcase oils [23].

5.4 Occupational and Nonoccupational Exposures

A review of the literature on toxic effects of used mineral-based crankcase oil made by ATSDR [23] has not identified specific populations that are known to be unusually susceptible to used oils.

People occupationally/repeatedly exposed to used motor oil would potentially be a very susceptible group:

- Workers involved in the remediation of environmental contamination from used oil re-refining or recycling facilities may inhale contaminated particulates or volatiles or be dermally exposed to contaminated soil, sludge, and sediment.
- Workers involved in collecting used mineral-based crankcase oil for recycling or re-refining may also be dermally exposed.
- Individuals who work or live in buildings that burn used mineral-based crankcase oil for heating fuel may be exposed to high levels of metal particulates in the respirable fraction [28–30].

General population may be exposed to used mineral-based crankcase oil when checking the engine oil level, but the resulting exposure to used oil is minimal. Individuals who live near roads that have been treated with used mineral-based crankcase oil as a dust suppressant may be exposed to dust particles that are contaminated with used oil.

5.5 Toxicity of Impurities and Additives

Focusing on the organic pollutants, heavy molecular incomplete combustion products such as PAHs and their analogs can accumulate in the used oil by a factor of up to 1,000 [22, 31]. In the used oil of crankcase-lubricated engines, 140 different PAHs have been found. PAHs are not created when lubricants are used: they are formed by the combustion of gasoline or gather in quenching oils after a long period of heavy-duty use. Moreover, PAHs can leave the engine in different ways, such as via particulates, oil leaks, and uncontrolled oil changes causing serious environmental damages.

PAHs are known to be highly toxic environmental contaminants with carcinogenic and mutagenic properties as previously established by the International Agency for Research on Cancer (IARC) in 1984 [32]. From a toxicological point of view applied to lubes, PAHs with three to seven rings such as benzo-α-pyrene, benzo-α-anthrachene, and chrysene can be considered the most important carcinogenic pollutants present in the used oil. It is important to underline that PAHs are also present in crude oil and in the unused base oil fluids. These compounds are often found in re-refined oil at a higher level than those found in new base oils.

Small amounts of PCBs were in the past incorporated into transmission fluids to control swelling of rubber seals [23] and could be detected in old used oil samples at ppm levels. It should be noted that PCBs have never been used in lubricant additives or in the creation of new motor oils. Because PCBs can form dioxins, which has severe health implications, production of PCBs for commercial purposes has almost ceased.

Concerning on metals, used oil contains high concentrations of lead, zinc, calcium, barium, and magnesium along with lower concentrations of iron, sodium, copper, aluminum, chromium, manganese, potassium, nickel, tin, silicon, boron, and molybdenum [23] originated as additive byproducts and derived from motor wear.

If considered as themselves, several of the oil additives, such as ZDTPs and phosphorus-containing chemicals, are toxic environmental contaminants due to the presence of heavy metals.

ZDTPs have a low acute systemic toxicity, and they can cause eye damage and skin irritation [33, 34]. Prolonged exposure to high concentrations of ZDTPs, calcium alkyl phenates, and magnesium, sodium, and calcium sulfonates had significant effects on the reproductive organs of male rabbits (testicular atrophy and reduction or absence of spermatozoa) which appeared to be species specific.

Some phosphorus-containing products were subjected to detailed investigation to clarify their toxicity [35, 36]. In particular, *tricresyl phosphate* (TCP) were deeply studied as a result of the *o*-cresol content in the feedstock as tri-ortho-cresyl phosphate (TOCP), a suspected neurotoxins. The absorption of tricresyl phosphates caused peripheral nervous system damage, leading to neuromuscular problems. Results from these studies also suggest that any isomer containing *o*-cresyl moiety are neurotoxic, and, for these reasons, the *o*-cresol content in the feedstock used for TCP production has been progressively reduced. Nowadays, levels of *o*-cresol are frequently less than 0.05% and TOCP levels are lower than ppb. However, in view of the past concerns, the use of TCP is now largely restricted to aviation gas turbine oils, and in most general applications this compound is substituted by other additives which are not able to induce organophosphorus-induced delayed neurotoxicity (OPIDN) after acute oral ingestion.

5.6 REACH and Lubricants

REACH stay for Registration, Evaluation, Authorisation and restriction of Chemicals. It is an European regulation (Regulation EC1907/2006) that came into force on 1 June 2007.

It impacts all substances manufactured or imported to several countries (Austria, Belgium, Bulgaria, Cyprus, Czech Rep., Denmark, Estonia, Finland, France, Germany, Greece, Hungary, Ireland, Italy, Latvia, Lithuania, Luxembourg, Malta, Netherlands, Poland, Portugal, Romania, Slovakia, Slovenia, Spain, Sweden, UK, Iceland, Liechtenstein, Norway, French Guiana, Guadeloupe, Martinique, Reunion, Channel Islands, Isle of Man, Dutch Antilles) in quantities greater than 1 ton/year.

Everyone in the lubricants supply chain has a role to play in the implementation of REACH. Each party (customers, importers, suppliers) needs to be aware of their respective obligations and prepared to communicate information on product usage and safe handling procedures, up and down the supply chain.

Base oils are defined as "substances" for REACH legislation, while blended lubricants are defined as "preparation" and they can also be used in articles. Both are affected by REACH legislation because they are produced/imported in quantity more than 1 ton/year.

For example, there are compounds such as alkanes, C10–13, and chlorinated compounds (such as short-chain chlorinated paraffins, SCCPs) that are used primarily as a cooling lubricant and hydraulic oil in metal working operations outside the EU and also as a flame retardant in rubber. In addition, they can be used as a plasticizer, in paints, varnishes, and sealants. These chemicals are of particular concern because they meet the persistent, bioaccumulative, and toxic criteria as well as the very persistent and very bioaccumulative criteria (vPvB) (Art. 57 d und e REACH Regulation). For these reasons they are included in the list of substances of very high concern (SVHC), which is considered a priority list for action to be taken within REACH.

Substance Information Exchange Fora (SIEF) are fora to help registrants who intend to register the same substance. SIEF facilitate data sharing between the companies, hence to avoid duplication of studies (unnecessary testing) and lead to agreed classification and labeling.

The most active SIEF regarding lubricants, is undertaken by CONservation of Clean Air and Water in Europe (CONCAWE – www.concawe.be). SIEF are the main responsible for the management and compilation of information for products registration under REACH. The lubricants registration is complicated, e.g., one of the product categories consists of 78 different base oil CAS numbers.

Access is granted by "license" to companies allowing use of the CONCAWE IUCLID5 files and Chemical Safety Report (CSR)for the registration of virgin base oils.

Furthermore, base oils and preparation are often UVCB substances (substances of Unknown or Variable composition, Complex reaction products or Biological materials). These substances have additional identification requirements due to their unknown or variable composition. Identifiers such as source, manufacturing process, and genetic code may be required to fully define the substance.

References

1. Mang T, Dresel W (2007) Lubricants and lubrification, 2nd edn. Wiley-VCH GmbH, Weinheim. ISBN 978-3-527-31497-3
2. UNEP (2002) Basel convention: technical guidelines on used oil re-refinining of other re-uses of previously used oil. Available from Secretariat of the Basel Convention, International Environment House, Chatelaine, Switzerland
3. Vazquez-Duhalt R (1989) Environmental impact of used motor oil. Sci Total Environ 79(1):1–23
4. Mang T, Lingg G (2007) Base oils. In: Mang T, Dresel W (eds) Lubricants and lubrification, 2nd edn. Wiley-VCH GmbH, Weinheim. ISBN:978-3-527-31497-3
5. Dresel W (2007) Syntethic base oils. In: Mang T, Dresel W (eds) Lubricants and lubrification, 2nd edn. Wiley-VCH GmbH, Weinheim. ISBN:978-3-527-31497-3
6. Rudnick LR (2009) Lubricant additives: chemistry and applications, 2nd edn. CRC Press, Taylor & Francis Group, Boca Raton, FL
7. Kline Group (2010) Global used oil 2009: market analysis and opportunities.
8. USEPA (1996) Managing used oil: advice for small businesses. EPA530-F-96-004
9. Braun J (2007) Additives. In: Mang T, Dresel W (2007) Lubricants and lubrification, 2nd edn. Wiley-VCH GmbH, Weinheim. ISBN:978-3-527-31497-3
10. International Agency for Research on Cancer (IARC) (1996) Carbon black (IARC Summary & Evaluation), vol 65
11. RTECS (2004) The registry of toxic effects of chemical substances. carbon black. http://www.cdc.gov/niosh/rtecs
12. World Health Organization (WHO) (1988) WHO Food Additives Series 22. Toxicological Evaluation of Certain Food Additives-Carbon Black. Prepared by the Joint FA O/WHO Expert Committee on Food Additives (JECFA). World Health Organization, Geneva
13. IPCS (2001) International Programme on Chemical Safety. Carbon Black. Chemical Safety and the Commission of the European Communities
14. McDonald RA (2009) Zinc dithiophosphates. In: Rudnick LR (eds) Lubricant additives: chemistry and applications. CRC Press Taylor & Francis Group, Boca Raton, FL. ISBN:978-1-4200-5964-9
15. Luther R (2007) Lubricants in the environment. In: Mang T, Dresel W (eds) Lubricants and lubrification, 2nd edn. Wiley-VCH GmbH, Weinheim. ISBN:978-3-527-31497-3
16. Bressi G (1997) Recupero degli oli usati. Collana monografie dell'Istituto per l'Ambiente. Milano, Italy
17. United States Environmental Protection Agency (USEPA) (1989) How to set up a local program to recycle used oil. Solid Waste and Emergency Response (OS-305), EPA530-SW-89-039A
18. Dean NL (1991) Encouraging consumers to buy re-refined oil. In: Proceedings 6th International Conference on used oil recovery and reuse "re-refining rebirth", San Francisco, May 28–3, pp 11–14
19. Wakeham SG (1977) A characterization of the sources of petroleum hydrocarbons in Lake Washington. J Water Pollut Control Fed 49:1680–1687
20. Hunter JV, Sabatino T, Gomperts R et al (1979) Contribution of urban runoff to hydrocarbon pollution. J Water Pollut Control Fed 51:2129–2138
21. Mackenzie MJ, Hunter JV (1979) Sources and fates of aromatic compounds in urban stormwater runoff. Environ Sci Technol 13(2):179–183
22. Van Donkelaar P (1990) Environmental effects of crankcase- and mixed-lubrication. Sci Total Environ 92:165–179
23. Agency for Toxic Substances and Disease Registry (ATSDR) (1997) Toxicological profile for used mineral-based crankcase oil. Available at: http://www.atsdr.cdc.gov/ToxProfiles/tp102.pdf
24. Fehrenbach H (2005) Ecological and energetic assessment of re-refining used oils to base oils: substitution of primarily produced base oils including semi-synthetic and synthetic compounds. Available at: http://www.noranews.org/attachments/files/106/IFEUReportLongVersion.pdf

25. Lu FC (1991) Basic toxicology: fundamentals, target organs, and risk assessment, 2nd edn. Hemisphere Publishing Corporation, New York, pp 93–115
26. Carmichael PL, Jacob J, Grimmer G et al (1990) Analysis of the polycyclic aromatic hydrocarbon content of petrol and diesel engine lubricating oils and determination of DNA adducts in topically treated mice by phosphorus-32-postlabelling. Carcinogenesis (London) 11(11):2025–2032
27. McKee RH, Plutnick RT (1989) Carcinogenic potential of gasoline and diesel engine oils. Fundam Appl Toxicol 13(3):545–553
28. Cooper JS (1977) The lead problem associated with recycled oil. National Bureau of Standards Special Publication No. 488. In: Proceedings of a Workshop on Measurements and Standards for Recycled Oil held at National Bureau of Standards, Gaithersburg, MD, November 22–23, 1976, pp 103–104
29. Hall RE, Cooke WM, Barbour RL (1983) Comparison of air pollutant emissions from vaporizing and air atomizing waste oil heaters. J Air Pollut Control Assoc 33(7):683–687
30. Mumford JL, Hatch GE, Hall RE et al (1986) Toxicity of particles emitted from combustion of waste crankcase oil: In vitro and in vivo studies. Fundam Appl Toxicol 7(1):49–57
31. Scheepers PTJ, Bos RP (1992) Combustion of diesel fuel from a toxicological perspective. I. Origin of incomplete combustion products. Int Arch Occup Environ Health 64:149–161
32. IARC (1984) Classification of mineral oils according to their carcinogenicity, vol 33
33. Hewstone RK (1994) Environmental health aspects of lubricant additives. Sci Total Environ 156(3):243–254
34. Hewstone RK (1994) Health, safety and environmental aspects of used crankcase lubricating oils. Sci Total Environ 156:255–268
35. Phillips WD (2000) Phosphate ester hydraulic fluids. In: Totten GE (ed) Handbook of hydraulic fluid technology. Marcel Dekker, New York
36. Daughtrey WR, Biles B, Jortner M (2006) Delayed neurotoxicity in chickens: 90 day study with Mobil Jet Oil 254. The Toxicologist, 90, Abstract 1467

Additives in the Plastics Industry

Lauran van Oers, Ester van der Voet, and Veit Grundmann

Abstract To obtain a picture of environmental and health impacts related to plastics additives, data are required with respect to the production and use of plastics and of their additives, to emissions of additives in all life cycle stages, and to the impacts resulting from those emissions. Statistics on plastics production, trade, and sales are available. Statistical data on plastics waste are scarce – waste trade data are available but most probably do not cover all relevant flows, plastics waste generation data are available only for the EU, and plastics waste treatment data are mostly absent. For the production and use of additives, reports from market analysts provide relevant and usable information. With regard to emissions to the environment, Life Cycle Inventory data are available for plastics production, but these do not include additives. Additive production data are scarce. It is, therefore, not possible to obtain a picture of plastics-related emissions of additives to the environment. For the use and waste treatment processes, LCI data with relevance for plastics additives are not available at all. Life Cycle Impact Assessment data are available for a number of additives, and could be supplied relatively easily for others. The most important data gaps, therefore, occur in the emissions data and in the data related to waste generation and waste treatment.

Keywords Additive, Environmental impact, LCA, Life cycle analysis, Plastic, Use, Waste treatment

L. van Oers (✉) and E. van der Voet
Institute of Environmental Sciences, Leiden University, Einsteinweg 2, 2333 CC, Leiden, The Netherlands
e-mail: oers@cml.leidenuniv.nl

V. Grundmann
Institute of Waste Management and Contaminated Sites Treatment, Dresden University of Technology, Pratzchwitzer Str. 15. 01796, Pirna, Germany

B. Bilitewski et al. (eds.), *Global Risk-Based Management of Chemical Additives I: Production, Usage and Environmental Occurrence*, Hdb Env Chem (2012) 18: 133–150,
DOI 10.1007/698_2011_112,
Published online: 17 August 2011

Contents

1 Introduction

Plastics are produced in large quantities and extensively used in society. Applications can be found in many, widely different fields: packaging, building and construction, electronics, furniture, agriculture, cars, and all kinds of consumer products. Almost all plastic materials contain additives, to create the material characteristics required for the application. These additives may enter the environment in various life cycle stages: as losses during production, leaching from products during use, or leaching from landfilled waste. Although plastics worldwide constitute a very large material flow, there seems to be little systematic information on the importance of the plastics life cycle as a source of emissions of additives. In order to provide a first idea of the magnitude and impacts of these emissions, we present some data in this chapter on worldwide plastics and additive flows. For an assessment like this, the following data are required:

- Data on the production, trade, and consumption of plastics in different regions in the world
- Data on the application of plastics in different products
- Data on plastics waste generation and treatment, in terms of recycling, incineration, and landfill
- Data on the production, trade and consumption of plastics additives
- Data on the application of additives in plastic materials
- Additives emission data with regard to production, use and waste treatment processes in the plastics life cycle
- Data on the environmental fate and impacts of emitted additives

In this chapter, we explore the availability and quality of these data.

2 Quantitative Data on Plastic Flows

2.1 Production, Trade, and Consumption of Plastics

Plastics production and consumption data are available, for the world, the continents, and for countries, in varying degrees of detail. Table 1 shows the production, import and export of plastics in several regions in the world. Production data are taken from PlasticsEurope [1,2]. Import and export data are based on Comtrade, Commodity Trade Statistics Database, of the UN [3]. The apparent consumption of plastics in the regions is calculated from those as production + import − export.

The world produces and consumes roughly 230.10^9 kg plastics per year, or 40 kg per capita on average. Consumption varies by region, from 500 kg/cap (NAFTA) to 15 kg/cap (Africa and Middle East). Asia, including China, is a net exporting region, while the NAFTA countries are the biggest importers. Data on plastics waste trade differ per region – it can be suspected that statistical reports are not as reliable on this issue as they are on the other flows. For most regions, they are one or two orders of magnitude smaller than the trade flows in primary material and plastic products.

Statistics show trade and production flows, and from these data an apparent consumption can be calculated. However, this does not represent the actual amounts of plastics in use: large stocks exist in society of plastic applications and products that have been accumulated in the past. No estimates exist of such stocks, but they will be larger in proportion to the life span of the application. It is to be expected that largest stocks can be found in buildings and infrastructure. These stocks could be important from the point of view of leaching of additives in the use phase.

2.2 Polymers and End-Uses

In order to be relevant for building up knowledge on additives, these statistics have to be broken down into different polymers, as well as different applications of the plastics. These data are not available for the world. However, they are available for the European Union. Figures 1 and 2 show the plastics sales in the EU-25 + 2, broken down into different polymers (Fig. 1) and different applications (Fig. 2) (Plastics Europe [2]). Both are relevant information for waste management and recycling: however, these data are not available for countries outside Europe.

Polymers produced in largest quantities are high density and low density polyethylene (HDPE/LDPE), polypropylene (PP), and polyvinylchloride (PVC). Together, they constitute ca. 60% of the sales. Packaging is the largest application of plastics. Building and construction is the second largest, the two together

Table 1 The production, import, and export of plastics and articles thereof in the world in 2005

Region	Production plastics	Import plastics and articles thereof				Export plastics and articles thereof				Apparent consumption[a]
		Import total	Import primary material	Import waste	Import articles thereof	Export total	Export primary material	Export waste	Export articles thereof	
EU27	5.75E + 10	9.43E + 09	5.58E + 09	2.32E + 08	3.62E + 09	1.54E + 10	9.83E + 09	1.67E + 09	3.91E + 09	5.30E + 10
NAFTA	5.29E + 10	3.86E + 10	1.31E + 10	7.27E + 08	2.47E + 10	2.43E + 10	1.63E + 10	1.37E + 09	6.67E + 09	6.78E + 10
Rest of Asia	3.80E + 10	1.68E + 10	1.34E + 10	4.02E + 08	2.99E + 09	3.21E + 10	2.49E + 10	8.37E + 08	6.33E + 09	2.32E + 10
China	3.45E + 10	2.46E + 10	1.82E + 10	4.96E + 09	1.44E + 09	1.03E + 10	1.89E + 09	4.47E + 07	8.41E + 09	4.39E + 10
Middle East, Africa	1.84E + 10	8.51E + 09	6.33E + 09	6.28E + 07	2.12E + 09	9.42E + 09	7.76E + 09	9.93E + 07	1.56E + 09	1.75E + 10
Middle East		4.66E + 09	3.49E + 09	4.27E + 07	1.13E + 09	8.37E + 09	7.11E + 09	6.95E + 07	1.19E + 09	
Africa		3.85E + 09	2.84E + 09	2.01E + 07	9.90E + 08	1.05E + 09	6.54E + 08	2.99E + 07	3.69E + 08	
Japan	1.27E + 10	2.94E + 09	1.47E + 09	3.08E + 06	1.47E + 09	5.89E + 09	4.10E + 09	1.06E + 09	7.29E + 08	1.08E + 10
Latin America	9.20E + 09	4.91E + 09	3.88E + 09	1.89E + 07	1.02E + 09	3.30E + 09	2.53E + 09	7.77E + 07	6.88E + 08	1.09E + 10
CIS	6.90E + 09	3.86E + 09	2.36E + 09	7.63E + 06	1.50E + 09	1.10E + 09	8.57E + 08	2.16E + 07	2.21E + 08	9.68E + 09
Unit	Kg	Kg	Kg	Kg	Kg	Kg	Kg	Kg	Kg	Kg

[a]Apparent consumption = production plastics + import primary material + import articles thereof − export primary material − export articles thereof

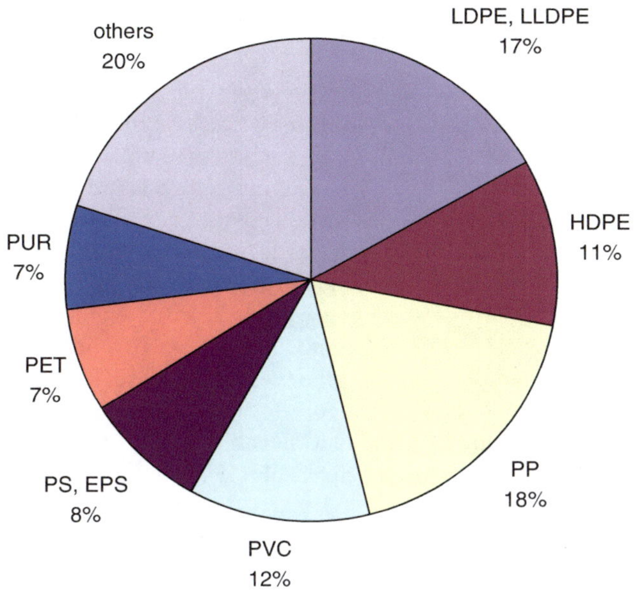

Fig. 1 Plastics demand per polymer, EU-25 + 2, 2008

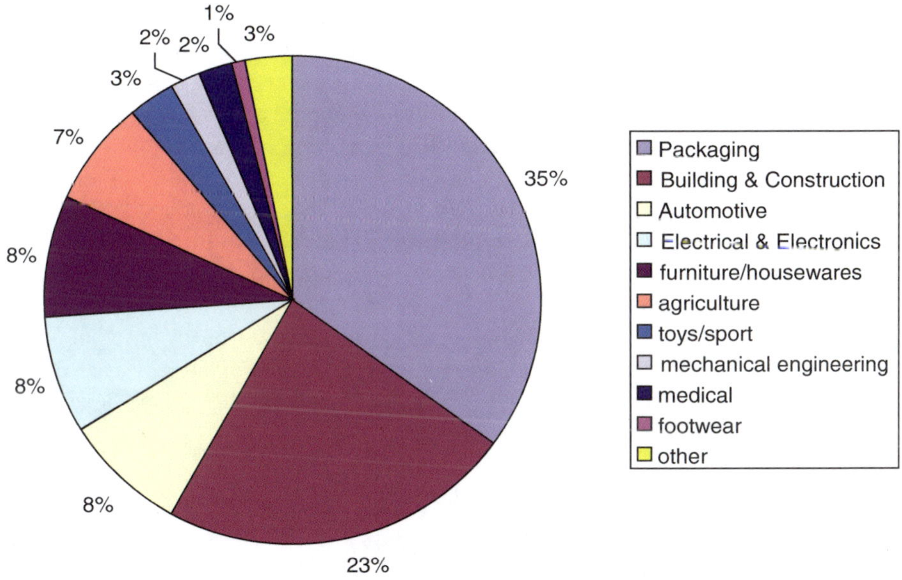

Fig. 2 Plastics demand per application, EU-25 + 2, 2008

represent approximately 60% of sales. Those two probably also represent extremes with regard to life span: packaging plastics will be discarded in the same year as they are sold, while life spans in infrastructure may be decades.

2.3 Plastics Waste Generation and Treatment

Data on plastics waste generation and especially treatment are scarce. Some data can be found for the EU25 and Norway and Switzerland in the year 2005 (Plastics Europe [1]), presented in Table 2. Note that these data refer to the European situation. For other parts of the world, data are more aggregate, incomplete or even absent. The OECD [4] provides some general data on waste generation and management for various regions of the world, and a fraction of plastics of total waste. Specific polymers are not identified. These data are presented in Table 3.

The total amount of plastics waste generated is 22 million tonnes, which is about 40% of the apparent consumption calculated in Table 1. This implies that the other 60% accumulates. Accumulation means the (yearly) increase of the stock of plastics in society. This value might be used as indicator for possible future discard flows of plastics to the waste treatment. Note that in Table 2, it is assumed that all waste is treated within the EU. According to Table 1, however, 1.67 tonnes (7%) is exported.

Slightly more than half of the generated plastics waste is landfilled. The remainder is recycled in one way or another. The most common form of recycling is energy recovery, or the combustion of plastics in municipal waste incinerators. Mechanical recycling involves the mechanical recovery of material from waste, during which the different types of plastics are usually shredded into small given particle sizes. After sorting the shredded material, it can be used to produce a high quality recyclate, substituting raw material as long as the properties of recyclates stay similar to the original raw materials. Industrial production waste and separately collected packaging waste have a more homogenous composition and make recycling more efficient. In comparison to that, recycling of plastics from inhomogeneous post-consumer plastic waste is more difficult, due to a more diverse and unpredictable composition. The presence of additives also hinders an efficient mechanical recycling. In most cases, secondary plastics are mixed and downcycled, which includes that they only can be used in applications where low quality materials will be sufficient.

An improved collection and sorting system for plastics from Municipal Solid Waste to accomplish an impurity amount less than 10% is necessary. Better labelling for more efficient disposal could also improve the users' behaviour.

Table 2 Post consumer plastics waste production and treatment in the EU25 + 2 in 2005

	Kg	%
Post consumer waste generated	2.2E + 10	
Landfilled	1.2E + 10	53
Recycled	1.0E + 10	47
Energy	0.64E + 10	29
Mechanical	0.36E + 10	16
Feedstock	0.04E + 10	2

Table 3 Waste generation and treatment in other OECD countries [4]

	Generation of municipal waste				Municipal waste treatment				
	Municipal waste (1,000 tonnes)	Of which : Household waste (1,000 tonnes)	Plastics % of municipal waste	(1,000 tonnes)	Recycling %	Composting %	Incineration %	Landfill %	Other %
Year	2005	2005	2005	2005	2005	2005	2005	2005	2005
Canada	13,375	13,375	3	359	27	12			
Mexico	36,088	27,785	6	2,205	3	0	0	97	0
USA	222,863	133,718	12	26,227	24	8	14	54	0
Japan	51,607	34,798	13	6,709	17		69	3	0
Korea	18,252	15,175	8	1,460	49	0	12	36	0
Australia	8,903	8,903	4	356	30			70	
New Zealand	1,541	1,541	8	123	15			85	6

Due to the missing reflection of the near infrared radiation (NIR), the sorting out of black plastic waste is impossible.

The most efficient way to recycle plastics to obtain pure polymers could be the transformation into original or similar molecules by using chemical reactions. Feedstock recycling refers to the reuse of plastics as a feedstock in other industrial processes, like reducing agent, fuel or even monomers to make new plastics. Plastic polymers are split into smaller molecules (e.g. CO or H_2), which can be used as a substitute for raw materials. In addition, originally included additives, which often cause unintended accumulation in recyclates, could be removed easier. The most common ways of chemical recycling are gasification, pyrolysis, hydrogenation, coking and for some plastics also the separation by using ionic liquids to depolymerize polymers into monomers. Until today the majority of feedstock recycling processes are not developed sufficient enough to be transformed into full-scale plants to be economically assessed.

Except for Japan and Korea, the largest share of waste management is landfill. Significant recycling percentages are achieved in various regions. After usage, plastics should undergo a recycling process and make their way into a recovered and monitored material with no unpredictable material properties. However, it is unclear if these percentages are also valid for plastics. Most probably, this is not the case.

The largest share of plastics worldwide is landfilled. This bears the risk that additives leach out of it. Due to their chemical stability, plastics degrade only very little in landfills. Chemical additives within the different types of landfilled plastics could produce adverse consequences to the environment if they were to be washed out with landfill leachate. Therefore, the management of the landfill is important. Unfortunately, no global data exist for that.

From an additive perspective, incineration is a good option. Plastics and enclosed organic additives are transferred into carbon dioxide and water, whereas metals do not degrade and end up in waste streams such as slag and ashes. These waste streams might be used as minerals for road construction and building materials. Necessary precautions should be taken to minimize leaching of heavy metals from these recycled waste streams. Incineration can be a recycling option when energy is recovered.

3 Quantitative Data on Flows of Plastics Additives

A plastic, randomly chosen, will contain about 20 additives. These additives have different functions in providing the plastic material with the desired characteristics. In Table 4, a number of additives is presented, grouped by their function, together with the polymers in which the additive is used [5]. It can be seen that some types of additives, like colourants or flame retardants, are used in all polymers. Others are more specific, such as heat stabilizers or plasticisers. The additives make out a significant share of the material, in some cases up to 60% of the weight.

Table 4 Selected additives, their concentration, and use in plastics [5]

Type	Example	Concentration [wt.% polymer]	Polymer in which applied: PVC	PE	PP	PUR	PS	ABS	PC	Cellulosics	Epoxy	Acrylics	Impact styrene	Engineering plastics	Unsaturated	% of market
Antioxidants	Phenolics, Phosphates, Amines, Thioesters	Low (<1)		X	X			X					X			1
Blowing agents	Chlorofluorocarbons, Azobisformamide, pentane	Moderate (0–5)	X		X	X	X	X								1
Colourants	Organic and inorganic pigments and dyes	Low (1–2)	X	X	X	X	X	X	X	X	X	X	X	X	X	4
Fillers/wood flours	Calcium carbonate, Polyester, PVC	High (10–50)													X	50
Flame retardants	Alumina trihydrate, Antimony oxide, Halogenated hydrocarbons, Organophosphates	High (10–20)	X	X	X	X	X	X	X	X	X	X	X	X	X	4
Initiators (free-radical)	Azo compounds, Peroxides	Low (<1)	X	X			X					X				1
Heat stabilizers	Organotin mercaptides, Lead compounds, Barium, Cadmium and Zinc soaps	Moderate (1–5)	X													4
Impact modifiers	Methacrylate butadiene styrene, Chlorinated PE, Acrylic polymers, Ethylene vinyl acetate	High (10–20)	X	X	X									X		2
Plasticizers	Phthalates, Aliphatic Di/Tri Esters, Polyesters, Phosphates, Trimellitates	High (20–60)	X							X						22
Reinforcers	Glass fibres, Polyester	High (10–40)									X				X	7
% of market			12	28	18	7	8									

Statistical data on the production of additives are scarce. However, market analysts make up reports of annual sales of additives. These sales can be used as an approximation of consumption, i.e. consumption by the manufacturing industry. Table 5 shows the sales of plastics additives, which is estimated at 22 million tonnes worldwide in 2005. The largest market is fillers with 50% of total world market. The second largest market is plasticizers with 22% of total world market, mostly phthalate plasticizers (18%).

Consistent growth in infrastructure, construction, and packaging industries place China as one of the fastest growing markets for plastics and thus plastic additives. The global manufacturing of plastics shifts to the Far East and so does the production of plastic additives. The majority of the plastic additives capacity expansion takes place in China, the Middle East, India, Russia, Eastern Europe, and other regions in Asia-Pacific.

Environmentally friendly products such as non-halogenated flame-retardants, organic heat stabilizers, conductive compounds, wood composites, nano composites, fibre-reinforced thermoplastics represent some of the potential growth areas for global additives. The market for plastic additives such as flame-retardants, antioxidants, and organic heat stabilizers, particularly for vinyl, is strong as Europe banned lead-based heat stabilizers. Consumer pressure and regulations spur demand for flame retardants, primarily in furniture foam and cable and wire insulation [5].

4 Process Emission Data on Plastics and Additives

An important source of emissions data are the databases collected to support Life Cycle Assessment (LCA) studies. LCA is used to compare different products on their cradle-to-grave environmental impacts (ISO 14040). Life Cycle Inventory (LCI) data are needed to specify the emissions, while Life Cycle Impact Assessment (LCIA) data are required to estimate the environmental impact of the emissions. In this section, we will explore the availability of LCI data.

An LCI database contains process data. Process data are quantified descriptions of the inputs and outputs of a process. Processes are understood as technical processes with physical inputs and outputs, and refer to the whole life cycle: production processes, use processes and waste treatment processes. Inputs can be raw materials extracted from the environment, but also outputs from other processes, such as intermediate materials, finished materials or products. Outputs are the products of the process, the emissions to the environment and waste flows to be disposed of.

At the moment there are many different LCI databases that contain process data related to the production, use and waste treatment of plastics and plastic additives. There are at least two initiatives that try to facilitate the overview of available LCI databases the Database Registry (the registry) by UNEP/SETAC [19], and the LCA Resources Directory by JRC-IES [20].

Table 5 Annual sales of plastic additives per segment and region in year 2005 [5]

Product segment	North America (1,000 tonnes)	Japan (1,000 tonnes)	Europe (1,000 tonnes)	Asia-Pacific (1,000 tonnes)	Rest of the World (1,000 tonnes)	Total World (1,000 tonnes)	%	%
Fillers	3,301	929	3,014	2,995	758	10,998	50	
Mineral fillers	1,980	576	1,740	1,729	448	6,473		29.4
Other fillers	1,321	353	1,274	1,266	310	4,524		20.6
Plasticizers	1,255	529	1,370	1,228	506	4,889	22	
Phthalate plasticizers	928	409	1,142	979	400	3,859		17.5
Other plasticizers	327	120	228	249	106	1,030		4.7
Reinforcing agents	447	141	435	439	109	1,571	7	7.1
Flame retardants	330	98	280	257	46	1,010	5	
Aluminium oxide hydrates	121	28	119	43	8	319		1.4
Antimony oxide	24	9	22	34	6	96		0.4
Bromine-based compounds	83	33	52	110	20	297		1.3
Chlorine-based compounds	17	4	11	36	6	73		0.3
Phosphorus-based compounds	49	18	57	31	5	160		0.7
Other flame retardants	35	7	19	4	1	66		0.3
Colourants	259	96	164	302	65	887	4	
Titanium dioxide	63	21	36	71	15	206		0.9
Iron oxides	35	14	22	37	9	116		0.5
Organic colourants	123	48	81	147	32	431		2.0
Other colourants	38	14	25	47	10	133		0.6
Performance additives	779	264	717	711	169	2,641	12	
Blowing/foaming agents	79	26	72	73	17	266		1.2
Impact Modifiers	154	54	146	139	40	533		2.4
Organic peroxides	48	17	48	49	12	174		0.8
Heat stabilizers	250	83	228	240	49	849		3.9
Light stabilizers	10	4	9	9	2	33		0.2
Antioxidants	78	27	71	66	17	259		1.2
Antistatic agents	8	3	7	8	1	28		0.1
Preservatives and biocides	34	12	32	34	8	120		0.5
Lubricating agents	77	25	68	63	15	248		1.1
Miscellaneous	42	13	38	30	8	130		0.6
Total	6,371	2,058	5,981	5,932	1,653	21,996		

The UNEP/SETAC Life Cycle Initiative has made an overview of available Life Cycle Inventory (LCI) databases around the world [6,7]. The overview is one of the deliverables of Task Force 1 (Database Registry) of the Life Cycle Inventory Programme. The aim is to provide a comprehensive, web-based listing of available LCI databases for the world LCA community. The UNEP/SETAC Database Registry is implemented and made available on web, since the end of 2009. The UNEP/SETAC Database Registry (the registry) has been queried on data on plastics and plastic additives. The registry is still in development and seems to be incomplete. In Curran and Notten [6], more databases are included.

In Table 6, the databases in the JRC-IES are presented. This is a rather comprehensive overview.

For some of the databases, a quick scan has been made on the availability of data regarding plastics and plastic additives.

Table 6 LCA resources directory by JRC-IES

Database + version	Supplier
CPM LCA Database	Center for Environmental Assessment of Product and Material Systems – CPM
DEAM™	Ecobilan – PricewaterhouseCoopers
DEAM™ Impact	Ecobilan – PricewaterhouseCoopers
DIM 1.0	ENEA – Italian National Agency for New Technology, Energy and the Environment
ECODESIGN X-Pro database V1.0	EcoMundo
Ecoinvent Data v1.3	Ecoinvent Centre
EIME V8.0	CODDE
EIME V9.0	CODDE
Esu-services database v1	ESU-services Ltd.
Eurofer data sets	EUROFER
Franklin U.S. LCI database	Franklin Associates, A Division of ERG
GaBi databases 2006	PE International GmbH
GEMIS 4.4	Oeko-Institut (Institute for applied Ecology), Darmstadt Office
IO-database for Denmark 1999	2.-0 LCA consultants
IVAM LCA Data 4.04	IVAM University of Amsterdam bv
KCL EcoData	Oy Keskuslaboratorio-Centrallaboratorium Ab, KCL
LC Data	Forschungszentrum Karlsruhe
LCA Database for the Forest Wood Sector	Bundesforschungsanstalt für Forst- und Holzwirtschaft (BFH)
LCA_sostenipra_v.1.0	Universitat Autònoma de Barcelona (UAB)
MFA_sostenipra_v.1.0	Universitat Autònoma de Barcelona (UAB)
Option data pack	National Institute of Advanced Industrial Science and Technology (AIST)
PlasticsEurope Eco-profiles	PlasticsEurope
ProBas	Umweltbundesamt
Sabento library 1.1	ifu Hamburg GmbH
SALCA 061	Agroscope Reckenholz-Tänikon Research Station ART
SALCA 071	Agroscope Reckenholz-Tänikon Research Station ART
SimaPro database	PRé Consultants B.V.
sirAdos 1.2.	LEGEP Software GmbH
The Boustead Model 5.0.12	Boustead Consulting Limited
Umberto library 5.5	ifu Hamburg GmbH
US Life Cycle Inventory Database	Athena Sustainable Materials Institute
Waste Technologies Data Centre	UK Environment Agency

4.1 Polymer Production Data

It appears that most of these databases include data on plastics production. Some only have data for "plastics", such as Umberto, others have production data for the individual polymers. However, the process data in all cases are aggregated data. This means that the process chain for plastics is not described in separate processes and materials. The aggregate process describes the extracted resources and emitted substances, along the cradle-to-gate chain, but does not distinguishing between intermediate materials in the chain, like additives, fuels, capital goods, etc. The database therefore does not allow for a detailed analysis of the individual processes of the total plastic chain and of the contribution of additives in the total environmental impact assessment of plastics. Only the ETH database [8], the precursor of the Ecoinvent database [17], does contain some disaggregated data. However, additives are not mentioned as specific inputs in the production of the plastics. An exception is the consumption of Bisphenol A which is an input in the production of PC. The emission data of plastics production does contain emissions of additives. This database, although quite old, contains more relevant data than most of the others.

On closer scrutiny, it appears that plastics production data in several of those databases (ECOINVENT, ELCD, GABI, IVAM, and possibly others as well) are based on industrial data from the PlasticsEurope database. This implies that these databases do not differ. NREL contains data for North America. As both Plastics Europe and NREL only provide aggregate data, it is not possible to distinguish additives, or any other individual inputs in these processes. The absence of any emissions of additives from the chain however is unexpected. On closer enquiry, it appears that the production data are excluding additives: the inventory refers to uncompounded plastic resins.

The properties of plastic materials depend on the addition of plastic additives and further processing. In order to compare plastic materials, the plastic resin data cannot be used as such, but should be combined with data on additive production as well as plastic conversion. The databases of PlasticsEurope and therefore also GABI, ECOINVENT, Umberto, and IVAM do contain some data of semi-manufactured products, like sheet, film, pipe, bottle, window frame. However, these conversion modules mainly contain transport, energy consumption, and packaging data. The modules need raw materials like resins and additives as input data.

4.2 Additive Production Data

In general, process data on the production of additives, like phthalates or brominated flame retardants, etc., are missing in the databases listed in Table 6. An exception is the production of some metals (compounds) and Bisphenol A. The IVAM database also contains data on the production of DEHP and lead stabilizer.

In view of this data gap, we suspect that many LCA case studies where plastics are part of the product do not include additives. The fact that the LCI data refer to uncompounded plastics is not obvious, and as additive production is rarely included in the databases one might assume these processes will be included in the production of the polymers.

4.3 Data on Plastics Use

LCI databases do not include data on the use phase of plastics. In most cases, the use phase is quite irrelevant for the emissions of additives – however, in some cases additives tend to leach out, corrode or evaporate from the plastic material during use. No data are available on those processes.

4.4 Polymer Waste Treatment

Most of the LCI databases contain some data on the incineration of (specific) plastics, like ECOINVENT, ELCD, GABI, and IVAM. A few databases contain data on landfill of (specific) plastics, like ECOINVENT and IVAM. However, data related to recycling of plastics are very poor. Furthermore, it is unclear whether emissions of additives in waste treatment processes are accounted for. For example, the ECOINVENT database reports no emissions of phthalates for PVC waste treatment. The background reports [9–11] lead to the suspicion that waste incineration is modelled in a material-specific way, and that the composition of the plastics includes additives. Organic additives of course degrade in the process of incineration; emissions of metals do appear. Landfill however seems to be modelled in a non-material specific manner.

In all, it can be concluded that the LCI databases on plastics contain relevant data for the production of plastic resins and the conversion of resins into materials or (half) products, but that data on the production of specific additives are missing. Furthermore, the conversion processes do not take into account the consumption, and thus possible emissions, of additives. The conversion processes are modules that mainly focus on energy consumption. Additional data about the consumption of additives and possible emissions should be added to the conversion module and will depend on specifications of the produced (half) product. Data related to recycling of plastics are missing or very limited.

It can also be concluded that data on waste treatment processes, and especially use processes, are scarce. LCA studies on plastics that rely on these general databases therefore are bound to be incomplete.

5 Data on Fate and Impacts of Plastics Additives

Since other chapters in this book are dedicated to these topics, we restrict the inventory to Life Cycle Impact Assessment (LCIA) data. LCIA uses data on environmental fate, exposure and toxic impacts from the field of risk assessment. It translates those data into so-called characterization factors [12], that can be used in LCA studies. Such factors provide a relative impact. For the impact category of Global Warming, for example, all emissions are expressed in kilogram CO2-equivalent. Likewise, the toxicity characterization factors are expressed in kilogram 1,4-dichlorobenzene equivalent.

For the environmental impacts of human toxicity and ecotoxicity there are numerous chemicals for which characterization factors are available, including chemicals that are used as additives in plastics, like heavy metals (zinc, lead, organotins) and phthalates. An overview of chemicals and their environmental impact factors can be downloaded from the CML website http://cml.leiden.edu/software/data-cmlia.html.

For some of the additives characterization factors for toxicity are not yet available. These cannot be included in the LCA Impact Assessment, unless characterization factors are developed for them. This can be done by using characterization models, like USES [13–15, 18], or UseTox [16]. These models use both the toxicity of a substance and their fate and exposure characteristics, like solubility, degradation, bio accumulation, etc., to arrive at characterization factors.

6 Conclusions

Statistics on plastics production, trade and sales are available and can be used to obtain a rough overview of the plastics consumption in regions of the world and sometimes in countries. Statistical data on plastics waste are scarce – waste trade data are available but most probably do not cover all relevant flows, plastics waste generation data are available only for the EU, and plastics waste treatment data are scarce. There are no statistics on additives; however, reports from market analysts provide relevant and usable information.

With regard to emissions to the environment, Life Cycle Inventory data are available for plastics production, although mostly at the aggregate level, and excluding additives. Additive production data are scarce. It is therefore not possible to obtain a picture of plastics-related emissions of additives to the environment. For the use and waste treatment processes, LCI data with relevance for plastics additives are not available at all. Life Cycle Impact Assessment data are available for a number of additives.

Most important data gaps, therefore, occur in the emissions data and in the data related to waste generation and waste treatment.

To estimate impacts of additives in plastics at a global level, especially with regard to recycling and other waste management options, the missing data must be supplemented. This can be done easily only for the LCIA data: impact factors can be derived from existing models such as UseTox. The impacts of plastics waste generation and management, especially with regard to additives, can be assessed at present only by collecting data on a case-by-case level. Missing emission data should be added to LCI databases to enable a more complete assessment of plastics applications. In Chapter 14 of this book, an approach is outlined to estimate such emissions data in the absence of measurement data.

References

1. Plastics, Europe (2007) The compelling facts about plastics. An analysis of plastic production, demand and recovery for 2005 in Europe. PlasticsEurope, Brussels, Belgium. www.plasticeurope.org
2. Plastics, Europe (2009) The compelling facts about plastics. An analysis of plastic production, demand and recovery for 2008 in Europe. PlasticsEurope, Brussels, Belgium. www.plasticeurope.org
3. UN (2010) Comtrade, commodity trade statistics database. http://comtrade.un.org/
4. OECD (2008) Environmental data compendium 2006–2008. Waste. http://www.oecd.org/statisticsdata/0,3381,en_2649_34395_1_119656_1_1_37465,00.html
5. GIA (2008) Plastic Additives. A global strategic business report. Global Industry Analysts, San Jose, USA. www.StrategyR.com
6. Curran MA, Notten P (2006) Summary of global life cycle inventory data resources. Prepared for: task force 1: database registry SETAC/UNEP life cycle initiative. http://www.epa.gov/NRMRL/lcaccess/pdfs/summary_of_global_lci_data_resources.pdf
7. Norris GA, Notten P (2002) Current availability of LCI databases in the World. http://lcinitiative.unep.fr/sites/lcinit/default.asp?site=lcinit&page_id=2F3B97E2-1697-4CD5-9D7E-D3CAD7214482#lcidel
8. Frischknecht (ed) (1996) Ökoinventare von Energiesystemen. Eidgenössische Technische Hochschule (ETH), Zürich
9. Doka G (2007a) Life cycle inventories of waste treatment services. Part I: general introduction, waste material compositions, municipal waste collection. Ecoinvent report no. 13, Dübendorf, Switzerland
10. Doka G (2007b) Life cycle inventories of waste treatment services. Part II: waste incineration. Ecoinvent report no. 13, Dübendorf, Switzerland
11. Doka G (2007c) Life cycle inventories of waste treatment services. Part III: landfills – underground deposits – landfarming. Ecoinvent report no. 13, Dübendorf, Switzerland
12. Guinée JB, Gorrée M, Heijungs R, Huppes G, Kleijn R, de Koning A, van Oers L, Wegener Sleeswijk A, Suh S, Udo de Haes HA, de Bruijn H, van Duin R, Huijbregts MAJ (2002) Handbook on life cycle assessment. Kluwer Academic, Dordrecht, Operational guide to the ISO standards
13. Huijbregts MAJ (1999) Priority assessment of toxic substances in LCA. Development and application of the multi-media fate, exposure and effect model USES-LCA. IVAM environmental research, University of Amsterdam, Amsterdam
14. Huijbregts MAJ, Guinée JB, Reijnders L (2000) Priority assessment of toxic substances in life cycle assessment, III: export of potential impact over time and space. Chemosphere 44 (1):59–65

15. Huijbregts MAJ, Thissen U, Guinée JB, Jager T, van de Meent D, Ragas AMJ, Wegener Sleeswijk A, Reijnders L (2000) Priority assessment of toxic substances in life cycle assessment, I: calculation of toxicity potentials for 181 substances with the nested multi-media fate, exposure and effects model USES-LCA. Chemosphere 41:541–573
16. Rosenbaum RK, Bachmann TM, Swirsky Gold L, Huijbregts MAJ, Jolliet O, Juraske R, Koehler A, Larsen HF, MacLeod M, Margni M, McKone TE, Payet J, Schumacher M, van de Meent D, Hauschild M (2008) Usetox – the UNEP-SETAC toxicity model: recommended characterisation factors for human toxicity and fresh water ecotoxicity in life cycle impact assessment. Int J LCA (2008) 13:532–546
17. Ecoinvent (2010) Ecoinvent database version 2.0 http://www.ecoinvent.ch/
18. Huijbregts MAJ (2000) Priority assessment of toxic substances in the frame of LCA. Time horizon dependency of toxicity potentials calculated with the multi-media fate, exposure and effects model USES-LCA. Institute for Biodiversity and Ecosystem Dynamics, University of Amsterdam, Amsterdam, The Netherlands. (http://www.leidenuniv.nl/interfac/cml/lca2/).
19. UNEP/SETAC (2011) Database registry of the UNEP/SETAC Life Cycle Initiative. http://lca-data.org:8080/web/guest/information
20. JRC (2011) LCA Tools, Services and Data. http://lca.jrc.ec.europa.eu/lcainfohub/index.vm

Part II
Case Studies

Problems in Paper Recycling of Vietnam

Nguyen Thi Diem Trang

Abstract Paper production can be considered as an important factor for the development of one country. On the way of development, Vietnam has increased steadily in paper consumption; however, the paper production did not get to the same speed. One of the reasons remains in input material which composes of wood or other agricultural fiber sources and recovered paper. Statistical data on pulp and paper production have been taken into consideration. Reasons for the low in recycling rate have been analyzed.

Moreover a picture of recovered paper in Vietnam has been drawn. In form of recycling villages, the recovered paper production is operated completely by hand or by using simple tools. Data in emissions of additives, and the impacts resulting from those emissions are lacking. Since recycled paper is used very popular in flow of domestic consumption, this product has shown a big risk to the human health and the production is a big risk to the environment.

Keywords Handicraft village, Paper production, Paper recycling, Policy, Pulp production, Risk-based management

Contents

N.T.D. Trang
Hanoi University of Science, Faculty of Chemistry, 19 Le Thanh Tong, Hanoi, Vietnam
e-mail: trang_ntd@yahoo.com

B. Bilitewski et al. (eds.), *Global Risk-Based Management of Chemical Additives I: Production, Usage and Environmental Occurrence*, Hdb Env Chem (2012) 18: 153–166,
DOI 10.1007/698_2011_116,
Published online: 14 September 2011

1 Introduction

Paper production can be considered as an important factor for the development of one country. Vietnam has increased steadily in paper consumption but it is not yet the case for production.

The paper quantity from domestic production satisfies only 60% of the demand (2008). The paper consumption was 1.8 million tons per year but domestic pulp production provided only 1.13 million tons, near 0.7 million tons must be imported. The recovered paper utilization meets 70% of the domestic paper production, but half of the recycled paper is imported.

From 1999 to 2007, the collection rate was 24–25%, but from the last 4 years until now this rate even has been reduced to 16–17%. The utilization rate remains the same like the number imported recovered paper (48–50%).

Figure 1 shows the paper recycling rate in Asian countries and US in 2007. Unfortunately, in this data, Vietnam stays in the lowest range.

The paper production of Vietnam in 2000–2008 is shown in Fig. 2. All parameters like paper-, pulp production, and paper consumption increased granulary, corresponding to the increase of the average paper consumption.

Most of the domestic generated wastepaper was destroyed, while the country has to spend a big budget for wastepaper importation from US, Japan, and other countries. In this connection, the *flow of input material* for pulp and paper production, which composes of wood or other agricultural fiber sources and recovered paper, may cause problems which need to be discussed.

The paper recycling with advantages in energy consumption, nature conservation as well as reduction of CO_2 emission and many others has not yet been well sustainably implemented. Clarifications for this situation are explained within the following sections.

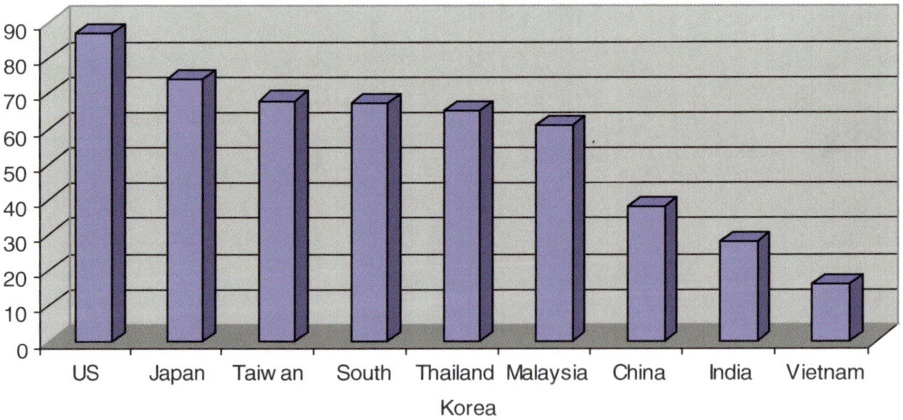

Fig. 1 Paper recycling rate in Asian countries and US in 2007. *Source*: Thoi bao Kinh te Sai Gon 7/12/2009

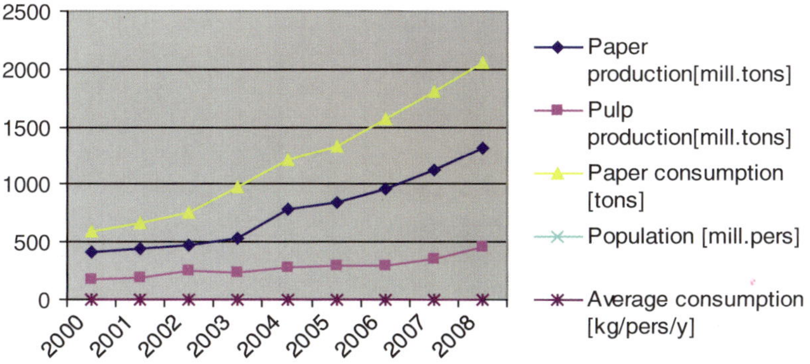

Fig. 2 Vietnamese pulp-, paper production, and paper consumption in 2000–2008. *Source*: Paper and Cellulose Industry Institute, 2007

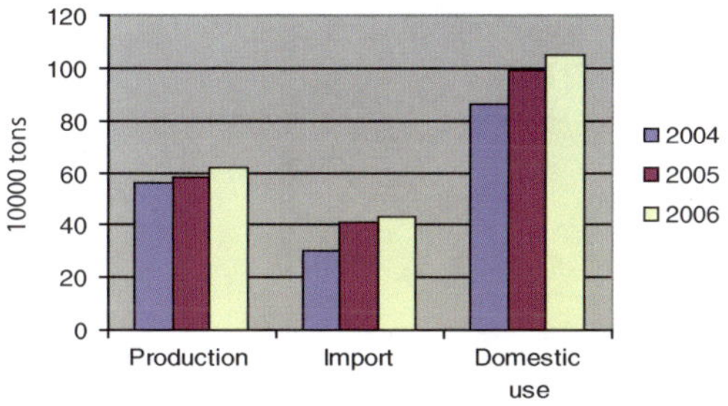

Fig. 3 Situation of pulp production from recovered paper (2004–2006)

2 Problem Definition

There are two figures showing the pulp production from recovered paper (Fig. 3) and the paper production (Fig. 4) from 2004 to 2006 [*source: www.vietpaper.com 2007)*[1]]. In these figures, the *domestic use increases* continually in all types of products. This situation also stays the same for the other period (2005–2007) included in Fig. 5.

According to the Paper and Cellulose Industry Institute (Ministry of Industry and Trade), the production of 1 ton of paper pulp requires 5 m^3 wood and 100 m^3 water. Meanwhile, 100,000–120,000 tons of recovered paper can be processed into at least 80,000 tons of paper pulp. With that amount, the country can save 400,000 m^3 wood/year and reduce CO_2 emissions if the used paper is well collected and recycled.

In fact, offices, schools, and households are the biggest sources of used paper discharge. People prefer using high quality paper for any purpose, and print less on both sides of a sheet. Therefore, *an unwanted amount of used paper goes into the waste stream.*

[1] www.dongcao.bacninh.com, www.vietnamnet.com, www.vietpaper.com (2007), www.thiennhien.net (23 sep 2007).

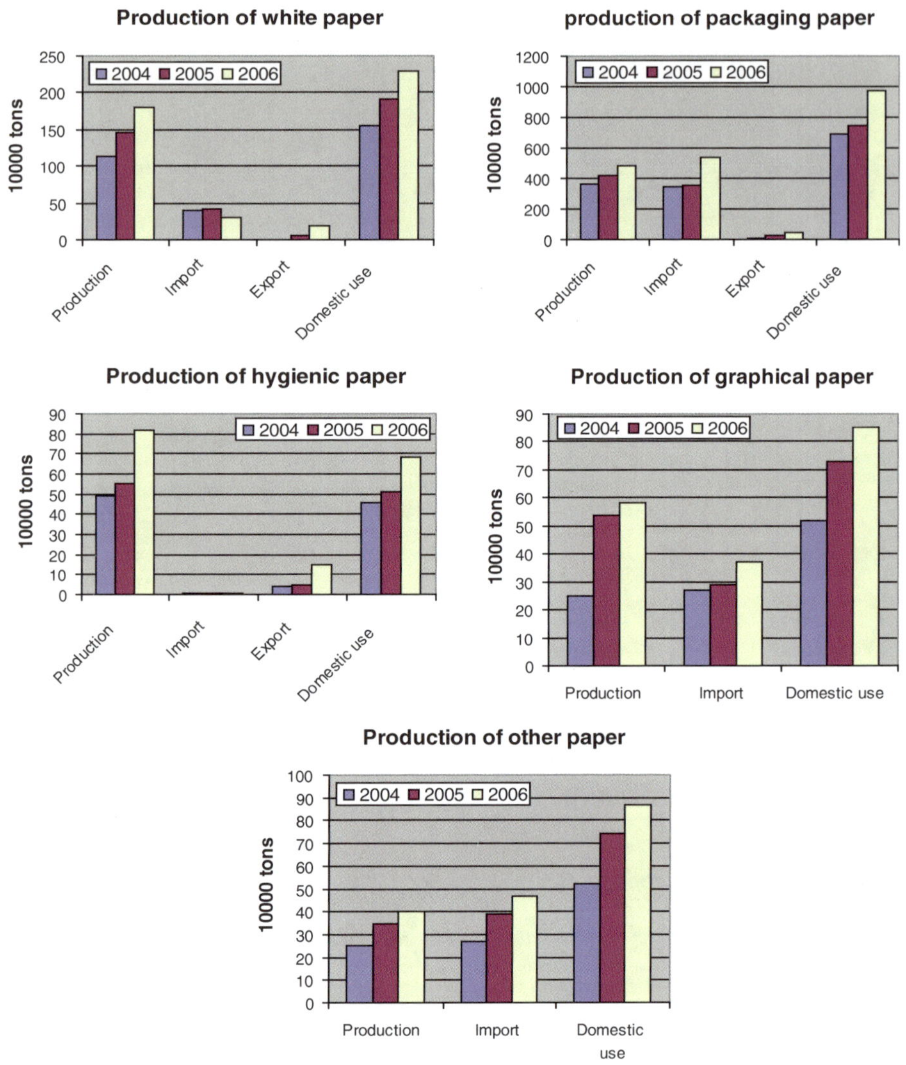

Fig. 4 Situation of different kinds of paper production from 2004 to 2006

Viet Tri Paper Company is one of the biggest pulp and paper producer in Vietnam. Last year (2008) this company needed 50,000 tons of recovered paper and had to import 20,000 tons because of *difficulties in sourcing it locally.*

Used paper *without segregation at source* was collected by the urban environmental companies with municipal solid waste. The quality of scrap paper, therefore, was low, subsequent to the low *recycling rate* (Fig. 1). From 1999 to 2007, the utilization parameter has almost not change (48% in 1999 to 50% in 2007).

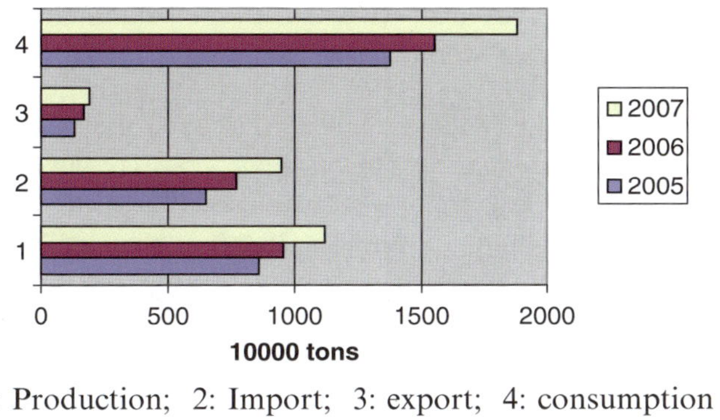

Fig. 5 Situation of production, export, import, and consumption of paper from 2005 to 2007

The waste paper collection has been done by waste pickers and scrap iron dealers; they become the main supplier of reused paper in recycling sector. From this source, however, *waste paper with low quality* (paper already being recovered several times and not separated at source) has come together with scrap paper of imported goods. Moreover, since *state invoice* is necessary from supplying source, the paper production plants do not like to buy reused paper from scrap iron dealers, they prefer the imported one. No paper collecting company has been created yet. In this context, the utilization rate is lower.

Every year Vietnam could produce 80,000 tons pulp from used paper, but it was lost from low collection rate. Vietnam still has not recognized the important role of wastepaper as a source of raw material. A *policy to encourage the paper recycling* needs to be issued.

Vietnam also needs to have a *Waste reduction policy* since this is the most important one in the waste management aspect (landfilling is still the common way of "treating" municipal solid waste here).

Accordingly, such problems have made difficulties for the flow of domestic used paper in the Vietnam paper production.

Paper recycling takes place mainly in household units. With low potential of investments in recycling technology, the discharge amount will remain high. How this activity could look like is described hereinafter within the next chapters.

3 Recycling

In Vietnam the recycling takes place mostly in recycling villages. Recycling villages are the same as handicraft villages, where the process of production is operated completely by hand or by using simple tools. Scrap becomes the raw material for manufacturing of new products.

There are about 80–90 recycling villages (from around 1,500 total handicraft villages with ten millions of workers) over the country, mainly for scrap paper, metal, and plastic (Fig. 6).

The majority of recycling villages is concentrated in the Northern part of the country and they are located in or next to the living areas. The development of these villages creates to a collecting network for waste, scraps, and raw materials, which improves the recycling pattern of the whole country. Figure 7 shows the common sight of scraps supply to recycling villages.

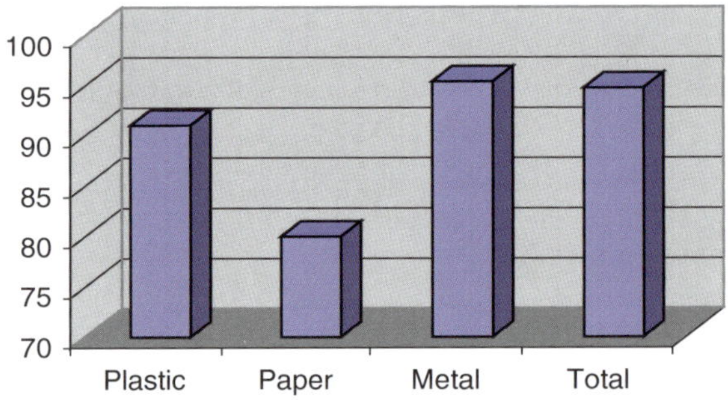

Fig. 6 Potential of recycling villages for plastic, paper, and metal 2003

Fig. 7 Common sight concerning recycling activity

Since 1986, together with the process of economic reform and industrialization of the country, the craft villages in the Red River delta have also gone through a period of rapid growth. According to recent reports, there are about 200 craft villages within the watershed of the river Cau. The river runs through the six provinces of Bac Kan, Thai Nguyen, Vinh Phuc, Bac Giang, Bac Ninh, and Hai Duong (Digregorio 1999). In craft villages, households have been divided into three types of different production activities: agricultural only, both agricultural and "extra-occupation", and "extra-occupation" only. However, "extra-occupation" has become the most important for local people, since it provides the main source of income for households. Therefore, they are accelerating investments to purchase machines to increase the labor productivity and to improve the product quality. In this way, many craft villages have become small industrial centers. This development, on the one hand has provided many employment opportunities and increased the income for local people. On the other hand, it has raised a number of problems, especially problems related to environmental pollution and its impacts on the health of the community (Digregorio 1999).

In general a craft village is a rural village, that has 35–40% of its households specialized in a certain occupation and which can live on the income from that occupation.[2]

Paper-recycling villages occupy 6.2% of the total recycling villages over the country. They concentrate in Hanoi, Bac Ninh, Thanh Hoa, Nam Dinh, and Hung Yen provinces in Northern Vietnam [1].

The typical villages are concentrated in two communes: Phu Lam (Tien Du) and Phong Khe (Yen Phong, Duong O craft village) of Bac Ninh province. Main products from these villages are packaging paper, low quality hygienic paper, or newspaper.

The small recycling units mainly use waste paper from domestic collection, whereas the bigger mills prefer imported waste paper, because of the higher quality of products.

Many advantages have been seen from paper recycling. It saves wood consumption, protects the forest, and contributes to nature conservation. Moreover, it reduces material consumption in pulp and paper production, lowers the material price in comparison to using wood and, therefore, reduces the price of paper products.

The big scale enterprise "Giay Sai Gon" started the used paper collection in schools in the beginning of 2009 by setting up three collection points in Ho Chi Minh City. They collect 4,500 up to 5,000 tons of used paper per month. This has brought significant advantage for the economy, environment protection, and also social aspects. However, this business was rather rare in Vietnam.

Breakdown the case study in recycling sector, one typical example can be clarified in the Phong Khe commune.

[2] Nguyen Van Ha, Doctor of Philosophy, Faculty of Forest, University of Toronto, 2005.

4 Paper Recycling Method in Phong Khe Commune

Phong Khe commune has 170 household recycling units, with a production capacity of 300 up to 10,000 tons/unit per year. The total production is about 400,000 tons paper/year. The way of collection route for providing input material is shown in Fig. 8.

The existence of these household-level paper recycling units has brought important income to the people. This activity also helps saving the cost of landfilling of used paper generated in the cities. The production activity here, however, mostly depends on experiences. Their equipments are old, home-made, and unsystematic, which leads to release of potential pollution. The raw materials are from all kinds of scrap paper. Chemicals like alkali, javen solution, and bleaching agents (with more content of chlorine for de-inking) have brought a big threat to the environment.

In Phong Khe commune, there is Duong O village, which is situated about 32 km northeast of the Vietnamese capital, Hanoi. According to statistics supplied by the People's Committee of Phong Khe commune (personal communication, June 2002), Duong O village currently has a population of 3,950, living in 700 households. Of many traditional paper-making craft villages that existed in the north of Vietnam for centuries, Duong O has become the most industrialized village.

Figure 9 illustrates the different steps of the recycling process: milling, bleaching, cooking, and final production.

In principal, the above illustrated steps demonstrate the process according to the following scheme (Fig. 10).

By using mostly waste paper as an input material, Duong O village produces about 200 tons of finished paper per day and employs about 1,900 workers working directly in the paper factories. Different categories of paper are produced in this village, such as toilet paper, tissues, votive paper, Kraft paper, and printing paper, and they are distributed throughout Vietnam. Duong O recycling village has created jobs for thousands of people in other localities as they participate in the process of trading waste paper, collecting, sorting, and cleaning wastepaper. They also do the transportation of waste paper, materials, or finished products.

Production process in such villages results in big threats to the environment. Biological oxygen demand (BOD), chemical oxygen demand (COD), and total suspended solids (SS) are the three most important parameters to categorize

Fig. 8 The collection route for the flow input of recycling process

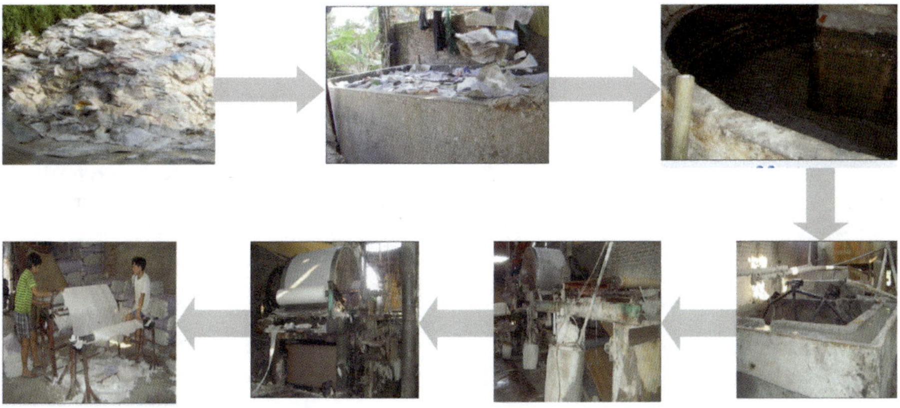

Fig. 9 Paper recycling process in Phong Khe

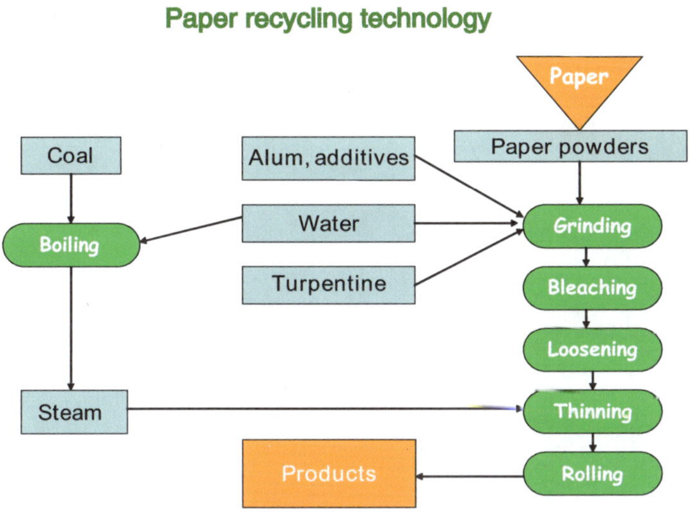

Fig. 10 A scheme of paper recycling technology

environmental outputs from the recycling units. The average levels of pollutants in the wastewater of recycling units in Duong O village have been much higher than the Vietnamese government's industrial wastewater standards (Table 1).

From the 170 recycling units in Phong Khe, ca. 3,500–4,000 m^3 of wastewater are discharged daily without treatment. The COD value was 8.1–8.9 times higher than the standard (2,500 mg/l), coliform 11–18.7 times, BOD 7–14 times (700 mg/l), and the pH value was also too high. Table 2 gives information about recharge flow of two symbolic recycling villages (village Pho Liem and Duong O) in Phong Khe commune.

Table 1 The Vietnamese drainage standard

Parameter	Unit	Limit values			References
		A	B		
			Drainage of pulp production	Drainage of paper production	
BOD_5(20°C)	mg/l	30	100	50	TCVN 6001 (ISO 5815)
COD	mg/l	50	300	200	TCVN 6491:1999 (ISO 6060: 1989)
TSS	mg/l	50	100	100	TCVN 6625: 2000 (ISO 11923: 1997)
AOX		7.5	15	15	

Table 2 Yearly recharge flow in two symbolic paper recycling villages

No.	Kind of recharge	Recharge on 1 ton of product	Amount of recharge in 1 year	
			Village Pho Liem (tons/year)	Village Duong O (tons/year)
1	Waste waster	8,2 m^3	100,923 (m^3/year)	129,938 (m^3/year)
2	Dust	1,83 kg	24,400	31,45
3	Air emission	6.802 kg	90.691	116.777,2
4	Solid waste	212,06 kg	2688,42	3461,34
	Waste of pulp and paper	81,80 kg	1.006,769	1.296,215
	Coal residues	76,5 kg	1019,99	1313,24
	Screw/brad, nylon	53,76 kg	661,661	851,889

Accordingly, this situation has become a concern of local people and researchers. For example, the BOD, COD, and SS values in wastewater in 2003 exceeded the standards by more than 10, 19.6, and 11 times, respectively. The BOD and COD have been increasing over time. For example, the level of BOD in 2003 was five times and the COD was twice as higher as in 1998. On the opposite, the SS parameter in 2003 was only half as high as in 1998. Major reasons for SS reduction might be the application of better technology with respect to recovery of paper fibers and enriching their experiences to increase their productivity.

As a consequence, untreated waste from recycling villages freely discharged into the environment has lead to many effects for this area. As water pollution occurs, it results in the increase of environmental parameters such as SS, COD, BOD, coliform...; high concentration of toxic heavy metals; pollution of surface water and underground water; exceed of organic matter (due to eutrophication); polluted by many chemicals such as alkali, javen solution, bleach agents, paper powder, fiber, oil, turpentine, and 40,000 tons coal/year. It brought to air pollution presented in dust, gas, smoke, hot steam (alkali), Cl_2, CO_2... and to solid pollution resulting in solid wastes. They are cinder, metal oxide, and pin, which include organic polymers like polyester, polystyrol, and glue (30%); solid waste like dust, sand (20%), and metal (30%). The soluble organic (20%) runs out with waste water; therefore, 80% sludge becomes of the not easy decompose form. Solid waste discarded freely on road, pond, and canal without hygienic care makes live available for mosquito, flies, rat (fermentation, biodegradation), and causes fetid smell.

Fig. 11 Polluted water flows into the Cau River in Bac Ninh Province, causing a threat to public health

The untreated waste has caused land degradation (soil pollution), which has been seen in the loss of organic components, barren, deserted, and uncultivable land. It also caused occupational diseases like deafness (noise), ENT (toxic substances). The most common illnesses are of gynecological nature, diarrhea, skin irritation, respiratory difficulties, and eye irritation. Birth defect and blood contained with heavy metals have also appeared. The resulting loss of control on the environmental aspect of production leads this area to polluted Ngu Khe river (surrounding Phong Khe district), which was gravely polluted with pitch-black water and paper fiber floating on it (Fig. 11). An area with three hectare field became uncultivatable and 30% local people have contacted with skin irritation, digestive problem, intestine, and respiration diseases.

5 Current Policy to Deal with These Problems

According to the Vietnam Pulp and Paper Association (VPPA), Vietnam does not have yet the policies and guidelines that encourage the wastepaper collection for paper recycling. VPPA has, therefore, proposed the government to issue these documents. It is necessary to increase public awareness about the value of saving and recycling paper in reducing deforestation and environmental pollution (VietnamPlus, Feb. 8, 2009).

Cleaner production (CP) throughout the whole production process needs also awareness. This is a preventive strategy to reduce waste generation, helping enterprises to gain benefits and to protect the environment. In the National Action Plan (2002) of CP, phase 1 includes the conduction of CP demonstration in some craft villages. However, CP has been applied very limited in paper recycling villages, mainly because of the lack of awareness and legal aspects [2].

The most important progress of recycling activity depends critically on the frame work conditions, which "guide" the behavior of the households and the business companies [3].

To enhance the recycling potential it is necessary to have a national policy encouraging this action by reducing tax for waste paper business, encouraging the investment on paper recycling technology, and issuing market tools to involve people in used paper recycling (Vietnam News TTX, 29/12/2009). The efforts made so far are confined to very small programs. For instance, only one program calls for teenagers to collect at least 1 kg of wastepaper per month from classrooms was done effectively in Ho Chi Minh City (2009). Even in the national development strategy for the paper industry, the government has emphasized producing paper pulp from wood and makes no specific plan for collecting and recycling wastepaper. To reduce paper consumption, in companies, offices, schools and households, there is a necessity to have a guideline for using paper category for suitable purpose, issue the white scale and amount for office paper, announce the environment friendly paper, etc.

To encourage paper recycling, some specific number should be announced, quite accurately, for example, to produce 1 ton of recycled paper can save 32 m^3 water, 4,200 kWh of electricity, reduce the cut down of 17 trees, etc. [4]. Hence, it might help to make people aware about many advantages of paper recycling and so to environmental protection.

In general, encouraging different components in the society to be involved with used paper collecting and recycling by "provide frame work conditions" might be the most important issue.

6 Conclusions

Pulp and paper production of Vietnam increases over years, but this development does not yet go on a sustainable way. The amount of imported recovered paper and pulp shows a higher percentage in comparison with those of the domestic supply. Therefore, it leads to problems in the flow of material input, which need to be taken into consideration.

There is also problem of used paper collection, since the collection rate is low. All factors like recovered paper utilization, utilization rate, and collection rate need to be considerately strongly.

The paper recycling activity in craft villages has contributed a remarkable sign in the recycling sector. However, there is still a problem in harmonization between production and environmental protection.

The recycled paper products with low quality, especially for hygienic paper delivered throughout the country may cause health problems for customers. An investigation concerning the risk-based management of chemicals in recycled paper and also in other products needs to be focused in the coming day.

Glossary and Definitions

Recovered paper (R.P.) utilization means use of recovered paper as raw material to produce new products.

Apparent collection means utilization plus exports minus imports of recovered paper.

Utilization rate means percentage of recovered paper utilization compared to the total paper production.

Collection rate means percentage of apparent collection compared to the total paper consumption.

Recycling rate means percentage of recovered paper utilization compared to the total paper consumption.

Utilization by sector means total use of recovered paper in a sector as a percentage of the overall recovered paper use.

Recovered paper mainly consists of cellulose pulp. EINECS identifies cellulose pulp as follow: "The fibrous substances obtained from the treatment of lignocelluloses substances (wood or other agricultural fiber sources) with one or more aqueous solutions of pulping and/or bleaching chemicals, composed of cellulose, hemi-cellulose, lignin and other minor components. The relative amounts of these components depend on the extent of the pulping and bleaching processes" (EINECS number 265-995-8, Waste and recovered substances, p. 14).[3]

Recovered paper may contain other constituents such as pigments, inks, glues, fillers, etc. Regarding the recovery and recycling process, constituents that have no specific function in the material (cellulose pulp), can, therefore, be considered as impurities.

Impurities are an unintended constituent present in a substance as produced. It may originate from the starting materials or be the result of secondary or uncompleted reactions during the production process. While it is present in the final substance it was not intentionally added (EINECS number 265-995-8, Waste and recovered substances, p. 8).

[3] Report of EINECS number 265-995-8.

References

1. Dang Kim Chi, Tuong Thi Hoi et al (2010) Guideline to improvement of the environment in paper recycling villages
2. Ngo Thi Nga (2006) Workshop "The way to circular economy for Vietnam", Hanoi University of Science, 2006
3. Wiesmeth H (2010) Workshop "Waste to energy for the city of tomorrow", Hanoi University of Science, Oct 2010
4. Tran Thi My Dieu (2009) Workshop "Management of used plastic and paper packaging in HCM City", University of Van Lang, HCM City, 18 Aug 2009
5. Diem Trang NT and Student Group K18, K22, Hanoi University of Science, Project study 2007–2008

Case Study on Printed Matter in Denmark

Henrik Fred Larsen

Abstract The implementation of the EU REACH regulation will most probably promote substitution within sectors handling a lot of different chemicals such as the printing industry. With the aim of being at the cutting edge of this development, the Danish printing industry started up a substitution project in 2006. A major part of the work has been mapping the presence of chemicals, which are the potential candidates for substitution (e.g., PBT, CMR, vPvB, EDS). The mapping comprises a combination of a literature study and an investigation of the actual (2007) presence of candidate substances at 15 Danish printing houses including the examination of almost 900 MSDSs (i.e., products). Furthermore, a focused search in the Danish Product Register has been included. More than 200 of the mapped substances are candidates for substitution according to Danish legislation (List of Undesirable Substances) and about 60 of these substances fulfill one or more of the criteria (e.g., CMR, EDS) for the REACH Annex XIV candidate list (Authorisation List). This case study presents the results of the mapping of chemical substitution candidates some of which may be relevant regarding accumulation of additives/impurities in globally recycled paper.

Keywords Chemical substitution candidates, Printing industry, REACH, Recycled paper, Substances of very high concern (SVHC)

Contents

H.F. Larsen
Department of Management Engineering, Technical University of Denmark, Produktionstorvet, Building 424, 2800 Lyngby, Denmark
e-mail: hfl@man.dtu.dk

B. Bilitewski et al. (eds.), *Global Risk-Based Management of Chemical Additives I: Production, Usage and Environmental Occurrence*, Hdb Env Chem (2012) 18: 167–182,
DOI 10.1007/698_2011_106,
Published online: 21 June 2011

Abbreviations and Symbols

AT	Danish Working Environment Authority
CMR	Carcinogenic, mutagenic, reprotoxic
DTU	Technical University of Denmark
ECHA	European Chemicals Agency
EDS	Endocrine disrupting substances
EFL	Effect list
EPA	Environmental Protection Agency
EU	European Union
IARC	International Agency for Research on Cancer
IPU	Institute for Product Development
LOFS	List of hazardous substances
LOUS	List of undesirable substances
MSDS	Material safety data sheets
NOEC	No observed effect concentration
PBT	Persistent, bioaccumulating and toxic
POP	Persistent organic pollutants
QSAR	Quantitative structure-activity relationship
REACH	Registration, Evaluation, Authorisation and Restriction of Chemicals
Rxx	Risk phrases (e.g., R45, R50/53)
SVHC	Substances of very high concern
vPvB	very Persistent, very Bioaccumulating

1 Introduction

The printing industry uses a high number of different chemicals in the production of printed matter. Based on the previous studies [1–5], it may be estimated that at least 500–1,000 single substances (e.g., hexane) and mixtures (e.g., gasoline) appearing in minimum 40 frequently used product groups, such as printing inks, cleaning agents, and fountain solutions, are used in the Danish printing sector. In addition, a higher number of other substances are part of less frequently used raw materials and ancillaries. The chance of identifying substances that meet one or more criteria for the EU REACH Annex XIV candidate list [6], the so-called "Substances of Very High Concern" (SVHC-substances), is therefore relatively high. These substances will most probably be liable to strict regulation on use (authorization/banning) leading to substitution.

A strategy of being prepared for the expected coming demand on substitution of specific chemicals was therefore formulated within the Danish printing industry in 2005. A substitution project was subsequently designed by the Graphic Association Denmark and the "Danish Newspapers' and Media Employers' Association" in cooperation with the Technical University of Denmark (DTU) and IPU. The project achieved funding by the Danish EPA. So, related to the now ongoing implementation of the EU REACH regulation (http://ecb.jrc.it/reach/) a substitution project within the Danish printing sector has been going on since 2006. The mapping of potential candidates for substitution has now come to an end and the work on actual substitution has started up. A substitution database has been created including all the mapped substances divided into groups of different levels of substitution need. In total, about 600 substances are included. Identification of prioritized candidates for substitution, their alternatives, and the actual substitution work together with printing houses and suppliers is now going on. In this chapter, the mapping of the chemical substitution candidates will be presented. Apart from updates included here, this work is also published in more detail in a recent report (in Danish) by the Danish EPA [7].

2 Methodology

The mapping of substitution candidates within the Danish printing industry is based on the following three main approaches:

1. Literature study (existing research)
2. Novel inventory within the Danish printing industry
3. Search in the Danish Product Register

As this study regards the Danish printing industry and because a lot of research has been done on chemical use at Danish printing houses, the literature study is primarily based on former Danish investigations [1–5, 8–16]. However, also German [17], Swedish [18–20], and American [21] literature has been included when relevant. A comparison of the substances appearing in the existing literature with those registered on the Danish EPA List of Undesirable Substances, that is LOUS 2004 [22], or meeting its criteria, has been done.

As a second approach, a novel inventory was performed within the Danish printing industry involving the use of chemicals in 2006 and 2007 at 15 Danish printing houses and including about 880 Material Safety Data Sheets (MSDSs) supplemented by technical data sheets. The data sheets were investigated and the appearing substances registered and compared with those appearing on LOUS 2004 [22] and on nine other "hazard" lists (see Sect. 2.2).

As the third approach, a focused search in the Danish Product Register (http://www.at.dk/sw52891.asp) has been performed on the Danish printing industry (Nace code 22: Printing industry, 2006). By law, producers, importers, and suppliers on the Danish market are obliged to notify hazardous substances and materials (including products/mixtures) to the Product Registry, that is the Danish Working Environment Authority. The tonnage limit for this duty is 100 kg hazardous product per year. Nonhazardous substances (e.g., as part of hazardous mixtures) are typically also reported. The search in the Product Register was focused on the so-called PBT/vPvB (Persistent, Bioaccumulating, Toxic/very Persistent, very Bioaccumulating) substances appearing on the EU online PBT system [23] and substances appearing on LOUS 2004 [22] or on the Danish "effect" list EFL 2004 [24] being a candidate list of LOUS 2004.

The reason for using three approaches in the mapping of substitution candidates is a wish of cross-checking and verifying the presence of the substances in the Danish printing industry today (i.e., actually 2006/2007). Experience from former investigations (e.g., [3]) shows that using only one mapping approach is not enough if reasonable certainty of the result is aimed for.

2.1 Printing Techniques Included

The main printing techniques included are offset printing supplemented by screen printing, digital printing, and book binding (finishing). The registered substances include primarily "printing industry specific chemicals" such as printing inks, cleaning agents, and fountain solutions, but nonspecific chemicals, such as hand cleaning agents and lubricants, are included to some degree. The coverage of the different printing techniques is shown in Table 1.

2.2 Included Hazard Lists, Criteria, and Substitution Levels

The main "hazard" lists used for classifying the identified substances are shown in Table 2. The associated priorities regarding substitution needs (only inherent properties) are also shown and further dealt with in Sect. 2.3.

Table 1 Printing technique coverage of the included printing houses in the inventory

Printing technique	Number of printing houses with technique
Sheet-fed offset	7
Cold-set offset (newspaper)	2
Heat-set offset	2
Screen printing	3
Digital printing	3
Book binding	4
Flexo printing (labels)	1

Table 2 Main "hazard" lists used in classifying the mapped substances

List	Substitution priority	Issuer	Ref.
LOUS 2004	1 (high)	Danish EPA	[22]
EFL 2004	1 (middle)	Danish EPA	[24]
LOFS	–	EC/EU	[25]
EDS-list	1 (high)	EC/EU	[22]
PBT-list	1 (very high)	EC/EU	[23]
REACH Annex XIV	1 (highest)	EC/EU/ECHA	[6, 26]
REACH Annex XVII	1 (very high)	EC/EU	[6]
ATs cancer list	2	The Danish Working Environment Authority	[27]
EFL 2000	2	Danish EPA	[28]
Self-classification list	–	Danish EPA	[29]

The REACH Annex XIV substances, the so-called Substances of Very High Concern (SVHC), are the most problematic ones. They will become part of the authorization under the REACH regulation [6], that is either forbidden or subject to restrictions. However, only 15 substances have yet (December 2010) been recommended by ECHA for inclusion in Annex XIV [26]. These substances are taken from the Annex XIV candidate list, which includes 46 substances today (December 2010) [26]. The substances appearing on the Annex XIV candidate list have to be recommended by EU member states and finally decided upon in ECHA's Member State Committee. Substances meeting one or more of the following criteria are potential candidates [6]:

1. Carcinogenic substances in category 1 (carc1) or category 2 (carc2): T (toxic); R45 (may cause cancer), R49 (may cause cancer by inhalation)
2. Mutagenic substances in category 1 (mut1) or category 2 (mut2): T (toxic); R46 (may cause heritable genetic damage)
3. Reprotoxic substances in category 1 (rep1) or category 2 (rep2): T (toxic); R60 (may impair fertility), R61 (may cause harm to the unborn child)
4. Substances meeting the criteria of PBT and/or vPvB substances [23]
5. Endocrine disrupters and substances not meeting the above-mentioned criteria but for which there is scientific evidence of possible alike serious effects on humans and/or the environment

As the REACH Annex XIV candidate list [26] was (and still is) quite limited (actually none substances was chosen at the beginning of this project in 2006), the Danish EPA List of Undesirable Substances, LOUS 2004 [22], was used in this project as the main list together with its associated candidate list, that is the Danish "effect" list EFL 2004 [24]. LOUS 2004 includes 270+ substances and the EFL 2004 list includes 6,400 substances. The criteria for these lists include the criteria for the REACH Annex XIV list as mentioned above, and furthermore stricter criteria on carcinogenicity (carc 3), ecotoxicity (R50/53), and more. The substances on EFL 2004 are mainly taken from the EU list of dangerous substances (29. Adaption), that is LOFS [25], and have to meet one or more of the following criteria (here expressed as risk phrases):

R33 Danger of cumulative effects
R39 Danger of very serious irreversible effects
R40 Limited evidence of a carcinogenic effect
R42 May cause sensitization by inhalation
R45 May cause cancer
R46 May cause heritable genetic damage
R48 Danger of serious damage to health by prolonged exposure
R49 May cause cancer by inhalation
R50/53 Very toxic to aquatic organisms, may cause long-term adverse effects in the aquatic environment
R58 May cause long-term adverse effects in the environment
R59 Dangerous for the ozone layer
R60 May impair fertility
R61 May cause harm to the unborn child
R62 Possible risk of impaired fertility
R63 Possible risk of harm to the unborn child
R64 May cause harm to breastfed babies
R68 Possible risk of irreversible effects

Furthermore, endocrine disrupters according to the EU EDS-list [22], most of the prioritized substances according to the EU Water framework directive [30] and substances registered on the EU PBT list [23] are also part of the Danish EFL 2004 list.

The 270+ substances on the Danish LOUS 2004 list was chosen by the Danish EPA, mainly by combining the EFL 2004 list with a search in the Danish Product Register on tonnage. CMR (Carcinogenic Mutagenic, Reprotoxic) substances in categories 1 and 2 (carc1, carc2, mut1, etc.) with a yearly consumption higher than 1 ton are all included. For the other substances on the EFL 2004 list, the tonnage limit was 100 ton. With certain exceptions [22] endocrine disrupters according to Appendix B of the LOUS publication [22] and substances that are especially problematic for other reasons, for example as related to ground water or waste streams, are also part of LOUS.

The PBT-list or PBT/vPvB-list includes substances that are under suspicion of being persistent (P), bioaccumulating (B) and toxic (T), or very persistent (vP) and very bioaccumulating (vB). The criteria are defined in the REACH regulation [6] and includes for P: half-life higher than 40–180 days (depending on media), for B: bioconcentration factors BCF higher than 2,000 (B) or higher than 5,000 (vB), and for T: chronic NOEC values below 0.01 mg/L and/or carcinogenic (carc1 or carc2) and/or mutagenic (mut1 or mut2) and/or reprotoxic (rep1, rep2, rep3) and/or classified by R48 (danger of serious damage to health by prolonged exposure). The list contains 127 substances of which 27 or found to be PBT or vPvB, 66 substances do not comply, 24 substances are still under evaluation, and 10 substances comply with screening criteria but final decision postponed [23].

REACH Annex XVII list substances that are all ready subject to restrictions and substances with known CMR properties (carc1, carc2, mut1, mut2, rep1, rep2) [6].

"ATs cancer list" is a list published every second year by The Danish Working Environment Authority. It contains substances that are either carcinogenic or mutagenic (CM). Besides the CM-substances appearing on the LOFS list [25] (and in REACH Annex XVII), it contains substances classified for carcinogenicity according to IARC (International Agency for Research on Cancer), that is as IARC 1, IARC 2a, and IARC 2b.

The former LOUS list, LOUS 2000 [31], together with its associated candidate list, that is the Danish "effect" list EFL 2000 [28], was based on stricter criteria than the main EFL-list used here, that is EFL 2004 [24]. Besides the R-phrases shown above for EFL 2004, the following was also included:

R23 Toxic by inhalation
R24 Toxic in contact with skin
R25 Toxic if swallowed
R26 Very toxic by inhalation
R27 Very toxic in contact with skin
R28 Very toxic if swallowed
R43 May cause sensitization by skin contact
R51/53 Toxic to aquatic organisms, may cause long-term adverse effects in the aquatic environment
R52/53 Harmful to aquatic organisms, may cause long-term adverse effects in the aquatic environment

Only the criteria (not the lists) for LOUS 2000 [31] and EFL 2000 list [28] are included in the project and only for substitution level 2 (see Sect. 2.3).

The self-classification list by the Danish EPA [29] contains 20,624 substances classified by one or more of the following Risk phrases: R22 (harmful if swallowed), R43, R40, R50 (very toxic to aquatic organisms), R50/53, R51/53, R52/53. It is a QSAR (Quantitative Structure-Activity Relationship)-based list created by running relevant QSARs on 47,000 substances with unknown toxicological properties. In this project mapped substances with unknown properties were checked against this list and if classified by R43, R40, R50/53, R51/53 or, R52/53, it was registered.

The appearance of a mapped substance on one or more of the lists described above is not mandatory for its registration in the project, that is the substitution database. If the substance meets one or more of the different criteria, for example according to the self-classification by the producer on the MSDS, it is registered.

2.3 The Substitution Database

All mapped substances have been compiled within an Access "Substitution database" (Microsoft Office Access 2003, SP3), which is freely available in Excel format at the homepage of the Graphic Association Denmark (www.ga.dk).

Based on the lists and criteria described in Sect. 2.2, the mapped substances are divided into three levels of substitution priorities:

- Level 1 (highest priority): Substances appearing on the following lists or meeting its criteria
  - Annex XIV candidate list
  - The PBT/vPvB list
  - The Danish EPA List of Undesirable Substances (criteria on tonnage excluded)
- Level 2 (second highest priority): Substances not meeting the criteria of level 1 but meeting the criteria of the former Danish EPA List of Undesirable Substances, that is R23, R24, R25, R26, R27, R28, R43, R51/53, and R52/53 (criteria on tonnage excluded)
- Level 0 (lowest priority): Environmental and health hazard unknown or known but not meeting the criteria of levels 1 and 2

These levels are only based on the inherent properties of the substances and needs to be combined with consumption, exposure, and socioeconomic issues when dealing with actual substitution, which is not included here.

The degree to which the substance appears on any of the lists shown in Table 2 (or complies with the list criteria to the degree known) is depicted for all included substances in the database. Furthermore, all registered substances are characterized by CAS number, name, etc. and with references to where it was found (literature search, printing house inventory, etc.). Known functions of the substance are shown (emulgator, pigment, binder...) and in which products (sheet-feed offset inks, fountain solutions, etc.) it was found. The name of the producer/supplier and the self-classification is also indicated. Regarding the search in the Danish Product Register, the range in percentage of the substance occurrence in different products, the number of products in which it occurs, and the amount (tonnage) of the substance on the Danish marked (Printing Industry, Nace code 22) are also shown in the database.

3 Results and Discussion

In total, 588 substances (including mixtures of hydrocarbons) are registered in the database (September 2008). 228 substances are allocated to substitution level 1 (including 31 mixtures of hydrocarbons and the like) and 89 substances to substitution level 2 (including 6 hydrocarbon mixtures). Substitution level 0 accounts for 245 substances. Furthermore, the database includes 26 hydrocarbon mixtures of which at least some most probably contains single substances allocated to substitution level 1 or 2.

The result of the mapping of substitution candidates at substitution level 1, divided into function groups is shown in Fig. 1. Regarding hydrogen mixtures, only the eight mixtures that were actually identified in the printing industry are included.

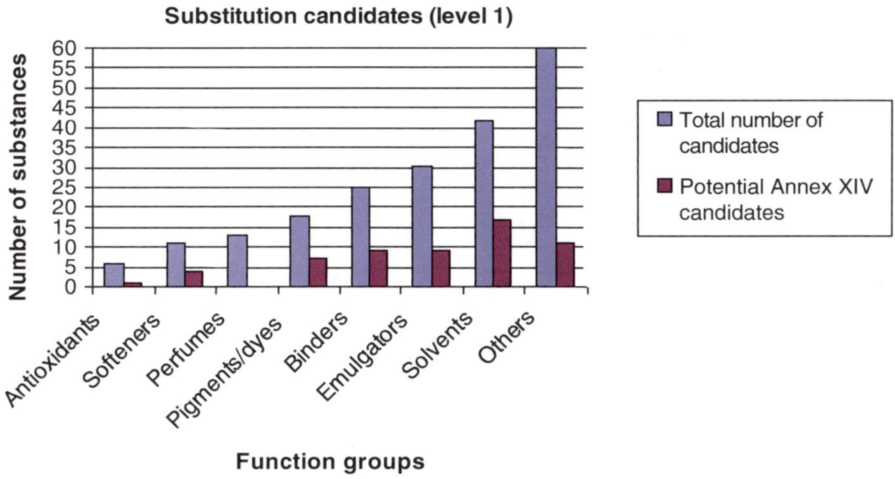

Fig. 1 The result of mapping substitution candidates within the Danish printing industry

The number of mapped substances appearing on the Danish EPA List of Undesirable Substances, LOUS 2004 (or meeting its criteria), in this way amounts to 205.

125 of the substances were found at least at one of the fifteen participating printing houses. Of the remaining 80 substances, 47 were found in the Danish Product Register and only 33 substances were solely found in the literature indicating that they may no longer be on the Danish market.

The substances occur in different product groups such as printing inks, cleaning agents, and fountain solutions. 58 (including 16 only found in the literature study) of the substances fulfill one or more (CMR or EDS) of the criteria for the REACH Annex XIV candidate list. Most substitution candidates are found within the function group on solvents, in total 42 substances. Seventeen of these are potential candidates for REACH Annex XIV (e.g., benzene, C14-17 chloro alkanes). In the group of pigments/dyes, seven out of eighteen are potential Annex XIV candidates (e.g., Pigment Yellow 34, Pigment Red 104).

It should be noticed that the total number of substances, that is 588, found in this study and registered in the database, far from covers the actual total consumption in the Danish printing sector. According to a search in the Danish Product Register, the printing industry in Denmark in 2006 consumed 2,067 substances, being components of 1,541 products and used in a total amount of 236,500 ton. Due to confidentiality, typically, only the identity of substances appearing in more than two products or notified by more than two producers/suppliers have been available to this project. Therefore, 1,404 substances appearing in the Product Register under Nace code 22 (Printing Industry) are unknown to this study. Furthermore, the searches in the Product Register have been focused on PBT substances, and "substitution level 1 substances" found in the literature search and the novel printing industry inventory.

The use of gray zone products (i.e., non-industry specific) in the printing sector is apparently significant. A search in the Product Register on 100 products identified by Product Register numbers on the MSDSs found in the novel printing industry inventory showed that only 48 were notified under Nace code 22 (Printing Industry). So, about half of the products were only notified under other industries, such as industries related to automobiles (Nace code 50) and mechanical engineering (Nace code 29).

The result of this study, as implemented in the database, may therefore be characterized as focused on hazardous substances and probably covering at the level of 10–20% of the total number of substances used in the Danish printing sector.

3.1 The PBT List

Regarding the twenty seven substances fulfilling PBT/vPvB/POP criteria according to the EU PBT/vPvB list [23], and therefore most probably candidates for Annex XIV, only one has been found in the Danish printing industry (a chlorinated organic substance, name confidential). Four substances, which are still under evaluation for PBT/vPvB properties (February 2009), were identified in thirty five products (especially printing inks and cleaning agents) in a total yearly substance consumption of 1.1 ton. Two of these have confidential names and the other three are decamethylcyclopentasiloxane, octamethylcyclopentasiloxane, and C14-C17 chloroalkanes.

3.2 The Annex XIV Candidate List

In Tables 3 and 4, the substances found in the Danish printing industry in 2006 and 2007 (i.e., the novel printing industry inventory and the searches in the Product Register), which meet one or more of the REACH Annex XVI criteria are shown. Eleven of these substances are now (December 2010) part of the Annex XIV candidate list [26], see Table 3. Regarding five out of these eleven substances, that is the lead-chromate pigments Pigment Yellow 34 and Pigment Red 104, and the phthalates DEHP, DBT, and BBT, inclusion in Annex XIV (Authorization List) is recommended by ECHA and adopted by the Member State Committee [32, 33].

3.3 Relevance for Recycled Paper

Regarding the three phthalates in Table 3, that is DEHP, dibutylphthalate, and benzylbutylphthalate, a total yearly consumption above 1 ton, an appearance in

Table 3 Substances appearing on the recently updated REACH Annex XIV candidate list [26] and found in the Danish printing industry

Name	CAS No.	Annex XIV criteria	Use
Chromtrioxide	1333-82-0	Carc 1, mut 2	Chrome plating (gravure)
Trichloroethylene	79-01-6	Carc 2	Inks
Cobalt-siccatives[a]	(10124-43-3)	(Carc 2)	Inks (off-set, screen printing)
Acrylamide	79-06-1	Carc 2, mut 2	Unknown (impurity?)
Pigment Yellow 34 (lead-chromate)[b]	1344-37-2	Rep 1	Inks (screen printing)
Pigment Red 104 (lead-chromate)[b]	12656-85-8	Rep 1	Inks (screen printing)
2-Methoxy ethanol	109-86-4	Rep 2	Photochemistry
Di(2-ethylhexyl)phthalate, DEHP[b]	117-81-7	Rep 2, EDS-list	Inks
Dibutylphthalate, DBT[b]	84-74-2	Rep 2, EDS-list	Inks (screen printing, flexo)
Benzylbutylphthalate, BBT[b]	85-68-7	Rep 2, EDS-list	Inks
Boric acid and borax	10043-35-3 and 1301-96-4	Rep 2[c]	Photochemistry

[a]Possible content of soluble cobalt(II)salts. Cobalt(II)sulfate, cobalt dichloride, cobalt(II)carbonate, cobalt(II)dinitrate, and cobalt(II)diacetate all appear on the recently updated REACH Annex XIV candidate list [26]. IARC categorizes all soluble cobalt(II)salts as possible carcinogenic, that is group 2B (http://monographs.iarc.fr/ENG/Monographs/vol86/mono86.pdf)
[b]Inclusion in Annex XIV (Authorisation List) recommended by ECHA and adopted by the Member State Committee [32, 33]
[c]"New" classification according to adaption 30 or 31 [34] and not implemented in the database (yet) as adaption 29 [25] was valid when the project was finished

about 40 products and a concentration range of 0.1–75% in the products are observed in the Danish printing industry. These substances are of interest as they are components of printing inks and remain in the ink after drying and therefore follow the substrate, that is paper, plastic, or textile, when recycled. They may therefore appear in the recycled material. Actually, according to a German investigation [35], dibutylphthalate has been found in recycled paper used for food packaging. Also, other substances mentioned in Tables 3 and 4 may be of interest as being components of printing inks such as the lead chromate pigments, the siloxanes, and bisphenol A. Furthermore, as mentioned earlier, 26 hydrocarbon mixtures, most probably containing hazardous single substances at substitution level 1 or 2 (e.g., hexane, heptane, naphthalene) are found in the Danish printing industry. Many of these are used as components in printing inks (and cleaning agents) and therefore may follow the printed substrate when recycled. Some of the hydrocarbon mixtures are used in relatively high amounts in the Danish printing sector such as "naphtha (petroleum), hydrodesulfurized (benzene <0.1%)" used at a total level of 1,500 ton/year, in 35 products with a content of 0.1–100%. Finally, it should be noted that highly toxic substances only found in the literature study, such as potassium dichromate and hydrocarbon mixtures with high benzene content (>>0.1%), are probably still in use at places on the world market with less strict

Table 4 Substances meeting Annex XIV candidate list criteria and found in the Danish printing industry. Not listed on the REACH Annex XIV candidate list but potential candidates that may be listed in the future

Name	CAS No.	Annex XIV criteria	Use
Benzene	71-43-2	Carc 1, mut 2	Inks, cleaning agents
Epichlorohydrin	106-89-8	Carc 2	Unknown (impurity?)
2-Methylaziridine	75-55-8	Carc 2	Inks (flexo)
Aziridine	151-56-4	Carc 2, mut 2	Inks (flexo, screen printing)
Propylenoxide	75-56-9	Carc 2, mut 2	Inks, cleaning agents
2-Methoxy propylacetate	70657-70-4	Rep 2	Inks (screen printing)
Triethylene glycol dimethylether	112-49-2	Rep 2	Brake fluid
2-Methoxypropan-1-ol	1589-47-5	Rep 2	Unknown
Alkylphenolethoxylates	(25154-52-3)	EDS-list	Inks, cleaning agents
Chloroalkanes, C14-17	85535-85-9	Possible PBT/vPvB-substance	Chain oil
Octamethylcyclotetrasiloxane (polydimethylsiloxane)	556-67-2 (9016-00-6)	Possible PBT/vPvB-substance	Inks
Bisphenol A	80-05-7	EDS-list	Inks, thermal paper
Resorcinol	108-46-3	EDS-list	Glue
Styrene	100-42-5	EDS-list	Inks, glue
Decamethyl-cyclopentasiloxane	541-02-6	Possible PBT/vPvB-substance	Inks
Stoddard solvent	8052-41-3	Carc 2	Unknown
Solventnaphtha (crude oil), hydrogen treated light naphthen- (benzene >= 0.1%)	92062-15-2	Carc 2	Cleaning agent

environment and health regulation (e.g., Asia), even though phased out on the Danish market. These substances may therefore be relevant when looking at globally recycled printing substrates such as paper, plastics, and textiles.

3.4 *Prioritization of Substances for Substitution*

In the ongoing second stage project, the actual prioritizations of substances for substitution are chosen among the candidates identified in this first stage project. Several criteria are included in the substitution campaigns within the Danish printing industry, that is severity of classification, relative high consumption/extended use and possibilities of successful implementation of the substitution. Substitution at the level of the printing house (e.g., use of existing alternative products) is prioritized. Examples of possible prioritized candidates for substitution are hydroquinone and heptane. Hydroquinone occurs in minimum 35 products (content: <0.1–64%) and the yearly consumption is minimum 8 ton. Heptane is used in more than 10 products (content 0.4–100%) and a minimum of 30 ton is yearly consumed.

4 Conclusion

The survey of chemicals which are potential candidates for substitution within the Danish printing industry resulted in about 200 substances/substance groups. About 60 of these substances fulfill one or more of the criteria for the EU REACH Annex XIV candidate list. Some of these, such as the phthalates and the lead chromate pigments, may be relevant when looking at the potential hazard of globally recycled paper based on printed matter. A substitution database has been created including about 600 substances characterized by their occurrence in identified products, function, environmental classification, and more. On this basis actual substitution is going on in the Danish printing industry.

Acknowledgement This study was financed by the Danish Environmental Protection Agency supplemented by in-kind contributions from the industry partners.

References

1. Lauritsen KB (1991) Renere teknologi i den grafiske branche (Cleaner technology in the printing industry). Miljøprojekt nr. 169 (book in Danish). Danish Ministry of the Environment, Environmental Protection Agency, Copenhagen
2. Sedorff L, Kjølholt J, Andersen HV, Jensen MM (1993) Miljøvurdering af fotokemikalier (Environmental assessment of photo chemicals). Miljøprojekt 218 (book in Danish). Danish Ministry of the Environment, Environmental Protection Agency, Copenhagen
3. Larsen HF, Tørsløv J, Damborg A (1995) Indsatsområder for renere teknologi i den grafiske branche – spildevandsvurdering. (Areas of intervention for cleaner technology in the printing industry – assessment of waste water). Miljøprojekt nr. 284 (book in Danish with English summary). Danish Ministry of the Environment, Environmental Protection Agency, Copenhagen
4. Larsen HF, Pedersen AR, Birch H, Rasmussen D, Hansen LE (1998) Miljøoptimering af rammevask ved serigrafi (Evaluation of Best Available Technology (BAT) in screen cleaning). Miljøprojekt nr. 381 (book in Danish with English summary). Danish Ministry of the Environment, Environmental Protection Agency, Copenhagen
5. Larsen HF, Helweg C, Pedersen AR, Andersen M, Wallström E, Hoffmann L (2002) Miljøoptimering af afvaskning ved tryk med vandfortyndbar flexofarve (Evaluation of Best Available Technology (BAT) in washing off water-based flexographic inks in the packaging printing industry). Miljøprojekt nr. 730 (book in Danish with English summary). Danish Ministry of the Environment, Environmental Protection Agency, Copenhagen
6. Regulation (EC) No 1907/2006 of the European Parliament of 18 December 2006 concerning the Registration, Evaluation, Authorisation and Restriction of Chemicals (REACH). (http://ecb.jrc.it/legislation/2006R1907EC.pdf)
7. Larsen H, Bøg C (2011) Kemikaliesubstitution i grafisk branche substitutionsdatabase med kortlægningsresultater (Substitution of chemicals in the printing industry – database and inventory results). Miljøprojekt nr. 1354 (book in Danish with English summary). Danish Ministry of the Environment. Environmental Protection Agency, Copenhagen
8. Havelund S (2002) Kortlægning af perfluoroktanylsulfonat og lignende stoffer i forbrugerprodukter – fase 2. Miljøprojekt 691 (book in Danish). Danish Ministry of the Environment, Environmental Protection Agency, Copenhagen

9. Hoffmann L, Poulsen P, Wallström E, Andersen K, Sørensen EH, Sørensen JB (2003) Substitution af alkylphenolthoxylater (APEO). Arbejdsrapport nr. 45 (book in Danish). Danish Ministry of the Environment, Environmental Protection Agency, Copenhagen
10. Larsen HF, Hansen MS, Hauschild M (2006) Ecolabelling of printed matter. Part II: Life cycle assessment of model sheet fed offset printed matter. Working Report No. 24. Danish Ministry of the Environment, Environmental Protection Agency, Copenhagen
11. Larsen HF, Helweg C, Pedersen AR, Boyd HB, Laursen SE, Hansen J (2000) Kemikalier i tekstiler (Chemicals in textiles). Miljøprojekt 534 (book in Danish with English summary). Danish Ministry of the Environment, Environmental Protection Agency, Copenhagen
12. Madsen T, Boyd HB, Nylén D, Pedersen AR, Pedersen GI, Simonsen F (2001) Environmental and health assessment of substances in household detergents and cosmetic detergents products. Environmental Project No. 615, 2001. Danish Ministry of the Environment, Environmental Protection Agency, Copenhagen
13. Pilemand C, Wallström E, Hoffmann L, Poulsen PB (2003) Substitution of Cobalt Driers and Methyl Ethyl Ketoxime. Environmental Project No. 884. Danish Ministry of the Environment, Environmental Protection Agency, Copenhagen
14. Rasmussen JO, Madsen T, Skak C, Nulén D (2003) Substitution af alkylphenolethoxylater (APE) i maling, træbeskyttelse, lime og fugemasser. Arbejdsrapport fra Miljøstyrelsen Nr. 46 (book in Danish with English summary). Danish Ministry of the Environment, Environmental Protection Agency, Copenhagen
15. Wallström E, Andersen BH, Jensen OK (1996) Videreudvikling af vandfortyndbare tekstiltrykfarver til Cool Flash trykning. Arbejdsrapport fra Miljøstyrelsen Nr. 79 (book in Danish). Danish Ministry of the Environment, Environmental Protection Agency, Copenhagen
16. Andersen P, Larsen HF, Trepile G, Trepile, L, Bøg C (1999) Anvendelse af vandfortyndbar farve til serigrafisk tryk (The use of water based inks in screen printing). Miljøprojekt nr. 508 (book in Danish). Danish Ministry of the Environment, Environmental Protection Agency, Copenhagen
17. Bauman W, Rothardt T (1999) Druckereichemikalien; Daten und Fakten zum Umweltschuts 2. Uberarbeitete and ergänzte Auflage (book in German). Springer, Heidelberg
18. Ludwiszewska J (1992) Grafiske branchen och dess kemikalieanvänding: Exempel från nogra företag inom Hälsingborgs kommun – Miljö- och Hälsoskyddskontoret Hälsingborg Kommun. Rapport 1/92 + Bilagsdel (book in Swedish). Hälsingborg Kommun, Hälsingborg
19. Brodin L, Korostenski J (1995) Miljøbelastningar från grafisk industri i Sverige. Grafiska Miljögruppen. Report version 2 (15-04-2003). Report to SFS (book in Swedish). Milgraf AB, Upplands Väsby
20. Brodin L, Korostenski J (1997) Miljøbelastningar från grafisk industri i Sverige: Screen-, Flexo-, Digitaltryck och Efterbearbetning. Preliminär rapport. Kompletterad och reviderad version. Grafiska Miljögruppen. Report version 3 (18-06-1997). Report to SFS (book in Swedish). Milgraf AB, Upplands Väsby
21. Leach RH, Pierce RJ, Hickman EP, Mackenzie MJ, Smith HG (eds) (1992) The Printing Ink Manual. Blueprint, Chapman & Hall, London
22. Miljøstyrelsen (2004) Listen over uønskede stoffer 2004. Orientering fra Miljøstyrelsen nr. 8, 2004, Miljøstyrelsen. + Opdatering af "Listen over uønskede stoffer 2004" og "Effektlisten 2004" af 17. juni 2005 (book in Danish). Danish Ministry of the Environment, Environmental Protection Agency, Copenhagen
23. EC (2008) Online PBT information system PBT. List of substances subject to evaluation of their PBT properties under the Interim Strategy for REACH and the ESR program ("an interim strategy for management of PBT and vPvB substances" (doc. ENV/D/432048/01, NOTIF/36/2001) was agreed by the Joint Meeting of the Competent Authorities for the Implementation of Council Directive 67/548/EEC (New Substances) and Council Regulation (EEC) 793/93 (Existing Substances) in June 2001). http://ecb.jrc.it/esis/index.php?PGM=pbt. Accessed Feb 2009

24. Miljøstyrelsen (2004) Effektlisten 2004. Orientering fra Miljøstyrelsen nr. 7, 2004, Miljøstyrelsen. + Opdatering af "Listen over uønskede stoffer 2004" og "Effektlisten 2004" af 17. juni 2005 (book in Danish). Danish Ministry of the Environment, Environmental Protection Agency, Copenhagen
25. Miljøstyrelsen (2005) Listen over farlige stoffer. Bekendtgørelse om listen over farlige stoffer, BEK nr. 923 af 28/09/2005 (29. tilpasning til den tekniske udvikling af Rådets direktiv 67/548/EØF). Danish Ministry of the Environment. Environmental Protection Agency, Copenhagen. Based on: COMMISSION DIRECTIVE 2004/73/EC of 29 April 2004 adapting to technical progress for the 29th time Council Directive 67/548/EEC on the approximation of the laws, regulations and administrative provisions relating to the classification, packaging and labelling of dangerous substances
26. ECHA (2010) ECHAs homepage: http://echa.europa.eu/home_en.asp. Recommendations on Annex XIV inclusions: http://echa.europa.eu/chem_data/authorisation_process/annex_xiv_rec_en.asp Candidate list: http://echa.europa.eu/chem_data/authorisation_process/candidate_list_table_en.asp. Accessed 21 Dec 2010
27. Arbejdstilsynet (2007) At-vejledning C.0.1, august 2007 (– erstatter april 2005). Grænseværdier for stoffer og materialer. Arbejdstilsynets liste over kræftfremkaldende stoffer og processer, bilag 3.6 (Book in Danish). The Danish Working Environment Authority, Copenhagen
28. Miljøstyrelsen (2000) Effektlisten 2000. Orientering fra Miljøstyrelsen nr. 6, 2000 (book in Danish). Danish Ministry of the Environment, Environmental Protection Agency, Copenhagen
29. Miljøstyrelsen (2001) Rapport om Vejledende liste til selvklassificering af farlige stoffer. Miljøprojekt Nr. 635, 2001 (book in Danish). Danish Ministry of the Environment, Environmental Protection Agency, Copenhagen
30. EC (2002) Decision No 2455/2001/EC of the European Parliament and of the Council of 20 November 2001. Establishing the list of priority substances in the field of water policy and amending Directive 2000/60/EC
31. Miljøstyrelsen (2000) Listen over uønskede stoffer 2000. En signalliste over kemikalier, hvor brugen på længere sigt bør reduceres eller stoppes. Orientering fra Miljøstyrelsen nr. 9, 2000 (book in Danish). Danish Ministry of the Environment, Environmental Protection Agency, Copenhagen
32. ECHA (2009) ECHA recommends strict control for seven substances of very high concern. Press Release: ECHA/PR/09/07. Helsinki, 02 June 2009. http://echa.europa.eu/
33. ECHA (2010) Sixteen Substances of Very High Concern closer to tighter control. Press Release: ECHA/PR/10/24. Helsinki, 03 December 2010. http://echa.europa.eu/
34. EC (2009) Commission Regulation (EC) No 790/2009 amending, for the purposes of its adaptation to technical and scientific progress, Regulation (EC) No 1272/2008 of the European Parliament and of the Council on classification, labelling and packaging of substances and mixtures
35. Brauer B (2010) Kontaminanten in Recyclingpapier – Problemstoffe, Problemlösungen. Chemisches und Veternäruntersuchungsamt Münsterland-Emscher-Lippe (CVUA-MEL). BfR-Forum Verbraucherschutz, Berlin, 28-10-2010 (slides in German)

Potential Reuse of Chemical Sludge from Textile Dyeing Processes

S. Pandey, H. Patel, and R. Johri

Abstract The textile industry plays a significant role in the Indian economy. The combined textile effluent from dyeing and printing clusters is treated in common effluent treatment plants (CETP) where the physicochemical treatment of the wastewater leads to the generation of chemical sludge in voluminous quantities. This sludge is considered a hazardous waste according to the Indian Hazardous Waste Management rules of 2008. Presently, the only option available to CETP operators for the disposal of this waste is a secure landfill, but this represents a costly option for them. The case study presented here attempts to find an environment-friendly and a cost-effective solution for the management of this chemical sludge. Sludge samples from various CETPs spread across the India were collected and they were physicochemically characterized; toxicity and microstructural aspects were also taken into consideration. To evaluate the suitability of the sludge as construction material, a solidification/stabilization (S/S) treatment of the chemical sludge was carried out using two binders: Ordinary Portland Cement (OPC) and Portland Pozzolona Cement (PPC). The evaluation of the solidified samples was carried out by considering their physical engineering properties, such as unconfined compressive strength and block density, and its chemical properties, such as leachability of heavy metals. A microstructural examination of the solidified samples was also performed using X-ray diffraction (XRD) and scanning electron microscopy (SEM). Characterization results revealed that the sludge samples were alkaline and had high electrical conductivity values. The concentration of heavy metals (Cr, Cu, Ni, Zn, Cd, and Pb) in the dried sludge, as well as in the leachate, was found to be less than that present in the prescribed limits (Indian Hazardous Waste Rules for sludge samples and US EPA limits for leachate). Oxides, such as SiO_2, Al_2O_3, Fe_2O_3, MgO, and SO_3, were present in a significant amount. Unconfined strength and block density data of the solidified blocks indicated that the chemical

S. Pandey (✉) and R. Johri
The Energy and Resources Institute, India Habitat Centre, Lodhi Road, New Delhi 110003, India
e-mail: spandey@teri.res.in

H. Patel
TERI University, 10 Institutional Area, Vasant Kunj, New Delhi 110070, India

B. Bilitewski et al. (eds.), *Global Risk-Based Management of Chemical Additives I: Production, Usage and Environmental Occurrence*, Hdb Env Chem (2012) 18: 183–206,
DOI 10.1007/698_2011_102,
Published online: 14 April 2011

sludge had potential to be used as a construction material for different kinds of applications. The microstructural examination of the solidified samples indicated a modification of the cement patterns.

Keywords Chemical sludge, Compressive strength, Leaching, Sludge reuse, Solidification, Stabilization

Contents

1 Introduction

Industrialization and urbanization have given rise to the production of a wide variety of industrial and consumer products, by-products, and waste. A major fraction of the waste generated constitutes a special or hazardous waste that possesses certain specific properties with respect to the impacts on the environment and the living world. Hazardous wastes are complex materials with properties that make them difficult to treat or dispose using conventional methods, and their persistence in the environment is higher than that of other wastes. This very nature of the hazardous wastes makes collection, handling, treatment, and disposal complicated, expensive, risk-laden, and difficult to manage.

Typical examples of hazardous wastes would include heavy metals, such as chromium, mercury, or nickel, solvents, cyanide wastes, pesticides or polychlorinated biphenyls (PCBs), or oil- and grease-laden wastes with toxic metals (e.g., sludge from leaded petrol storage tanks).

Large-scale production of a variety of chemicals, energy use, and other developmental activities such as agriculture, urbanization, and healthcare in India have led to the generation of huge quantities of waste in the form of solid, liquid, and gases. The improper disposal of these wastes may cause contamination of air, surface water, groundwater, soils, sediments, and biota. Contamination of groundwater by landfill leachate poses risk to downstream surface water and wells, and is a major environmental concern. Air emissions and liquid waste are treated on-site according to legal requirements. However, the situation is more complex with hazardous waste because this type of waste requires special collection, storage, treatment, and disposal. The problem of treatment is complicated due to the heterogeneity of the waste in terms of chemistry, bioavailability, and toxicity [1].

The amount of hazardous waste generated in India is estimated to be around 6.23 million tonnes (Mill T) according to the estimates of Central Pollution Control Board [2]. Part of this waste is recyclable (49.5%), a 6.7% can be incinerated, and the remaining 43.8% is disposable in secured landfills [2]. Twelve states of the country (Maharashtra, Gujarat, Tamil Nadu, Orissa, Madhya Pradesh, Assam, Uttar Pradesh, West Bengal, Kerala, Andhra Pradesh, Karnataka, and Rajasthan) account for 97% of the total hazardous waste generation. The top four waste-generating states are Maharashtra, Gujarat, Andhra Pradesh, and Tamil Nadu. On the other hand, states such as Himachal Pradesh, Jammu and Kashmir, and all northeastern states except Assam generate less than 20,000 tonnes per annum.

Hazardous waste regulations have prompted the awareness of this type of waste. However, effective practices for a safe management are still to be enforced in India. There are currently several methods for the treatment and disposal of the hazardous waste. Nevertheless, technical issues associated with the waste treatment and disposal need to be strengthened, and technology input requirements have to be worked out to achieve a sustainable management of waste. Although expensive and advanced methods can be adopted to treat hazardous waste, they still result in potentially hazardous residues and are no longer amenable to carry out effective treatments. Thus, waste minimization is the best approach to avoid the problem of treatment and disposal. There exists also the need to develop various clean, environment-friendly, and economic technologies for the safe disposal of hazardous waste in India.

The effectiveness of a hazardous waste management system refers to how well it reduces the quantities and hazards of waste, ideally approaching zero for both. The various options available are:

(a) To take measures that prevent the generation of waste
(b) To recover and recycle waste constituents
(c) To destruct and treat the waste (conversion to nonhazardous forms)
(d) To dispose the waste (storage and landfill)

Hazardous waste treatment and disposal technologies have been developed to handle several types of wastes, and new technologies are still emerging. The objective of waste treatment is to modify the physical and/or chemical properties of the waste to make it suitable for a safe disposal. There are several treatment technologies (e.g., acid–base neutralization, incineration, chemical fixation/solidification, etc.) which may be used before the ultimate disposal of the hazardous wastes. Treatment technologies convert waste into a more innocuous form, immobilize toxic components, or reduce the quantity of the waste. They are characterized by their effectiveness to treat specific waste types, their waste residue generation factor, and their cost and risk considerations [3]. Physical and chemical treatment processes include activated carbon adsorption, distillation, hydrolysis, ion exchange, solvent extraction, membrane separation technologies, air stripping and steam stripping, evaporation, catalytic extraction processing, filtration and separation, chemical precipitation, solidification and stabilization, oxidation, reduction and ozonation. Thermal treatment processes include liquid injection incinerators, fluidized-bed incineration, plasma systems, and pyrolysis processes. Examples of biological processes include activated sludge, aerobic lagoons, anaerobic lagoons, spray irrigation, tricking filters, and waste stabilization ponds.

Since even the most advanced treatment methods result in residues that are no longer amenable to a cost-effective treatment, the disposal of the waste is the final process and represents a key issue in the overall hazardous waste management program. Disposal facilities act as a permanent repository for the waste residues generated at the treatment facility. The options for disposal include deep well disposal, dilution and dispersal, ocean dumping, sanitary landfill, and land burial. The economics of waste disposal determines, ultimately, the amount and type of waste that is moved to distant disposal sites. The choice of disposal must be based on an evaluation of the economics and the potential pollution risks.

Solidification/stabilization (S/S) is a technology used for treating industrial solid wastes containing toxic constituents to prevent their dissolution and release to the environment [4]. It has been used for decades as a final treatment step before the disposal of both radioactive and chemical hazardous wastes. Solidification refers to those techniques that encapsulate the waste in a monolithic mass of high structural integrity. Solidification does not necessarily involve a chemical interaction between the waste and the solidifying agents, but may mechanically bind the waste into a monolith. Contaminant migration is restricted by decreasing the surface area exposed to leaching and/or by isolating the surface within the impervious capsule. Stabilization refers to those techniques that reduce the hazard potential of the waste by converting the contaminants into their least soluble, mobile, or toxic form. It is used extensively for the treatment of heavy metal containing inorganic sludges, contaminated soils containing toxic metals and organic constituents, incinerator residues, and bulk wastes from power plants. It is widely used because of its simple operation, relatively lower cost, and the desirable characteristics of treated waste reached [5, 6]. The most common binder systems used in S/S systems are alkaline matrices, such as lime and Portland cement. These are inexpensive, incorporate wet wastes and their alkalinity greatly reduces the solubility of many inorganic toxic or hazardous metals. The S/S matrix can also be modified using additives such as fly ash, soluble silicates, slag, and clay.

In an S/S process with cement, water reacts chemically with cement to form hydrated silicates and aluminates resulting in a solid monolithic mass. Hydration of dicalcium silicates and tricalcium silicates forms C-S-H gel, also known as tobermorite, and crystalline calcium hydroxide. The C-S-H gel is responsible for the strength development after the initial setting of the mixture.

Real wastes are complex mixtures with contrasting physical and chemical properties that are difficult to characterize. The synergistic or antagonistic effects of multicomponent mixtures can influence both hydraulic and pozzolonic[1] activity. Several heavy metal salts have been implicated in the inhibition of the hydration

[1]A pozzolan is a material which, when combined with calcium hydroxide, exhibits cementations properties.

process with a consequent retardation in the setting and curing of the cement/waste matrix, and thus reduction in strength development [7, 8].

Solidification/stabilization systems can be evaluated using physical, chemical, and microstructural methods [9]. Setting and strength development are often used as indicators of solidification, and the leaching test is often used to assess the extent of fixation [10–12].

Several researchers have attempted to use the S/S technology for the treatment of different types of industrial wastes, mostly heavy metal containing inorganic/organic sludges:

- Arsenic-containing wastes [13–16]
- Stainless steel factory sludge containing Cr and Ni [12, 17]
- Electroplating sludge containing Cr, Pb, Ni, and Zn [18, 19]
- Foundry sludge [20]
- Steel foundry electric arc dust [21–24]
- Nickel hydroxide sludge [25]
- Oil refinery sludge [26, 27]
- Asbestos waste [28]

S/S-treated waste can safely be disposed to landfills and can also be reused in construction materials depending on its physical engineering properties and chemical properties. The demand for construction material is escalating due to increasing urbanization. Therefore, the use of waste material in construction can not only save on cost of construction but also gainfully use it. Several researchers have attempted to reuse waste material into various types of construction materials, such as:

- Blocks from sewage sludge using fly ash, lime, and gypsum [29, 30]
- Arsenic and iron sludge in brick making [31]
- Sewage and sludge ash in brick and tile making [32, 33]
- Sludge ash for light weight aggregates [34, 35]

2 Problem Definition

The production pattern of clothes in the textile industry shows a wide variation with respect to quantity and type of fabrics produced. It includes the production of cotton fibers blended with man-made fibers such as polyester, nylon and rayon, and it consists of a series of unit processes that are complex and highly specialized. The unit processes basically fall into two categories: (1) dry processes and (2) wet processes.

The dry processes consist of the following steps:

1. Opening, blending, and mixing
2. Carding
3. Combing
4. Spinning

5. Slashing
6. Weaving and knitting

The wet processes comprise the following steps:

1. Desizing
2. Scouring
3. Bleaching
4. Mercerizing
5. Dyeing
6. Printing
7. Finishing

Wet processes are the main source of wastewater generation in the textile industry. Next they are described in detail:

- *Desizing*: This process involves treating the gray cloth with synthetic detergents and caustic soda and/or acids or enzymes. This removes the sizing material so that the dyes can penetrate and develop in a uniform manner.
- *Scouring*: The natural fat wax nitrogenous impurities are removed in this process by boiling the fabric with a 1% caustic soda solution and other chemicals.
- *Bleaching*: In this process, impurities present in the fabric are removed using oxidizing agents, such as bleaching powder, sodium hypochlorite, and hydrogen peroxide. After washing the cloth, which becomes highly alkaline, it is treated with diluted acid and sodium bisulfate, and then it is washed and dried.
- *Mercerizing*: It is a physical phenomenon by which the fabric is impregnated with a concentrated caustic soda solution so as to provide luster, strength, and dye affinity to the cloth.
- *Dyeing*: It is a process by which the rolled cloth is passed through a tray of dyes, called a jigger, for color fixation. This process is repeated several times, and chemicals and heat are sometimes added to fix the color.
- *Printing*: Printing is a localized dyeing which is carried out on wax tables with the help of screens. After printing, the cloth goes through a sodium silicate or steam process for color fixation.
- *Finishing*: In this process, starch, softeners, and other chemicals are applied on the fabric before drying and packing. The process can use either starch for hard finish or softeners for soft finish.

A schematic of the wet processes is shown in Fig. 1.

The textile industry involves a large amount of water consumption, i.e., the specific water intake for the textile industries varies from 95 to 400 L/kg of fabric depending on the type of processes applied and water use efficiency.

Wastewater is the dominating waste stream in the textile industry due to processing in aqueous solutions or suspensions. The quality and the quantity of the wastewater vary depending on the type of cloth to be processed (i.e., cotton, synthetic, wool, etc.) and also from mill to mill [36]. The specific wastewater discharge also varies according to the production method, ranging between 40 and 300 L/kg of finished product. General characteristics of the textile wastewater can be described as follows:

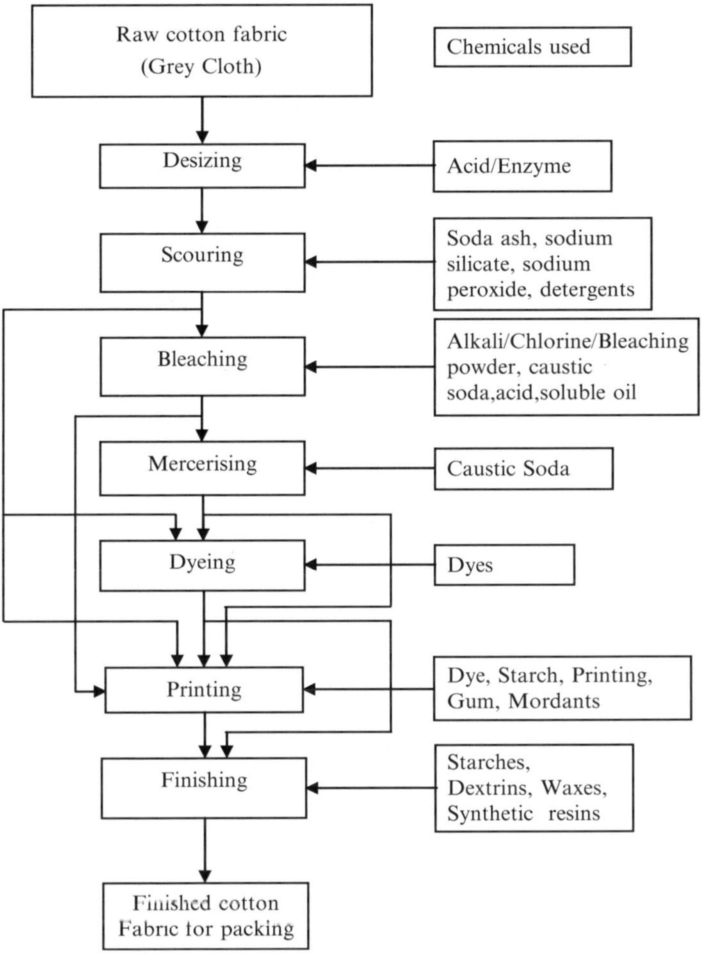

Fig. 1 Flow chart indicating the various processes involved in the wet processing of cotton fabric

- Intense coloring caused by a large number of unfixed remaining dyes
- High temperatures since many reactions require high temperatures
- Pollution with organic material, such as starch, carboxy methyl cellulose, surfactants, etc., in comparison with the pollution levels of domestic water
- High biochemical oxygen demand (BOD) values and high solid contents
- A large amount of refractory chemical oxygen demand (COD) caused by high-molecular synthetic textile auxiliaries (TA) and dyes
- High alkalinity
- High conductivity due to the salt character of many TAs
- High absorbable organic halogen (AOX) sulfide and heavy metal concentrations due to chlorinated bleaching agents, and halogens, sulfurs, or heavy metals contained in dyes in some cases
- High phosphate contents when pyrophosphates are added because of the use of hard process water

The textile industry in India is one of the oldest industries in the country and has a major presence in the Indian economy, contributing to about 14% of manufacturing value addition to gross domestic product and is one-third of the India's gross export earnings.

In India, the textile industry is generally established in the form of small and medium enterprises. They face specific problems when dealing with the wastes generated in their industrial processes. Often, they lack the space needed for setting up an effluents treatment plant. Besides having limited resources, they also have a limited understanding of the problem and a lack of technical ability for the wastewater management. However, during the last decade, several common effluent treatment plants (CETPs) have been established to treat the wastewater in small- and medium-scale industrial clusters [37].

The treatment scheme in typical CETPs for textile wastewater consists of the following steps: pretreatment (screening and grit chamber, oil and grease separation), physicochemical treatment using lime and ferrous sulfate in the presence of suitable coagulation aids, biological treatment in the form of activated sludge process (ASP), trickling filters, or rotating biological contractors (Fig. 2).

In a CETP, pretreated wastewater from individual textile units is collected, homogenized, and then pumped to a flash mixer where flocculants and lime solutions are added. Flocculants lead to particle destabilization and are very effective in removing the color. Since the addition of flocculants leads to a decrease in pH, lime is also dosed to prevent any abnormal decrease in pH, which could affect flocculation. The effluent from the flash mixer flows then to a clariflocculator where a polyelectrolyte is dosed to help in particle bridging and compaction of the sludge. The sludge is then transferred to a thickener to concentrate the solids and then it is dewatered in a centrifuge. The dewatered sludge is then dried in sludge drying beds.

The physicochemical treatment of the textile effluents leads to the generation of a large quantity of chemical sludge which is inorganic and non-biodegradable in nature. The dyestuff present in the wastewater gets precipitated in the form of settleable sludge slurry in the presence of coagulants, such as lime and ferrous sulfate. The chemical sludge contains inert solids, polymer solids, precipitated dye products, metal salts, and other chemicals. The sludge is considered hazardous as per the Hazardous Waste Management, Handling and Transboundary Movement Rules of 2008, listed in schedule I under category no. 34.[2] The amount and properties of the sludge depend on many factors, such as the composition of the wastewater, the chemicals used, and the treatment process.

Currently, large quantities of chemical sludge are lying untreated in CETP premises and await a suitable disposal method. The improper management of the sludge and its dumping on the ground leads to many adverse effects on environment and people, directly or indirectly. It can have many impacts, such as surface and

[2]http://www.moef.nic.in/legis/hsm/HAZMAT_2265_eng.pdf.

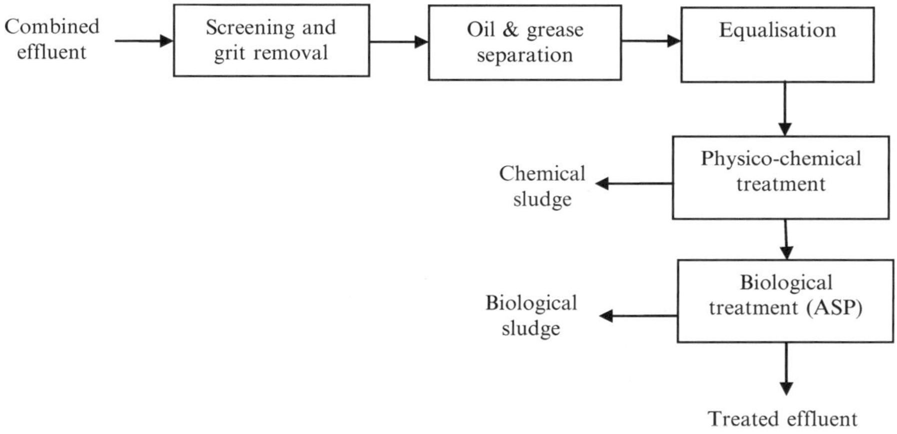

Fig. 2 Wastewater treatment and sludge generation

groundwater contamination, effects on public health, effects on the atmosphere due to odor problems, land degradation, etc. However, waste treatment may involve a considerable financial burden. This represents an important problem for the textile industry in India, because it generally comprises small- and medium-scale units, which are not very strong as far as financial and management aspects are considered. Conventional waste treatment/disposal methods, such as landfilling and incineration, may not be suitable because the leachate from the landfilling sites and the residues from the incinerators can cause secondary pollution. Moreover, they are also not an economically viable option. Textile chemical sludge cannot be used for land application because of its chemical contents.

Scarce literature is currently available about the characterization and reuse of chemical sludge from textile CETPs in India. Ansari and Thakur [38], Balasubramanian et al. [39], and Baskar et al. [40] have all attempted to characterize the chemical sludge and the options of reusing it in construction materials with varying degrees of success. Efforts are still going on to find out a suitable economic and environment-friendly solution for the safe reuse/disposal of this chemical sludge. This research work is an attempt toward the stabilization of this type of chemical sludge, which can result in a matrix where the mobility of the pollutants of concern is considerably reduced so that it is rendered harmless and can be safely disposed of or reused.

3 Adopted Approach

Chemical sludge samples were collected from the CETPs of four sampling sites in India, i.e., Balotra and Pali CETPs in Rajasthan, and Manickapuram and Mannarai CETPs in Tirupur (Fig. 3). The sludge samples were characterized for various

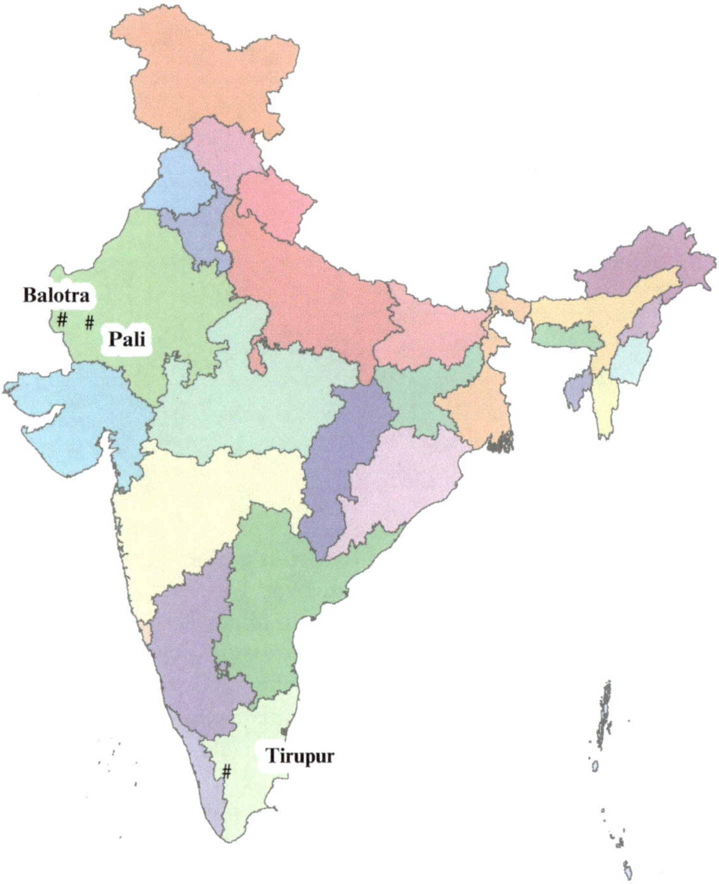

Fig. 3 Location of chemical sludge sampling sites

physicochemical parameters and for the concentrations of several heavy metals of concern. Standardized procedures were used to know about the nature of the sludge, so that appropriate treatment option(s) could be selected. Due to the limitations associated with other management options such as landfilling and incineration, the S/S option was selected for the treatment/reuse of the sludge samples. Treatability studies were carried out using different sludge binder systems, such as Ordinary Portland Cement (OPC) and Portland Pozzolona Cement (PPC). The evaluation of the S/S treatment was carried out in terms of physical engineering properties and leachability of heavy metals. A toxicity assessment was also carried out for the untreated and treated sludge samples. The solidification/stabilization phenomenon was also studied using XRD and scanning electron microscopy (SEM) techniques. Finally, the treatment technology was also evaluated in terms of costs and benefits, and it was compared with the present mode of disposal.

3.1 Sludge Sampling and Casting of Molds for S/S Tests

Grab samples were randomly obtained from sludge drying yards in the CETPs premises. Samples of 5 kg were collected in wide mouth polypropylene jars and they were refrigerated at 4°C for preservation.

After sludge characterization, the sludge obtained from the CETP at Balotra was selected for detailed S/S studies. The sludge was dried in a hot air oven for 24 h at 105°C. After drying, the sludge was powdered in a ball mill and then it was passed through a 212 μ BSS sieve. Two types of commercially available cements were used as binder systems: OPC and PPC. OPC was of 43 grade according to BIS IS: 12269 [41] and PPC was also of 43 grade according to IS 1489 (part 1): 1991 [42]. Mortar cube steel molds of 70.6 mm × 70.6 mm × 70.6 mm were used for the casting of the cement-sludge pastes. The general procedure for the casting of molds is described in the next paragraph.

The cubes were prepared according to the BIS standard IS 4031(6) [43]. The mixture for each cube was placed on a nonporous plate. The recipe was mixed with a trowel for 1 min until the mixture had a uniform color. Distilled water was slowly added into the dry mixture to promote hydration. It was then mixed at high speed for 3 min until the attainment of a predetermined water–cement ratio and a uniform color of the mixture. In assembling the mold ready for use, the joints between the halves of the mold were covered with a thin film of petroleum jelly, and a similar coating of petroleum jelly was applied between the contact surfaces of the bottom of the mold and its base plate to ensure that no water escaped during vibration. The interior faces of the mold were treated with a thin coating of mold oil. The mixture was hand-compacted for the removal of entrained air with the help of a prodding rod after casting each layer. The filled molds were kept undisturbed for 24 h in moist conditions (relative humidity 92%) and at room temperature (27–29°C). At the end of this period, the specimens were removed from the molds and were subjected to an experimental set of curing conditions. Three samples were prepared for each ratio.

4 Results and Discussion

4.1 Characterization of Chemical Sludge

The results of the physicochemical analysis of the different chemical sludges dried at 105°C are given in Tables 1 and 2. The chemical sludge was generally light green to dark green in color and odorous in nature. The sludge was found to be alkaline in nature, with pH varying from 8.02 to 9.00. It is comparable to the average value range given in the literature [38, 44, 45]. Electrical conductivity values were varying from 0.21 to 0.66 S/m. Ansari and Thakur [38] have reported electrical conductivity values ranging from 1.15 to 1.16 S/m in a characterization study of the textile sludge of effluent treatment plants. The electrical conductivity values

Table 1 Physicochemical characterization of the textile sludge at four sampling sites

Parameter	Sampling site			
	Balotra	Pali	Manickapuram	Mannarai
pH	8.78 ± 0.16	8.02 ± 0.06	8.85 ± 0.21	9.00 ± 0.10
Electrical conductivity (S/m)	0.66 ± 0.02	0.39 ± 0.01	0.21 ± 0.02	0.38 ± 0.01
Moisture content (%)	5.40 ± 0.28	10.33 ± 0.12	66.65 ± 2.13	57.31 ± 1.61
Dry solids (%)	94.60 ± 0.43	89.67 ± 2.25	33.35 ± 3.50	42.69 ± 2.24
Volatile solids (%)	38.60 ± 2.55	48.37 ± 1.29	35.63 ± 4.04	34.30 ± 1.58
Fixed solids (% of dry solids)	61.40 ± 1.79	51.63 ± 3.22	64.37 ± 3.68	65.70 ± 1.48
TOC (%)	610.40 ± 0.42	17.82 ± 0.83	1.23 ± 0.14	1.70 ± 0.06
Calorific value (kJ/kg)	4109.64 ± 20.61	8651.31 ± 53.13	14.65 ± 0.64	0
Specific gravity	1.11 ± 0.02	1.06 ± 0.01	1.07 ± 0.01	1.14 ± 0.04
Density (kg/m^3)	1116.00 ± 1.06	1061.31 ± 8.75	1,065.40 ± 5.44	1,143.83 ± 6.61

Note: All values are expressed as mean ± standard deviation and three values were determined in each case

Table 2 Heavy metals concentration (mg/kg on dry weight basis) of the chemical sludge at four sampling sites

Heavy metal	Sampling site			
	Balotra	Pali	Manickapuram	Mannarai
Cd	4.56 ± 0.15	4.59 ± 0.10	4.25 ± 0.04	5.41 ± 0.10
Cu	223.38 ± 14.69	389.83 ± 10.05	43.38 ± 5.21	39.83 ± 1.59
Zn	182.47 ± 1.53	386.94 ± 2.27	90.88 ± 2.01	73.48 ± 0.94
Ni	88.75 ± 2.23	33.62 ± 1.48	78.44 ± 1.93	23.73 ± 2.90
Co	12.70 ± 0.95	13.46 ± 0.53	12.12 ± 1.09	13.34 ± 0.02
Cr (III)	218.48 ± 0.94	316.33 ± 9.23	79.83 ± 0.99	32.00 ± 1.74
Cr (VI)	BDL	BDL	BDL	BDL
Pb	43.65 ± 1.24	52.04 ± 0.96	20.31 ± 1.03	33.15 ± 1.26

Note: All values are expressed as mean ± standard deviation and three values were determined in each case
BDL below detection limit

indicate the presence of ionic compounds in the sludge. The total amount of dry solids in the sludge varied from 33.35% to 94.60% and volatile solids from 34.30% to 48.37% of the total solids. These values provide a rough approximation of the amount of organic matter present in the sludge samples. According to the literature, the volatiles content has been found to be in the range of 24.2–80% [38, 44, 45]. This is roughly in agreement with the range of values obtained in this study.

The calorific value was found to be high for the sludge from Balotra and Pali CETPs (4109.64 kJ/kg and 8651.31 kJ/kg, respectively), whereas it was found to be very low for the Manickapuram CETP sludge (14.65 kJ/kg) and zero for the Mannarai CETP sludge. The high calorific value of the Balotra and Pali CETP sludges might be due to a high concentration of certain organic chemicals in different wet processing stages. The range of values mentioned in the literature is varying from negligible to 2072.5–2085.0 kJ/kg as reported by Bhalerao et al. [44] and Ansari and Thakur [38].

Table 3 Heavy metals concentration (mg/L) in the leachate of four sludge samples

Heavy metal	Sampling site				Regulatory limit
	Balotra	Pali	Manickapuram	Mannarai	
Cd	0.07	0.08	0.05	0.04	1.0[a]
Cu	0.49	0.79	0.51	0.44	130[b]
Zn	0.52	0.60	0.43	0.41	500[b]
Ni	0.43	0.78	0.53	0.65	2.0[c]
Co	0.34	0.40	0.29	0.33	N/av
Cr	0.56	0.65	0.46	0.51	5.0[a]
Pb	0.76	0.54	0.45	0.60	5.0[a]

[a]US EPA TCLP limits (http://www.ehso.com/cssepa/TCLP.htm)
[b]100 times the drinking water standards assuming that 100 times dilution would occur before the leachate could reach the drinking water (http://www.osti.gov/bridge/purl.cover.jsp;jsessionid=A824A52E29907179605AAB00ED6A768A?purl=/7147587-bG4MwM/)
[c]German leachate quality standards [46]

The concentration range of heavy metals on a dry weight basis was: Cu (39.81–389.83 mg/kg), Ni (23.72–88.75 mg/kg), total chromium (32.00–316.33 mg/kg), hexavalent chromium (below detection limit), Cd (4.25–5.41 mg/kg), Co (12.12–13.46 mg/kg), Pb (20.31–52.04 mg/kg), and Zn (73.48–386.94 mg/kg). The heavy metals Cu, Ni, Cr, and Zn were found to be in the highest concentrations, followed by relatively lower concentrations of Pb, Co, and Cd. The heavy metals concentration values found in the present study are relatively high when compared with the literature, except for the case of hexavalent chromium which is found to be below the detection limit, whereas in literature it was found to be in the range of 4.7–14.00 mg/kg on a dry weight basis [38]. The concentration of all heavy metals, including chromium species (Cr (III) and Cr (VI), the latter being more toxic), are within the regulatory limits for all the four sampling sites, according to the Schedule II of Hazardous Wastes (Management, Handling and Transboundary Movement) Rules 2008.[3]

The heavy metal concentrations of the toxicity characteristic leaching procedure (TCLP) leachate are given in Table 3. The regulatory limits are also mentioned in the table.

When compared with the regulatory levels, the heavy metals concentrations in the TCLP leachate at all the four sampling sites were found to be within the prescribed limits. In the literature, Palanivelu and Kumar [45] reported very low levels of heavy metals, which were also below the TCLP regulatory limits. This indicates that the risk of leaching of heavy metals from the chemical sludge is very minimal and controlled.

[3]http://www.moef.nic.in/legis/hsm/HAZMAT_2265_eng.pdf.

4.2 Structural Analysis Results

X-ray diffraction (XRD) examination of sludge revealed that it had high calcium oxide contents, owing to the use of lime in the physicochemical treatment of the sludge, but the content was lower than that of OPC. Calcium is the most essential substance required to provide strength in cement structures since it is the key element required in the formation of the calcium silicate hydrate, which is responsible for the hardening of the cementitious mixtures. The CaO content was greater than that found by Baskar et al. [40], who reported a value of 28.4%. Other oxides, such as SiO_2, Al_2O_3, Fe_2O_3, MgO, and SO_3, were also present in significant amounts. The content of oxides of silica was comparable to that of OPC, whereas the contents of oxides of aluminum, iron, magnesium, and sulfur were found in higher percentages when compared with that of OPC. Alkali oxides (Na_2O and K_2O) contents were comparable with that of OPC. The chloride concentration was quite high in the sludge in comparison with OPC, which is not desirable for its use in manufacturing of cement.

The content of Fe_2O_3 was comparable to the literature value reported by Baskar et al. [40], which was found to be 9.1%. Content values of SiO_2 were higher than that reported by Baskar et al. [40], which was 7.1%. However, the SiO_2 content value of the Pali sludge was comparable. The values of Al_2O_3 content in the sludge of the four sampling sites were also much higher than that reported by Baskar et al. [40] (0.698%).

The presence of significant amounts of calcium, silica, aluminum, and magnesium oxides in the chemical sludge collected from the four textile CETPs suggests that the chemical sludge may have some potential for use in construction materials.

The XRD patterns of the chemical sludge from the Balotra CETP are given in Fig. 4, while the scanning electron micrograph is presented in Fig. 5. This XRD pattern reveals the presence of main compounds: $CaCO_3$ (calcite), $CaCO_3$ (vaterite), and small amounts of calcium hydroxide $Ca(OH)_2$.

According to SEM analysis, the sludge particles were in the form of agglomerates with a diameter varying from 1 to 40 μm showing the predominance of spherical particles.

4.3 Solidification/Stabilization Studies Results

The chemical sludge had roughly around 5% moisture content. The sludge was dried in a hot air oven for 24 h at 105 ± 0.5°C. After drying, the sludge was powdered in a ball mill and then was passed through 212 μ BSS sieve. OPC and PPC were used as the binder systems. OPC was of 43 grade according to BIS IS: 12269 [41], and PPC was also of 43 grade according to IS 1489 (part 1): 1991 [42].

The unconfined compressive strength and the block density of the blocks solidified with OPC and PPC are presented in Figs. 6 and 7. For blocks with OPC, the compressive strength varied between 2.63 and 22.54 MPa after 14 days

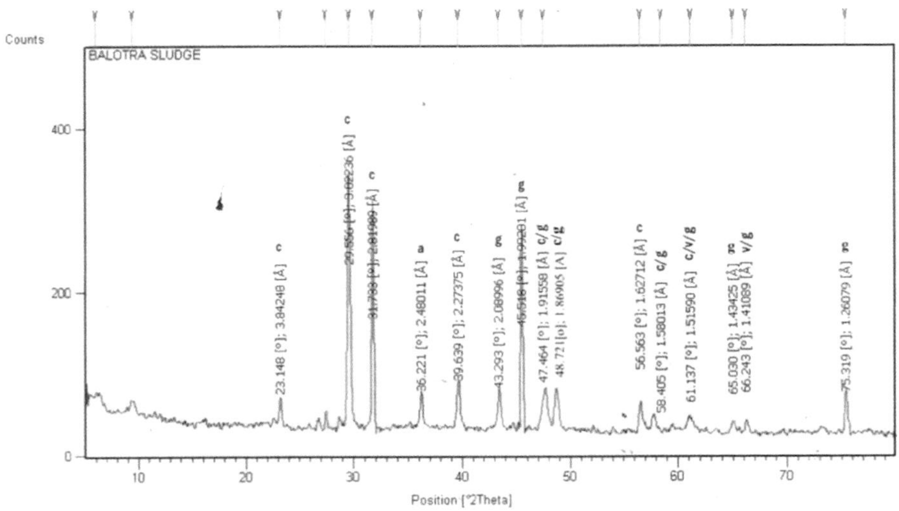

Fig. 4 XRD pattern of the chemical sludge sampled in Balotra CETP

Fig. 5 SEM image of the chemical sludge sampled in Balotra CETP

of water curing and between 6.48 and 24.89 MPa after 28 days of water curing. The maximum compressive strength was found for the control sample (53.46 MPa after 14 days of water curing and 65.80 MPa after 28 days of water curing). The compressive strength of PPC-sludge blocks varied between 2.78 and 17.42 MPa for 30, 40, 50, 60, and 70% sludge addition after 14 days of water curing and between 3.62 and 33.37 MPa after 28 days of water curing for the same sludge additions. The maximum compressive strength was also found for the

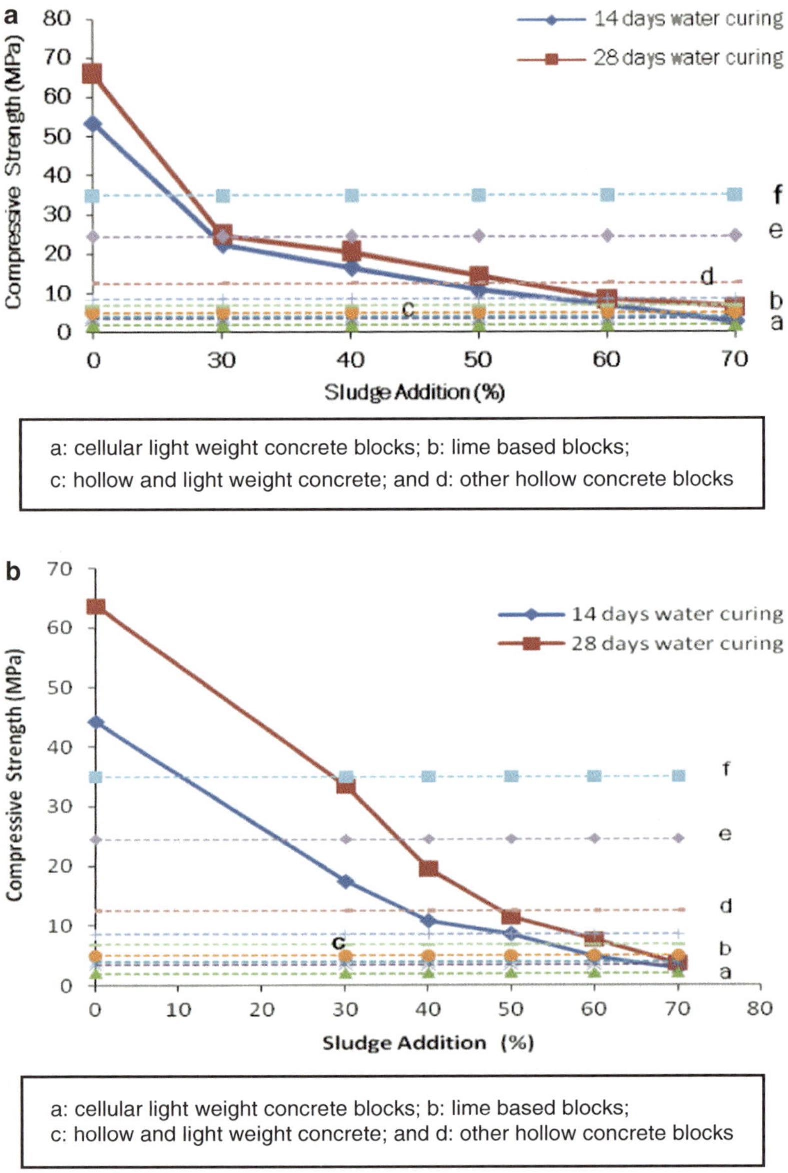

Fig. 6 (**a**) Compressive strength of OPC-sludge blocks; (**b**) Compressive strength of PPC-sludge blocks

control sample (44.26 MPa after 14 days of water curing and 63.69 MPa after 28 days of water curing). The block density of OPC-sludge blocks varied between 1,361.34 and 1,813.90 kg/m^3 after 14 days of water curing and between 1,386.40

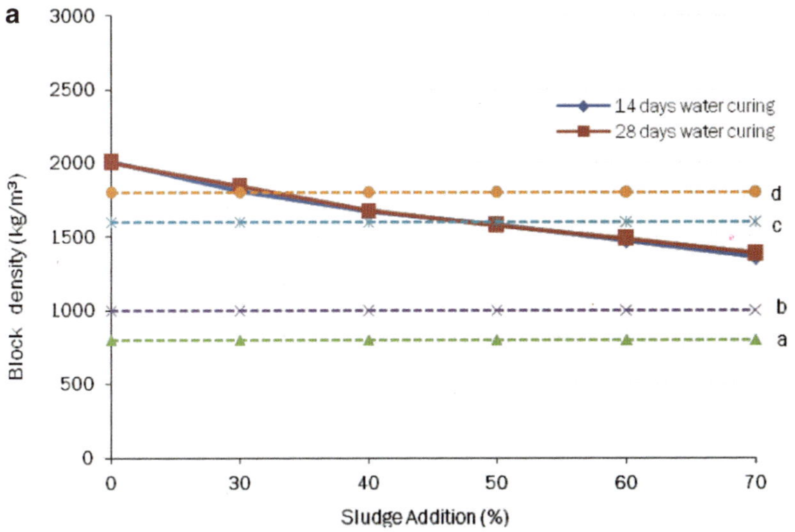

a: cellular light weight concrete blocks; b: lime based blocks;
c: hollowand light weight concrete; and d: other hollow concrete blocks

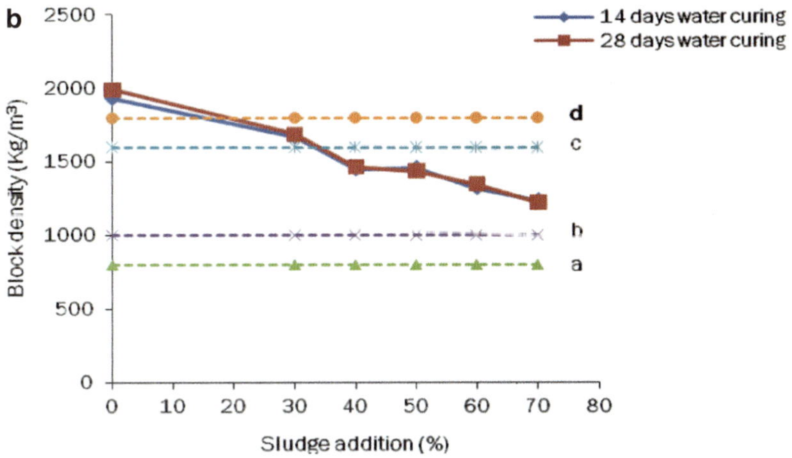

a: cellular light weight concrete blocks; b: lime based blocks;
c: hollow and light weight concrete; and d: other hollow concrete blocks

Fig. 7 (**a**) Block density of OPC-sludge blocks; (**b**) Block density of OPC-sludge blocks

and 1,842.34 kg/m^3 after 28 days of water curing. The block density of PPC-sludge blocks varied between 1,236.38 and 1,669.60 kg/m^3 after 14 days of water curing and between 1,222.17 and 1,688.72 kg/m^3 after 28 days of water curing.

The maximum block density was found for the control sample (1,930.90 kg/m^3 after 14 days of water curing and 1,991.77 kg/m^3 after 28 days of water curing).

Several strength requirements are solicited for different building materials:

- Nine classes of bricks and fly ash-bricks: 6.86–24.5 MPa [47]
- Hollow and solid concrete blocks: 4 and 5 MPa (IS 2185 (Part I) – 1979) [48]
- Hollow and light weight concrete blocks: 8.5 and 12.5 MPa (IS 2185 (Part II) – 1983) [49]
- Soil-based blocks: 2 MPa (IS: 1725–1982) [50]
- Lime-pozzolona concrete blocks for paving: 3.5 MPa (IS: 10360–1982) [51]

Considering the strength requirements of these construction materials, the cement-sludge blocks prepared in this work fulfill the requirements for most of the materials with non-structural purposes.

Block density requirements for several building materials are the following ones:

- Hollow and light weight concrete blocks: 1,600 kg/m^3 [49]
- Hollow and solid concrete blocks: 1,800 kg/m^3 [48]
- Lime-based blocks: 1,000 kg/m^3 [51]
- Cellular light weight concrete blocks: 800–1,800 kg/m^3 [52]

Therefore, when compared with these standards, cement-sludge blocks from this experimental study (which have from 1,386.40 to 1,842.34 kg/m^3) fulfill the density requirements of different types of construction material.

As the percentage addition of sludge increased, the compressive strength decreased. There might be several reasons for the decrease in compressive strength. Sludge is a complex mixture containing several types of chemical species which can have deleterious effects on the hydration process leading to a decrease in the compressive strength [53, 54]. Several metal salts have been implicated to have an adverse effect on the hydration process. The presence of Zn and Pb salts might be responsible for a decrease in the compressive strength. The formation of a membrane around cement particles with the precipitation of calcium hydroxyzincate ($CaZn_2\ (OH)_6{\cdot}2H_2O$) can prevent the required water and ion transport for cement hydration [7, 8]. Thus, the suppression or retardation of cement hydration adversely affects the strength development. Hydration occurs at a much slower rate than the reference with presence of Pb, which causes a decline in the unconfined compressive strength. Coating of tricalcium silicate component of cement with $Pb_3CaO{\cdot}SiO_2$ (C_3S) or complexation of Pb might also cause retardation in cement hydration [55, 56]. Moreover, the water content of the samples might have also contributed to the decrease in the unconfined compressive strength. It might have been excessive for the cement hydration. It has been reported by Conner (1990) that as the water– cement ratio increases, the percentage of larger pores increases, and hence the permeability of the product increases [57].

For exploring the further applications of the sludge, it was also used for the fabrication of micro concrete roofing (MCR) tiles and pavement tiles. The percentage of sludge used and the resulting strength in the two cases are presented in Tables 4 and 5 for two different water curing periods (14 and 28 days).

In case of MCR tiles, the maximum strength was found for the control samples, i.e., 60 kg after 14 days of water curing and 70 kg after 28 days of water curing. The strength obtained for all the combinations of sludge replacement was almost comparable to each other. It was around 40% of the strength control sample (Fig. 8). This indicates that other mix ratios can be designed to achieve the higher strengths. The interfering species in the sludge had a retarding effect on the hydration reactions of cement, thus lowering the strength of the sludge-replaced samples. In the case of pavement tiles, the strength of the control sample was found to be 140 kg after 14 days of water curing and 290 kg after 28 days of water curing (Fig. 9). It was evident from the results that sludge replacement in cement gave better results in comparison with the other combinations. For this combination, the

Table 4 Results of testing of sludge-based MCR tiles

Block ID	Load (in kg)	
	14 days	28 days
Control	60	70
50% Sludge replacement in cement	20	30
50% Sludge replacement in coarse sand	25	25
50% Sludge replacement in aggregate	20	25

Table 5 Results of testing of sludge-based pavement tiles

Block ID	Load (in kg)	
	14 days	28 days
Control	140	290
50% Sludge replacement in cement	130	230
50% Sludge replacement in coarse sand	25	25
50% Sludge replacement in aggregate	50	50

Fig. 8 Micro concrete roofing tiles made using chemical sludge

Fig. 9 Roofing tiles made using chemical sludge

strength after 28 days of curing was found to be nearly 80% the strength of the control sample.

5 Conclusions

The results from the study have led to following conclusions:

– The concentrations of heavy metals in the sludge as well as in the leachate generated as per the TCLP were lower than the regulatory levels prescribed in the Indian Hazardous Waste Management Rules and as prescribed under the TCLP for the sludge to be categorized as hazardous.
– The presence of a significant amount of calcium, magnesium, silica, and aluminum oxides in the sludge indicated that it had a potential to be used as construction material. This was confirmed by unconfined strength and block density values of solidified blocks.
– Experiments with cement and sludge for the fabrication of MCR tiles and pavement tiles also confirm that the sludge can be safely used as a construction material.

References

1. Misra V, Pandey SD (2005) Hazardous waste, impact on health and environment for development of better waste management strategies in future in India. Environ Int 31:417–431
2. CPCB (2009) National inventory of hazardous waste generating industries & hazardous waste management system in India. http://www.indiaenvironmentportal.org.in/files/hazardous-waste_inventory_final_report_2009.pdf. Accessed 22 July 2009
3. Blackman WC Jr (1996) Basic hazardous waste management. CRC Press Lewis Publishers, Boca Raton
4. Idachaba MA, Nyavor K, Egiebor NO, Rogers RD (2001) Stability evaluation of a cement based waste form to microbially induced degradation. Waste Manag Res 19:284–291

5. Soundararajan R (1990) An overview of present day immobilization technologies. J Hazard Mater 24:199–212
6. Baker PG, Bishop PL (1997) Prediction of metal leaching rates from solidified/stabilized wastes using the shrinking unreacted core leaching procedure. J Hazard Mater 52:311–333
7. Gervais C, Ouki SK (2002) Performance study of cementitious systems containing zeolite and silica fume: effects of four metal nitrates on the setting time, strength and leaching characteristics. J Hazard Mater B93:187–200
8. Olmo IF, Chacon E, Irabien A (2001) Influence of lead, zinc, iron (III) and chromium (III) oxides on the setting time and strength development of Portland cement. Cement Conc Res 31:1213–1219
9. Stegemann JA, Cote PL (1990) Summary of an investigation of test methods for solidified waste evaluation. Waste Manag 10:41–52
10. Bishop PL (1988) Leaching of inorganic hazardous constituents from stabilized/solidified hazardous wastes. Hazard Waste Hazard Mater 5:129–143
11. Hills CD, Pollard SJT (1997) The influence of interference effects on the mechanical, microstructural and fixation characteristics of cement-solidified hazardous waste form. J Hazard Mater 52:171–191
12. Malviya R, Chaudhary R (2006) Study of treatment effectiveness of solidification/stabilization process for waste bearing heavy metals. J Mater Cycles Waste Manag 6:147–153
13. Dutre V, Vandecasteele C (1996) An evaluation of the solidification/stabilization of industrial arsenic containing waste using extraction and semi-dynamic leach tests. Waste Manag 16:625–631
14. Leist M, Casey RJ, Caridi D (2003) The fixation and leaching of cement stabilized arsenic. Waste Manag 23:353–359
15. Singh TS, Pant KK (2006) Solidification/stabilization of arsenic containing solid wastes using Portland cement, fly ash and polymeric materials. J Hazard Mater B31:29–36
16. Kundu S, Gupta AK (2008) Immobilization and leaching characteristics of arsenic from cement and/or lime solidified/stabilized spent adsorbent containing arsenic. J Hazard Mater 153:434–443
17. Singhal A, Tiwari VK, Prakash S (2008) Characterisation of stainless steel pickling bath sludge its solidification/stabilization. Build Environ 43(6):1010–1015
18. Asavapisit S, Chotklang D (2004) Solidification of electroplating sludge using alkali-pulverised fuel ash as cementitious binder. Cement Conc Res 34:349–353
19. Sophia AC, Swaminathan K (2005) Assessment of the mechanical stability and chemical leachability of immobilized electroplating waste. Chemosphere 58:75–82
20. Andres A, Velasco FM, Coz A, Ruiz C, Viguri JR, Irabien JA (2002) Treatment of foundry sludges by stabilization/solidification with cement and siliceous binders. Fresenius Environ Bull 11:849–853
21. Hamilton IE, Sammes NM (1999) Encapsulation of steel foundry bag house dusts in cement mortar. Cement Concr Res 29:55–61
22. Skvara F, Kastanek F, Pavelkova I, Solcova O, Maleterova Y, Schneider P (2002) Solidification of waste steel foundry dust with Portland cement. J Hazard Mater B 89:67–81
23. Fuessle RW, Taylor MA (2004) Long-term solidification/stabilization and toxicity characteristics leaching procedure for an electric arc furnace dust. J Environ Eng – ASCE 130:492–498
24. Salihoglu G, Pinarli V, Salihoglu NK, Karaca G (2007) Properties of steel foundry electric arc furnace dust solidified/stabilized with Portland cement. J Environ Manage 85:190–197
25. Yin CY, Ali WSW, Lim YP (2008) Oil palm ash as partial replacement of cement for solidkfikcation/ stabilisation of nickel hydroxide sludge. J Hazard Mater 150:413–418
26. Montogomery DM, Sollars CJ, Perry R, Tariing SE, Barnes P, Henderson E (1991) Treatment of organic-contaminated industrial wastes using cement-based solidification/ stabilization – microstructural analysis of cement – organic interaction. Waste Manag Res 9:103–111
27. Athanasios KK, Voudrias EA (2007) Cement-based stabilization/solidification of oil refinery sludge: leaching behavior of alkanes and PAHs. J Hazard Mater 148:122–135

28. Chan YM, Agamuthu P, Mahalingam R (2000) Solidification and stabilization of asbestos waste from an automobile brake manufacturing facility using cement. J Hazard Mater B77:209–226
29. Shieh CS, Roethel FJ (1989) Physical and chemical behavior of stabilized sewage sludge blocks in sea water. Environ Sci Technol 23:121–125
30. Bednarik V, Vondruska M, Sild M, Vondruska E (2000) Characterisation of products from fluidized-bed combustion of coal. J Air Waste Manag Assoc 50:1920–1926
31. Rouf MA, Hossain MD (2003) Effects of using arsenic-iron sludge in brick making. In: Fate of arsenic in the Environment, Proceedings of the BUET –UNU International Symposium, Dhaka, Bangladesh, February 5–6
32. Okuno N, Takahashi S (1997) Full scale application of manufacturing bricks from sewage. Water Sci Technol 36(11):243–250
33. Wiebusch B, Seyfried CF (1997) Utilization of sewage sludge ashes in the brick and tile industry. Water Sci Technol 26(11):251–258
34. Bhatty JI, Reid KJ (1989) Moderate strength concrete from lightweight sludge ash aggregates. Cement Compos Lightweight Concrete 11:179–187
35. Bhatty JI, Mallisci A, Iwasaki I, Reid KJ (1992) Sludge as pellets as coarse aggregates in concrete. Cement, Concrete and Aggregates, CCAGDP 14(1):55–61
36. Correia VM, Stephenson T, Judd SJ (1994) Characterisation of textile wastewaters – a review. Environ Technol 15:917–929
37. Ministry of Environment and Forests (MoEF) (2008–2009) Abatment of pollution. Ministry of Environment and Forests annual report 2008–2009. http://www.envfor.nic.in/report/0708/chap04.pdf. Accessed 26 July 2009
38. Ansari AA, Thakur BD (2001) Sludge management in textile industry; disposal, recycling and reuse of primary sludge. Asian Textil J 10:55
39. Balasubramanian J, Sabumon PC, Lazar UL, Ilangovan R (2005) Reuse of textile effluent treatment plant sludge in building materials. Waste Manag 26(1):22–28
40. Baskar R, Meera Sheriffa Begum KM, Sundaram S (2006) Characterisation and reuse of textile effluent treatment plant waste sludge in clay bricks. J Univ Chem Technol Metallurgy 41(4):473–478
41. Bureau of Indian Standard (1987) Specification for 43-grade ordinary Portland cement IS 12269:1987. BIS, New Delhi
42. Bureau of Indian Standard (1991) Specification for Portland pozzolona cement Part I Flyash based (Third Revision) IS 1489: Part 1:1991. BIS, New Delhi
43. Bureau of Indian Standard (1980) Methods of physical test for hydraulic cement –determination of compressive strength of hydraulic cement IS 4031 (6). BIS, New Delhi
44. Bhalerao NA, Moghe CA, Bhavasar RS, Kharat RB (1997) Sludge management process – a review of its application to industrial sludges. Indian J Environ Prot 17(7):535–541
45. Palanivelu K, Kumar R (2001) Characterisation and leachability studies in textile effluent treatment plant sludge. Environ Pollut Control J 5(1):38–40
46. Ehrig H-J (1989) Leachate quality. In: Christensen TH, Cossu R, Stegmann R (eds) Sanitary landfilling: process, technology and environmental impact. Academic Press, London, pp 213–230
47. Bureau of Indian Standard (1992) Common burnt clay building bricks – specifications (fifth revision). BIS, New Delhi
48. Bureau of Indian Standard (2005) Methods of testing for cement masonry units- hollow bricks and solid blocks IS: 2185 (part-I). BIS, New Delhi
49. Bureau of Indian Standard (2005) Methods of testing for cement masonry units- hollow bricks and solid blocks IS: 2185 (part-II). BIS, New Delhi
50. Bureau of Indian Standard (1982) Specifications for soil based blocks used in general building construction IS: 1725. BIS, New Delhi
51. Bureau of Indian Standard (1982) Specifications for lime pozzolona concrete blocks for paving IS: 10360. BIS, New Delhi

52. Bureau of Indian Standard (1972) (Revised 2001) Methods of test for autoclaved cellular concrete products (Determination of unit weight or bulk density & moisture) Part 1,2,3,4,5,6 & 8. IS:6441
53. Trussell S, Spence RD (1994) A review of solidification/stabilization interferences. Waste Management, vol. 14, no. 6, pp. 507–519
54. Batchelor B, Overview of waste stabilization with cement, Waste Management (2006) 26:689–698
55. Thevenin G, Pera J Interactions between lead and different binders, Cem. Concr. Res. 29 (1999) 1605–1610.
56. Palomo A, Palacios M Alkali-activated cementitious materials: Alternative matrices for the immobilisation of hazardous wastes Part II. Stabilisation of chromium and lead, Cement and Concrete Research 33 (2003) 289–295
57. Conner J, Chemical fixation and solidification of hazardous waste, Van Nostrand Reinhold, New York, 1990.

Risk Management of Chemicals in the Leather Sector: A Case Study from Sweden

Stefan Rydin

Abstract The leather industry is a traditional industrial sector. The industry uses both a high variety and high amount of chemicals during the production of leather from raw hides and skins. The chemicals used will end up in the product, in the environment (wastewater, solid waste, air) and in by-products. This chapter describes how a modern tannery in Sweden is working with the risk management of chemicals in order to reduce the health and safety risks at the company and also to reduce the environmental impact of the company and avoiding hazardous chemicals in the product. The tannery is using 350 different chemicals in the different processes and in general the tannery adds 3 kg of chemicals for every kilogram of leather that is produced. The tannery has implemented environmental management systems and has a very good control of all processes in the tannery. The chapter focuses on how the tannery assesses new chemicals before they are introduced, how the tannery substitutes potential dangerous chemicals by less dangerous alternatives, how the tannery reduces the impact of chemicals to the environment by the use of best available techniques complemented by treatment facilities such as a newly built wastewater treatment plant and finally how the company monitors the content of certain chemicals in the finished leather.

Keywords Additives, BAT, Case study, Chemicals, Eco-label, GADSL, Leather, REACH, Substitution, Wastewater treatment, Water

Contents

S. Rydin
Nordeconsult Sweden AB, Sofiaparken 3C, 222 41 Lund, Sweden
e-mail: stefan.rydin@nordeconsult.com

B. Bilitewski et al. (eds.), *Global Risk-Based Management of Chemical Additives I: Production, Usage and Environmental Occurrence*, Hdb Env Chem (2012) 18: 207–224,
DOI 10.1007/698_2011_108, © Springer-Verlag Berlin Heidelberg 2011,
Published online: 21 June 2011

Abbreviations

BAT	Best available techniques
BBP	Butylbenzylphtalate
BDE	Bromodiphenyl ether
BREF	BAT reference document
CAS	Chemical abstracts service
CLP	Classification, labelling and packaging of substances and mixtures
DBP	Dibutylphtalate
DBT	Dibutyl tin
DEHP	Di-(2-ethylhexyl)-phtalate
DIBP	Di-iso-butylphtalate
ECHA	European Chemicals Agency
EINECS	European inventory of existing chemical substances
EU	European Union
GADSL	Global automotive declarable substance list
GASG	Global automotive stakeholders group
GHS	Globally harmonized system of classification and labelling of chemicals
HBCDD	Hexabromocyclododecane
NP	Nonylphenol
NPE	Nonylphenol ethoxylate
LGR	Lederinstitut Gerberschule Reutlingen
PAH	Polycyclic aromatic hydrocarbons
PBB	Polybrominated biphenyls
PCP	Pentachlorophenol
PFOS	Perfluorooctane sulphonate
REACH	Registration, evaluation, authorization and restriction of chemicals
SCCP	Short chained chlorinated paraffins

SVHC	Substance of very high concern
TBT	Tributyl tin
TCEP	*Tris*(2-chloroethyl)phosphate
TCMTB	2-(Thiocyanomethylthio)benzothiazole
TeCP	Tetrachlorophenol
TEPA	*Tris*-/aziridinyl)-phosphinoxide
TPhT	Triphenyltin
TRIS	Tri-(2,3-dibromopropyl)-phosphate
VOC	Volatile organic compound

1 Introduction

The leather industry uses in general a high number of chemicals during the production processes. The Swedish tannery that is used as a case study in this chapter is no exception, and the tannery was using 350 different chemicals in 2009 (a large part consisting of different types of dyes).

A practical problem faced by many tanneries is that many of the chemical products purchased are preparations (mixture or solution composed of two or more substances). Many suppliers do not specify the chemical composition of the products, so the tannery has to seek additional information from the chemical suppliers in order to determine the environmental impact of the chemical preparations they are using. Safety data sheets are the main information source but in many cases the information in the sheets is scarce. It is expected that more information on the chemicals that is used will be given in the future as a result of the new European Chemical Legislation called registration, evaluation, authorization and restriction of chemicals (REACH) [1] which entered into force on 1 June 2007.

This chapter describes a case study from Sweden and is illustrating how one European tannery has decided to approach the potential problem with the use of a high amount of chemicals. The tannery is Elmo Leather AB in Sweden. The tannery was founded in 1931. The production started up as 100% shoe leather tannery and during the 1960s and 1970s Elmo produced also leather for the garments-, furniture- and accessories-industries. In the late 1970s, the production was concentrated mainly on the high-quality upholstery leather and today the tannery produces high-quality upholstery leather for mainly the automotive and furniture industries. The production capacity of the company is 2,500 hides/day.

Elmo Leather has for many years been active in reducing the environmental impact from the leather production and as part of the environmental policy Elmo is seeking a continuous improvement of the environmental performance.

The environmental policy of Elmo Leather AB state "We will continuously improve our company according to predetermined environmental objectives by applying a holistic view of our activities' impact on the environment and by contributing to long-term sustainable activities by developing and offering high-quality

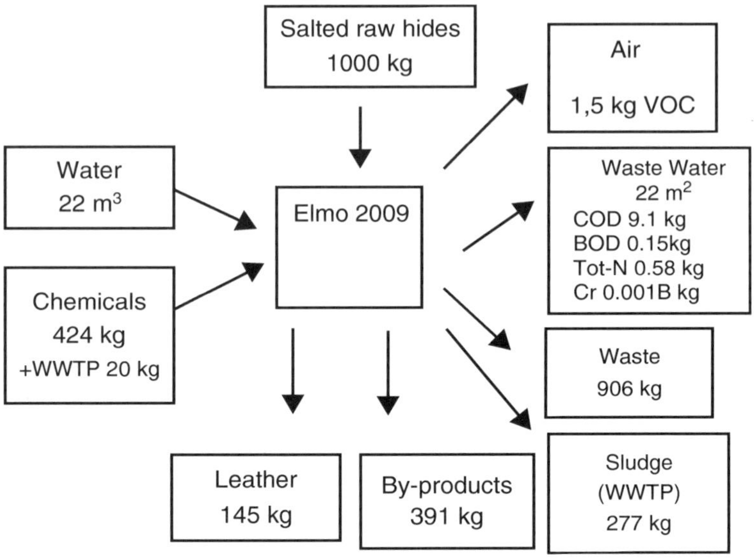

Fig. 1 Massbalance at Elmo Leather in 2009

leather products that have an increasingly low impact on the environment. Our activities mean that transportation, the use of chemicals, emissions into water and air as well as the dumping of waste have a significant effect on the environment and it is our task to continuously decrease all negative environmental consequences".

As a result of the environmental policy, the tannery has for many years been leading in environmental considerations and is therefore more active in this area than most other tanneries. The tannery should therefore not be considered as representative for the leather industry on how chemicals are managed but instead be seen as an example of a tannery using best practise in the leather sector regarding chemical management in the tannery.

It should be noted that less than 20% of the weight of the hides that will be processed will actually end up as leather. The remaining part will become waste or by-products. Furthermore, the tannery adds in average 3 kg chemicals for every kilogram of leather that is produced. A mass balance from the production at Elmo Leather is given in Fig. 1.

2 Introduction of New Chemicals

Elmo like many other tanneries is using approximately 350 chemicals in their production processes. The number of new chemicals introduced by the company every year varies substantially from year to year. As an average during the last 10 years it could be estimated that approximately 25–30 new chemicals/year have been introduced in the company (at the same time around the same number of chemicals have been withdrawn from the company). However, every year between 50 and 150 new chemicals are assessed and evaluated. There are many reasons for introducing new chemicals in the production and examples are: improvement of leather properties, new colours, lower cost, more environmentally friendly or already existing chemicals are withdrawn from the market (e.g. in connection with the close down of a chemical supplier).

One important tool in the assessment of new chemical substances is the PRIO tool developed by the Swedish Chemical Agency. PRIO [2] is a web-based tool intended to preventively reduce risks to human health and the environment from chemicals. The aim of PRIO is to facilitate in the assessment of health and environmental risks of chemicals so that people who work as environmental managers, purchasers and product developers can identify the need for risk reduction. To achieve this PRIO provides a guide for decision-making that can be used in setting risk-reduction priorities.

The recommendations on which chemicals are prioritized for risk-reduction measures are based on the environmental quality objective “a non-toxic environment” adopted by the Swedish parliament, work towards sustainable development and are in line with the objectives in the EU chemicals legislation, REACH. Throughout PRIO reference is made to Swedish legislation and other Swedish considerations.

Linked with PRIO are a number of environmental and health criteria for the substances to be prioritized in the risk-reduction work and a database containing examples of such substances. The PRIO tool applies to chemical substances of high concern as regards effects on health or on the environment. These high-priority substances are in PRIO divided into two levels of prioritization: phase-out substances and priority risk-reduction substances.

Furthermore, any new chemical will be assessed with reference to the so-called candidate list. The candidate list is part of the European Chemical legislation on REACH [1]. Substances that are included in the candidate list have been identified as substances of very high concern (SVHC). The list is therefore also sometimes called the SVHC list. Substances on the list may have very serious and often irreversible effects on humans and the environment.

From this list, chemical substances will be chosen to be included in Annex XIV in the REACH legislation (the annex includes substances that requires a special authorization to be used). The substances in the candidate list are hence candidates to be included in the list of chemicals where special authorization will be necessary for use. There are today (February 2011) 46 chemicals included in the candidate list and the list is continuously updated with new chemicals.

Substances are added to the candidate list by the European Chemicals Agency (ECHA). The inclusion of a substance in the list may have legal obligations on companies. The obligations include the following:

- Everybody who manufactures, imports or sells articles that contain more than 0.1% of the substance is obliged to give his/her customers information in order for the article to be handled in a safe way. This information should include the name of the substance.

- Information should always be given to industrial customers who use the article in their industrial operations.
- Private consumers are entitled to receive similar information from the suppliers of the article on request and without cost.

Examples of chemicals which are or have been relevant for the leather industry today on the list are: short-chained chlorinated paraffins (SCCPs), boric acid, disodium tetraborate, chromates and phthalates (DIBP, DBP, BBP and DEHP). As soon as a chemical substance is put on the candidate list, many brands and eco-labels react immediately restricting the use of the candidate substance.

One of the objectives with the SVHC list is that the substances gradually should be replaced by alternative substances and techniques when this is financially and technically possible.

Furthermore, all new chemicals to be introduced by Elmo are also checked according to the chemicals included in the EU Directive 2008/105/EC on environmental quality standards in the field of water policy [3]. The European Commission has established a list of substances and the list contains currently 33 chemical substances and any new chemical preparations at Elmo should not contain any of the listed chemicals.

The chemical will also be checked according to requirements from the customers of the company. In particular, the automotive industry present detailed specifications on many different chemicals and substrates. In the past, most of the different automotive industries had their own specifications and different demands were present from different customers. However recently, the automotive industry has started to cooperate regarding which chemicals that should be restricted and declared. The industry has developed the so-called global automotive declarable substance list (GADSL) document [4]. The GADSL is the result of a year-long global effort of representatives from the automotive, automotive parts suppliers and chemical/plastics industries who have organized the global automotive stakeholders group (GASG). The GASG's purpose is to facilitate communication and exchange of information regarding the use of certain substances in automotive products throughout the supply chain. The GADSL document contains 2,732 chemical substances in 2011 from 139 categories of chemicals. The decision to list a substance on the GADSL is based on the following criteria.

The substance should be expected to be present in a material or part in the vehicle. Either of the following conditions should apply:

- The substance is regulated, or is projected to be regulated by a governmental agency or authority.
- It is demonstrated, by testing under Organization for Economic Cooperation & Development (OECD) guidelines for testing chemicals, conducted under good laboratory practice (according to the OECD principles on good laboratory practice as revised in 1997), that the substance may be associated with a significant hazard to human health and/or the environment, and its presence in a material or part in a vehicle may create a significant risk to human health and/or the environment. Other scientifically valid methodology, based on the weight of evidence, may also be considered.

A substance that causes a functional problem in vehicle design may be included if its presence in a vehicle part exceeds a level shown to be problematic by an international industry standard test.

Reportable threshold levels will be based on the lowest level required by regulation or reasonably required by scientific evaluation.

A reportable substance when present in a material or part in a vehicle will be shown on the GADL with a classification "P" or "D" defined as follows.

A substance designated "P" is either prohibited by regulation for use in certain applications or may not exceed regulated threshold limits. A substance designated "D" must be declared if it exceeds the defined threshold limits.

As an additional source of information, the chemicals are also assessed according to information in different databases such as chemical substances, CLP/GHS, H-Class, N-Class and the list of substances database. A description of these different databases is given below.

Chemical substances database is an interactive register of 32,183 chemical substances. It contains about 2,974,000 facts of which 335,685 are searchable names. Those are divided into English, Swedish, German and French, as well as trade names. The chemical substances database supplies information about physical data including structural formula, risk and safety phrases, hazard symbols, directions for handling and storage, and transport directions for senders and transport organizers. This includes the possibility to print transport instructions in writing.

The CLP/GHS database includes lists of harmonized classification and labelling for certain substances or groups of substances which are legally binding within the European Union.

The N-Class database contains information on the classifications of more than 7,000 substances. The data primarily concern environmental effects constituting the basis for classifications and classification proposals. Classifications of fire and health hazards are also included.

H-Class is a database developed by the Nordic Council of Ministers. It includes dangerous substances/groups of substances, which have been discussed or are presently under discussion by the European Commission. Even substances not classified because of lack of evidence or insufficient information are included. The database primarily concern classification and labelling of health effects but physical–chemical properties and classifications of environmental hazards are also described to provide an overview of the complete classification of substances included in the EU harmonized "list of dangerous substances" contained in Annex I of the Council Directive 67/548/EEC. Documents concerning classification proposals and comments, agendas and reports from working-group meetings for substance entries are available.

The list of substances database contains more than 130,000 chemical substances with CAS numbers. Approximately 100,000 of these also have an EINECS number. The register of the list contains 170,000 synonyms.

It is hence a substantial work that is carried out in order to assess new chemicals at Elmo Leather AB. The chemicals are assessed according to information in several databases although it should be noted that many of the databases actually contain similar information about the chemicals.

The extensive assessment plays a very important part as support for decisions for the tannery when the tannery finally selects new chemicals. Other important factors are the cost for the chemical and the eventual improvement of the leather properties.

3 Substitution of Chemicals

Besides the evaluation of new chemicals at the company, existing chemicals are continuously controlled and assessed. Elmo Leather AB is continuously replacing chemicals which may be hazardous with other more environmental friendly chemicals. The same information sources that are used for the introduction of new chemicals are used for the continuous assessment of existing chemicals. In particular, it is important if chemicals that are used by the company will be placed on one of the lists mentioned in the previous chapter (e.g. candidate list, water framework list or GADSL). Elmo has developed their own database comprising all chemical substances that are used by the company in order to facilitate this work. The database also includes information on the final destiny of the chemical substances (water, air or product).

As a result of this, the use of several chemical substances has been substantially reduced during recent years and also preparations containing some selected priority chemicals have been phased out during the last years (e.g. preparations containing acrylamide and nonylphenol (NP) have been phased out).

Examples of the ongoing work with substitution of existing chemicals are given below for some chemicals.

3.1 Reduction of Nitrogen-Containing Chemicals

One high priority during recent years has been to replace nitrogen-containing chemicals in the production, since the reduction of nitrogen discharge to the environment is a high priority in Sweden. Discharges of nitrogen from agriculture, industrial facilities, cleared forest areas and municipal wastewater treatment plants, along with nitrogen deposition on lakes and seas, bring about marine eutrophication. Elevated concentrations of nitrogen in the water boost production of many organisms and cause changes in species composition. The results are reduced depth of visibility in the water, massive algal blooms and impaired oxygenation in bottom water.

Examples of measures are the introduction of best available techniques (BAT) such as hair removal during liming/unhairing and deliming using carbon dioxide. Besides the introduction of these techniques, a review of all chemicals containing nitrogen has been done and replacement taken place where possible. The reduction of the nitrogen discharge for the last years can be found in the Fig. 2.

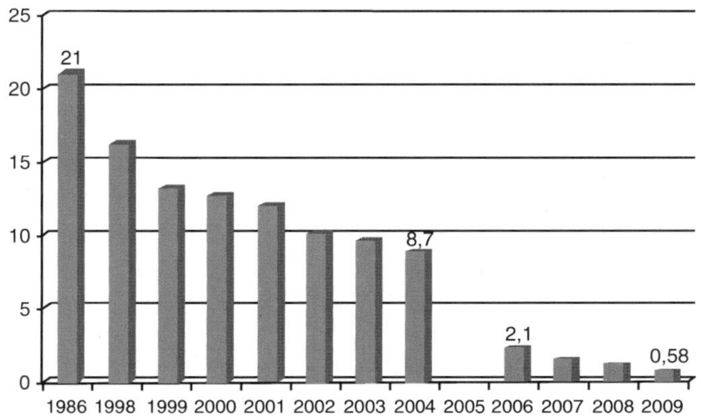

Fig. 2 Nitrogen discharge in kg/salted raw hide at Elmo

3.2 *Reduction of Chemicals Containing Phosphorus*

As part of the Swedish strategy to reduce the marine eutrophication, attention has also been made regarding discharge of phosphorus from the tannery. Nitrogen and phosphorus that are discharged from sewers and leaching from farmland cause severe environmental problems in water and forest areas.

The tannery therefore initiated an internal project in order to reduce the use of phosphorus-containing chemicals in the tannery. As a result, the concentration of phosphorus in water to the wastewater treatment plant was reduced from 31 to 14 mg/l in 2007 (as average values).

3.3 *Reduction of Organic Solvents (Air Emissions)*

Elmo Leather AB has put considerable efforts in reducing the emission of volatile organic compounds (VOCs) due to very strong demands from the regional authorities. The emission of VOCs from the company is far below the limit values in the EU Industrial Emissions Directive [5]. Several measures have been taken during the last years. Only water-based dyes are used in the finishing unit. One key point has been to reduce the amount of solvents in the water-based products that are used (it should be noted that also water-based finishing products contain solvents). Furthermore, the spray guns have been changed and a new control system has recently been installed.

In order to remove dye-particles from the spraying cabinets a water scrubbing system has been installed. The system is using a specifically developed filling material to improve the separation of particles.

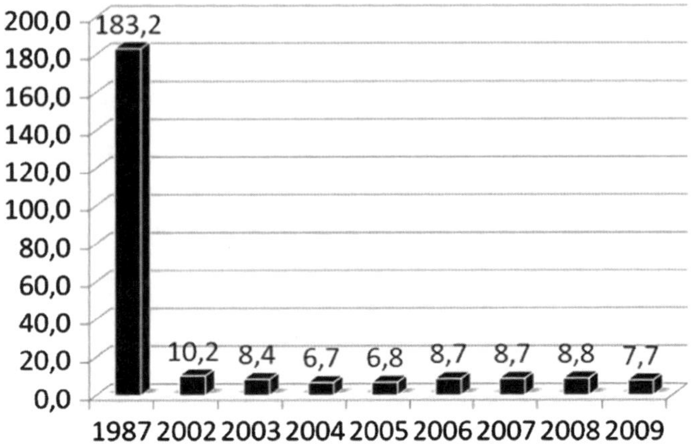

Fig. 3 Discharge of solvents (VOC) from Elmo Leather AB in g/m² leather produced

The specific emission of solvents from Elmo Leather during the last years compared to the level in 1987 is given in Fig. 3. The figure can be compared to the EU Industrial Emissions Directive which state that the specific emission should be lower than 75 g/m² produced leather (for a company with the size of Elmo Leather AB).

It should also be noted that the type of solvents used are changed and today mainly alcohols and glycols are used which should be compared with the more problematic solvents such as toluene and cyclohexanone, which were used in the past.

One consequence of using water-based finishing is that hardeners have to be used in order to obtain the necessary qualities of the leather. The company has taken several measures in order to reduce the health effects of using hardeners.

3.4 Replacement of Chemicals Containing Nonylphenol Ethoxylates

Nonylphenolethoxylates (NPEs) can degrade releasing NP in anaerobic conditions. NP is persistent to the aquatic environment, moderately bioaccumulative, extremely toxic to aquatic organisms and has showed endocrine-disrupting properties.

The discussions about the endocrine-disrupting properties of NP started in the 1980s. Elmo Leather used at that time large amounts of NPEs as a wetting agent and approximately 50 tonnes were used every year by the company. The substitution of NPEs as wetting agent was given high priority and contacts were made to the chemical suppliers. Elmo Leather succeeded within 6 months to replace NPE as wetting agent with other chemicals.

Another use of NPE was as a component in some commercial finishing products (although the amount of NPE used in these products was much lower). This was a more difficult and complex task and took a much longer time. It was first in the beginning of 2000 that the company had managed to replace all products containing nonlylphenol ethoxylates (two products used in small amounts were difficult to replace).

3.5 Replacement of Chemicals Containing Acrylamide (Prop-2-Enamide)

Acrylamide is a chemical that is a neurotoxin and potential carcinogenic chemical. The chemical was much discussed in Sweden in the late 1990s. The chemical was used as a sealant in connection with the building of a tunnel in the South of Sweden (Hallandsåsen). The chemical leaked out from the tunnel and poisoned the surroundings causing the death of fish and cattle were seriously affected.

The chemical was also used by the leather industry as an additive in some chemical products used in the finishing processes. The company started quickly in cooperation with chemical suppliers to substitute acrylamide with other chemicals. The process was successfully done and finally in 2004 all chemicals including acrylamide had been replaced by the company.

The acrylamide case illustrates that chemicals quickly may come into focus due to alarming reports in the press which will make it necessary for the industrial companies to substitute the chemicals rapidly.

3.6 Replacement of Selected Azo Dyes

Around 90% of all leather and textile dyes are azo dyes. The azo dyes break down during reductive conditions to form aromatic amines. Twenty two (22) aromatic amines are forbidden in the EU Regulation 1907/2006 [1] since these 22 amines are known to be carcinogens or potential carcinogens.

The focus on these aromatic acids started in Germany and in 1994 it became forbidden according to the German Consumer Goods Regulation to produce, import and sell garments and fabrics dyes with certain azo colourants (those azo dyes which can break down under reductive conditions to release any of the 22 aromatic amines). Since the German leather market is the most important market in Europe, all major chemical suppliers of dyes very quickly substituted dyes containing prohibited azo colourants. In 2002, the European Commission also prohibited the use of azo dyes which by reductive cleavage of one or more azo groups may release one or more of the 22 amines. The dyes were prohibited according to the EU Directive 2002/61/EC from July 2002 amending for the 19th time the Council

Directive 76/769/EEC. This Directive was later replaced by Regulation EC 1907/2006 (REACH) according to an EU Regulation from 2009 [6].

The leather industry and the chemical suppliers had already in 1994 started to remove those azo dyes from the market as a result of the German legislation. It was therefore no problem for Elmo Leather to comply with the European legislation in 2002 since the problematic azo dyes had been removed from the market several years before.

3.7 Biocides

Pesticides are used in farm animal husbandry (e.g. ectoparasiticides) to prevent animal pests, fly infestation, and beetle attacks on the animals. Furthermore, biocides can also be used to preserve the hides before they arrive into the tannery. It is hence possible that biocides are introduced to tannery through their main raw material (hides and skins). However, Elmo is primarily using Scandinavian hides which are only salted without the use of pesticides. The tannery has hence avoided the risk of unwanted biocides in the raw material by the choice of raw material (Scandinavian salted hides).

The use of biocides is in Europe controlled and regulated by the EU Biocidal Products Directive [7]. Registration of a biocide is expensive and only a few active biocide substances are available for the leather industry in the EU. Besides this, the biocides also have to be registered and approved at a national level. As a consequence, only one type of pesticide is allowed to be used by the tannery. It is a fungicide based on TCMTB [2(thiocyanomethylthio)benzothiazole]. The addition of TCMTB is minimized at the company and they are continuously following trends and developments regarding biocides.

4 Best Available Techniques

The company has implemented a number of measures to reduce the environmental impact and examples of the measures are given below. The company is often referred to and used as a good example in the EU BAT reference document which is produced as part of the EU Industrial Emissions Directive [5].

4.1 Unhairing

The function of the liming and unhairing process is to remove hair, interfibrillary components and to open up the hide structure. Normally, the process is carried out in alkali (in most cases lime) and sulphide where the hairs are dissolved; however, this brings a heavy organic load of the tanning wastewater.

The unhairing process at Elmo is carried out in mixers. In order to reduce the environmental impact of, for example COD and nitrogen by the wastewater treatment plant, equipment for filtration of the process water has been installed [8]. In every mixer for the beamhouse processes (seven mixers), filtering equipment has been installed and the process water is recirculated over the filters and the hairs are filtered. The hairs are transported by a conveyor belt to a tank where they are stored before agricultural use.

4.2 Chromium Recovery

Chromium is used worldwide as the most common tanning agent in the leather industry and around 80–90% of all leather is chrome tanned.

Elmo has during the last years reduced the chrome tanning and today less than 50% of the production is tanned by chromium (however this change mainly depends on customer requirements). The chromium uptake has been increased by process measures such as elevated temperature and pH in combination with the adding of certain chemicals to facilitate the uptake of chromium. Furthermore, the chromium containing process waters are treated at source before the water is discharged to wastewater treatment. The wastewater from the chrome tanning, sammying, wet aftertreatments and some other waste streams containing chromium are collected in a tank. The waste chrome liquors are precipitated with sodium carbonate (alkali) at a temperature of about 50–60°C. The chromium content in the water after precipitation is below 1 mg/l [8].

4.3 Use of Water-Based Finishing

As mentioned in another place in this chapter, Elmo is only using water-based finishing since several years ago instead of using finishing with solvents.

4.4 Energy Production from By-Products

Elmo is delivering so-called fleshings for biogas-production in nearby located biogas plants. Fleshing is a mechanical operation and the process of cutting away the subcutaneous fat and connective tissue from the flesh side of the hide. The waste is called fleshings and is the largest waste product at Elmo Leather. More than 50% of the waste from the company consists of fleshings.

The fleshings from Elmo Leather is transported to a nearby biogasplant and utilized to produce biogas through anaerobic digestion. Due to its fat content, fleshings generates a high gas yield and is therefore very attractive for biogas-producers. Several biogas plants in the vicinity have showed interest to utilize fleshings in their plants. After the anaerobic treatment, the remaining biomass will be used for agricultural purposes and thereby reduce the need for artificial fertilizers. Since the biomass from the biogas process is used for agricultural purposes, the fleshings had to be approved before it could be accepted by biogas plants. As an example, the content of a number of heavy metals was controlled prior to the acceptance of fleshings for biogas production.

5 Release of Chemicals into the Environment

A main priority for Elmo Leather is to reduce the environmental impact from the tannery. The tannery has an environmental permit from 2004 which regulates the amount of pollutants that are allowed to be released. The main environmental impact from the tannery is the wastewater. The company is discharging around 20 m^3 for every tonne rawhide that is processed. The permit from 2004 has been supplemented by a permit from 2007 regulating the discharge of emissions to the air.

The first priority of Elmo Leather AB is as mentioned before to prevent the pollution at source by using best available techniques (BAT) or even techniques better than BAT (as defined in the BREF-document for tanneries). However, there will always be an environmental impact from tanning of hides and therefore it is necessary to also use end-of-pipe-solutions to minimize the environmental impact from the industrial activities.

5.1 Wastewater Treatment Plant

The tannery Elmo Leather AB in Sweden finalized in 2005 the construction of a new wastewater treatment plant using an innovative system for nitrogen removal [9]. The tannery had in the past discharged their wastewater to the municipal treatment plant. The innovation of the new plant is the implementation of a nitrification and denitrification step in the treatment of tannery wastewater. The technology had not before been considered feasible in wastewater treatment plants for the leather industry, due to the composition of the tannery wastewater.

The total cost for the plant was around five million EUR and the project was supported by the EU LIFE financial instrument, since it is using an innovative technology for nitrogen removal from the leather industry. The plant was taken into operation in April 2005 and the official inauguration took place on the 1 June 2005.

There is an extensive monitoring programme of the water discharged from the wastewater treatment plant. The monitoring programme was finally defined and

Table 1 Performance of the wastewater treatment plant at Elmo Leather AB

	Incoming 2009–2010 (mg/l)	Outgoing 2009 (mg/l)	Outgoing 2010 (mg/l)	Reduction 2009 (%)	Reduction 2010 (%)
BOD	4,300	5.8	7.7	99.9	99.8
COD	9,000	357	443	96.0	95.1
Nitrogen	630	23	16	96.3	97.5
Chromium	7	0.072	0.037	99.0	99.5
Phosphorus	21	0.28	0.25	98.7	98.8

Table 2 Outgoing concentration of selected parameters from the wastewater treatment plant at Elmo Leather AB

	Outgoing 2010 (mg/l)
Copper	Below detection limit
Zinc	0.07
Lead	Below detection limit
Nickel	0.02
Arsenic	Below detection limit
Chromium VI	Below detection limit
PAH	Below detection limit

approved by the regional environmental authorities on 7 June 2005 and valid from 1 June 2005. The following parameters are monitored monthly: flow, pH, nitrogen-total, nitrogen-ammonium, COD, chromium, phosphorus, BOD, SS and TOC. Furthermore, parameters like zinc, nickel, copper, lead, arsenic, oil-index, PAH, EOX, chromium VI, toxicity and biodegradability are monitored once every year. Furthermore, tests of the sludge are carried out six times every year on mixtures of sludge. In total, 56 parameters are analysed and controlled in the sludge.

Some values for the performance of the treatment plant on key parameters are given below in Table 1 and the outgoing concentration of other parameters is given in Table 2.

It should be noted that the plant now has been running during several years and the performance of the plant has been very stable and can be mentioned that the reduction of nitrogen in the plant has never been below 85% despite the Scandinavian winter conditions.

5.2 *Air Emissions*

The company has taken measures to reduce the release of chemicals to the air. The main measure is the switch from solvent-based to water-based chemicals during the finishing processes, which was presented in more detail previously in this chapter. Furthermore, the company has installed water scrubbers on all stacks from the finishing department. These scrubbers are mainly installed to reduce the amount of small particles in the exhaust air but will also have a positive effect on the solvents that are released into the environment (since the solvents are water soluble).

Measurements have shown that around 50% of the volatile organic solvents that are present in the exhaust air are removed by the water scrubbers. The used water from the scrubbers is transported to the wastewater treatment plant.

6 Chemicals in Products

Besides monitoring the amount of chemicals released to the environment, the content of some chemicals in the final products is also measured. Some of the customers make requirements on some specific compounds that have to be declared or not present into the finished leather product.

The European automotive industry has to comply with the End of Life Vehicle Directive requirements and has set up a database to record each component used in cars. To ensure that information on restricted substances is collected worldwide, the GADSL has (as has been reported earlier in this chapter) been established and requires those in the supply chain to provide information for the listed substances. The list contains 2,732 substances in 2011 and the GADSL list is normally updated each year.

There are an increasing number of eco-labels for leather products available on the market. In most cases, behind each eco-label is a test institute or a group of test institutes, who are responsible for monitoring the compliance and issuing the eco-labels. Elmo is testing part of their production using two different eco-labels. The first eco-label that is used by the Elmo is the "Leather produced in an environmentally acceptable way and tested for hazardous substance" from the German Lederinstitut Gerberschule Reutlingen (LGR). LGR is a training, testing and research centre for the leather industry located in Baden Württemberg in Germany. This eco-label puts down criteria both for the production of leather and for the content of hazardous substances in leather and is quite widely used in the leather industry.

The chemicals that are restricted according to this German eco-label are PCP, azo dyes forbidden according to the legislation and hexavalent chromium. The limit values are set according to the European legislation.

Furthermore, Elmo has part of the production approved according to OeKO-TEX 100 [10]. This part of production is made for one major customer requiring leather without chromium as a tanning agent. This eco-label controls a number of substances and has developed limit values for the chemical substances. This eco-label includes criteria for a high number of substances. The following substances are regulated and therefore necessary to control in the leather: formaldehyde (released), extractable heavy metals (antimony, arsenic, lead, cadmium, chromium, hexavalent chromium, cobalt, copper, nickel and mercury), heavy metals in digested sample (lead and cadmium), pesticides (sum including PCP), chlorinated phenols [pentachlorophenol (PCP) and tetrachlorophenol (TeCP)], phthalates (sum of DEHP, BBP, DBP and DIBP), organic tin compounds (TBT, TPhT and DBT), other chemical residues [ortophenylphenol, arylamides, PFOS, Short Chain Chlorinated Paraffins (SCCP) and TCEP], chlorinated benzenes and toluens, emissions of volatiles (e.g.

formaldehyde, toluene and several more). Furthermore it is not allowed to use a number of colourants specified in the criteria such as cleavable arylamines having carcinogenic properties and dyes categorized as carcinogenic or allergenic and furthermore the use of a number of flame-retardant products are prohibited (PBB, TRIS, TEPA, pentaBDE, octaBDE, deca-BDE, HBCDD, SCCP and TECP).

In addition, some customers have established their own "restricted substance" specifications which have to be followed by the company (in the past before the establishment of the GADSL in particular the different automotive industries their own lists). The tendency during the last years has been that more and more customers are requiring copies of analytical protocols regarding restricted substances and not only the eco-label certificate (e.g. eco-label from LGR).

7 Conclusions

The leather industry is using a high number of chemicals in the production and it is not uncommon that tanneries use 300–400 different chemicals in their production. A careful and accurate management of chemicals is therefore necessary for the industry to minimize the health and safety aspects by being a chemically intensive industry.

This chapter has described the chemical management of a modern and well-developed tannery in Sweden. The tannery has done substantial work in reducing the environmental impact from leather production and also to have a control of the chemicals that are used. The tannery in this case study should not be considered as an average tannery but instead as an outstanding tannery in a traditional industrial sector. The case study (Elmo Leather AB) could be seen as one example of what is possible to do and therefore act as an example for other tanneries.

The main incentives for the tannery to implement this chemical management are the environmental policies of the company, environmental legislation and customer demands. Their main customer is the automotive industry which very often put strict demands on the use of chemicals by their suppliers.

One barrier for a tannery to get information about the chemical they are using is the lack of sufficient information given by chemical suppliers in, for example safety data sheets. However, due to the chemical legislation in Europe (REACH), it can be expected that more information about the chemical composition of different mixtures will be given in the future.

Another barrier is that tanneries very often are small and medium-sized enterprises and therefore do not have the resources to carry out an efficient chemical management at the company nor have the capacity to assess new or existing chemicals. The tannery used as a case study has one environmental manager who is responsible for the environmental issues at the company. In many small tanneries, the responsibility is taken by the Managing Director or the Head of the Production and these issues sometimes will have a quite low priority in many tanneries.

However, in the future it can be expected that more tanneries will implement a management of the chemicals used in the tannery. The main drivers for this development can be predicted to be customer demands and also future legislation. It is not only the automotive industry that restricts the use of certain chemicals but also many of the global brands publish their own lists of restricted substances and update them at regular intervals. In most cases the global brands base their specification limit values on the regulations that exist.

References

1. European Commission (2006) Regulation No 1907/2006 of the European Parliament and of the Council of 18 December 2006 concerning the Registration, Evaluation, Authorisation and Restriction of Chemicals (REACH)
2. PRIO – A tool for risk reduction of chemicals. Swedish Chemical Agency Web. Available at: http://www.kemi.se. Accessed 25 February 2011
3. European Commission (2008) Directive 2008/105/EC of the European Parliament and of the Council of 16 December 2008 on environmental quality standards in the field of water policy
4. Global Automotive Declarable Substance List website. Available at: http://www.gadsl.org. Accessed 25 February 2011
5. European Commission (2010) Directive 2010/75/EU of the European Parliament and of the Council of 24 November 2010 on industrial emissions (integrated pollution prevention and control)
6. European commission (2009) Regulation no 552/2009 of 22 June 2009 amending Regulation (EC) No 1907/2006 of the European Parliament and of the Council on the Registration, Evaluation, Authorisation and Restriction of Chemicals (REACH) as regards Annex XVII
7. European Commission (1998) Directive 98/7/EC of the European Parliament and of the Council of 16 February 1998 concerning the placing of biocidal products on the market (Biocidal Products Directive)
8. Rydin S (2006) Environment in focus at Elmo Leather. Leather International August/September 2006
9. Rydin S, Svensson L (2006) Innovative waste water treatment plant at Elmo Leather AB. In: Proceedings of the IULTCS II. Eurocongress Istanbul, Turkey
10. Oeko-Tex 100 Eco-label website. Available at: http://www.oeko-Tex100_PUBLIC/index.asp?cls=21. Accessed 25 February 2011

Lubricants Recycling – A Case Study: How Italy Managed to Become an Excellence and an Example for the Other EU's Member States

Fabio Dalla Giovanna

Abstract It is relatively common for people to recycle paper, glass and metals (aluminium and steel). The infrastructure and services for this recycling activity are available and relatively well understood. However, the recycling of petroleum products is less well known and sometimes confusing. Just 1 L of used oil can contaminate *1 million* litres of water. In Italy, used oil re-refining practice is common since the fifties, and it is today a reality thanks to an excellent organization system and thanks to the priority given to the used oil recycling vs. burning. This practice has now been strengthened by the European Union with the recent Waste Framework Directive, which imposes a priority order hierarchy for waste management, where recycling is at an upper level than burning with energy recovery.

This chapter describes how Italy, within the past EU15, was the only Member State which did not receive an infringement procedure by the Court of Justice of Luxembourg for not having respected the priority of regeneration according to Directive 75/439/EEC.

Legislative framework, collection system, incentives, good communication policies and technology are the key points that made Italy the most advanced European Country in the used oil re-refining sector. Today, advanced technologies have been developed to obtain recycled lubricants with quality and performances equivalent to the virgin products and, for those producing Gr.II/II+ base oils even better.

This case study can be taken as an example of the excellent practice for the other EU27 Member States for conserving a non-renewable fossil resource, for protecting the environment and saving money by reducing the dependence from the crude oil producing countries.

Keywords Recycling, Re-refining, Used oil

F. Dalla Giovanna
Viscolube Spa, Via Tavernelle, 19, 26854 Pieve Fissiraga (LO), Italy
e-mail: f.dallagiovanna@viscolube.it

B. Bilitewski et al. (eds.), *Global Risk-Based Management of Chemical Additives I: Production, Usage and Environmental Occurrence*, Hdb Env Chem (2012) 18: 225–252,
DOI 10.1007/698_2011_100,
Published online: 30 March 2011

Contents

1 Legislative Framework

Re-refining was born in Italy during the Second World War where, for autarchic reasons, waste lubricating oil was collected and recycled in order to be less dependent on the crude oil producing countries and suppliers of raw materials/commodities.

Over the years, re-refining has become an activity having environmental values and recycling has become a need related both to the exhaustion of the fossil resources and to ecological reasons.

1.1 Past and Previous Legislative Framework

Dir 75/439/EEC (amended by Dir 87/101/EECand Dir 91/692/EEC), implemented by D.Lgs 27th January 1992, n. 95

Objective: management

It provides for measures aimed at assuring waste oil collection and disposal without harmful effects upon man and the environment. It introduces several definitions (waste oil, disposal without harmful effects upon man or the environment, priority of re-refining) and also a specific regulation for waste oils, to create an efficient system of waste oil collection, storage and re-use (or, as an ultimate option, destruction), privileging re-refining rather than other more impacting ways of disposal. D.Lgs 95/1992 has cleared up nationally the competences, authorizations and ways of management of the activities of collection and disposal of used oils.

Dir 91/156/EEC implemented by D.Lgs 5th February1997, n. 22
Objective: Regulation of waste management

Legislation concerning waste. It obliges the Member States to pursue some objectives in accordance with certain standards and principles.

It introduces the notion of waste and a hierarchy in the targets concerning waste (reduction at source, material recovery, energy recovery, disposal in dump). The framework here outlined coexists and is perfectly consistent with what defined by the former D.Lgs 95/1992.

Dir 91/689/EEC implemented by D.Lgs 5th February 1997, n. 22
Objective: Regulation of hazardous waste management

It prescribes the necessary instruments to monitor and manage hazardous waste. It gives a definition of hazardous waste in which also the oils and oily and mineral substances are included.

European Directive 2008/98/EC
Objective: waste management

It establishes the measures necessary to protect the environment and human health promoting the waste re-use and recycling.

Legislative Decree n. 95 of 1992
Objective: Technical provisions concerning waste oil disposal

It regulates the methods of analysis and the technical requirements for the holder's storage sites; moreover, it regulates the necessary requirements for granting the authorization to collection and disposal.

Legislative Decree n. 209 of 22nd May 1999
Objective: regulation of PCB–PCT disposal

It integrates the rules and regulations governing polychlorinated biphenyls and polychlorinated terphenyls (PCB–PCT) disposal; particularly important is article 8 concerning waste mineral oil mixing.

Legislative Decree n. 133 of 11th May 2005
Aim: Provisions concerning the hazardous waste incineration plants

It transposes Directive 2000/76/EC and defines the technical provisions concerning the running of hazardous waste incineration and co-incineration plants.

Legislative Decree n. 59 of 18th February 2005
Objective: integrated reduction of pollution

It transposes Directive 1996/61/EC concerning the integrated pollution prevention and reduction IPPC. In attachment I, the plants regulated by this Decree are listed including those devoted to waste management; they must conform themselves to the new authorization system named "AIA" (Environmental Integrated Authorization).

Law 16 of 27th February 2002
Objective: urgent measures on excises, gas oil, and waste oils

It lays down urgent measures concerning waste oil disposal. Nevertheless until today, the relevant implementing regulation has not been issued yet; therefore, it is very likely that this law will never become effective.

Legislative Decree 152/2006 ***ENVIRONMENTAL CODE***
Aims: Reorganization and updating of the national environmental legislation

In comparison with the statutory consortiums for waste management, the new code provides for this sector to be open to "free competition", an aspect which could create enforcing problems in the management of hazardous waste.

Legislative Decree n. 4 of 16th January 2008
Objective: amendment to the contents of D.Lgs 152/06 "Testo Unico Ambientale"

It restores the Consortium's unicity and defines the reference bases for the New Statute.

Ministerial Decree of 8th April 2008
Objective: ***regulation of the municipal waste collection centres***

Separate collection, as provided for by article 183, paragraph 1, letter c) of D.Lgs 3rd April 2006, n. 152, and subsequent amendments.

Ministerial Decree of 13th May 2009

It modifies the contents of DM 8th April 2008 with reference to some aspects related to the regulation of the collection centres of the municipal waste collected separately.

Law n. 166 of 20th January 2009

It's the Conversion into law of the Legislative Decree n. 135 of 25th September 2009, taking urgent measures for the implementation of EC obligations and for the enforcement of judgements of the EC Court of Justice.

1.2 2009 Legislative Framework

The new legislation in 2009 was:

D.Lgs 135/09 – Law 166/09: provisions for re-refined oils: D.Lgs 135/09, subsequently become Law 166/09, aims at amending what is contested by the European Commission to the Italian Government (infringement procedure nr. 2004/2190) concerning the tax relief in favour of the lubricating oils produced through used oil re-refining. This facilitation consisted in a 50% reduction of the consumption tax on re-refined lubricating oils. The European Commission contested to the Italian Authorities to keep a system of tax reliefs on re-refined lubricating oils favouring the national productions to the detriment of those coming from the other Member States, openly infringing the non-discrimination principle established, in fiscal matters, by art. 90 of the EC Treaty. The Law 166/09 repeals the provision according which the name "used oils" refers to the used oils collected in Italy and, as regards the above-mentioned relief, it modifies the tax provisions concerning the consumption tax on re-refined lubricants. Besides article 13 of the decree remits to COOU, the task of paying to the re-refining companies a compensation for the disposal of the waste oil collected, abolishing the tax relief on waste lubricants; moreover, the companies putting lubricating oils on the market must pay to COOU a contribution sufficient to cover the cost of the aforesaid compensation.

Another interesting element in Law 166/09 is that it entrusts the Consortium with the duty to charge itself with the possible higher costs of re-refining (costs not recoverable from the market), and it provides for the used oil collected to be shared out among the refineries according to specific criteria and that the compensation paid evolves over time. The directives concerning the Consortium in the course of its activities are the following:

- To agree with the re-refining companies upon the technical parameters for the selection of the waste oils suitable for re-refining.
- To support the regenerable waste oil collection.
- To give the regenerable used oils to the re-refining companies, which apply for them in proportion to the ratio between quantities collected, quantities requested and authorized production capacities.
- To pay a compensation to the re-refining companies for the disposal, depending on the market situation, the re-refining costs and the possible revenue that the Consortium itself could get should the waste oils collected be sent to combustion. This fee is paid with reference to the quantity of lubricating base produced per tonne of waste oil and to its quality.
- To ensure that the waste oil which is not regenerable but re-usable is sent to combustion and that the waste oil which is not re-usable is disposed of in accordance with the anti-pollution provisions.

The choice to entrust the Consortium with the task of paying to the re-refining companies a fee for disposal through re-refining is for COOU a further acknowledgment of the Institutions' reliance.

DM 13th May 2009: new provisions for the waste collection centres

In the Ministerial Decree 13th May 2009, some of the measures introduced by the Ministerial Decree 8th April 2008 on the technical–managerial requirements concerning the waste collection centres were reformulated. Among its main measures, DM 13th May 2009 extended the time period in which the plants must conform to the new provisions, allowing the subjects not in line yet with these new technical–managerial provisions but authorized by the Regions (or by other local authorities) to go on operating until 18th January 2010, after which it is necessary to comply with the new environmental parameters. Furthermore, DM 13th May 2009 increased the typologies of waste that can be sent to the collection centres, adding 13 categories to those already foreseen in the original version of DM 8th April 2008. The Consortium has developed a programme to map the collection centres already operating on the national territory through the collaboration of the operators of this sector. Once the mapping is made, it is COOU's target to develop a software at the citizens' disposal so that they can easily trace the collection centres (green spots) located on the national territory.

DM 17th December 2009 SISTRI: a new system for the traceability of waste

The new control system of waste management, named "SISTRI", was designed with the aim of replacing the stock book of waste, the forms and the annual declaration (Modello Unico di Dichiarazione Ambientale MUD) with electronic devices.

All information will be concentrated on a software to which every operator of the waste sector will have to make reference to communicate the quality and quantities of the waste managed; a tracking system is foreseen through detection systems (black boxes) installed on the trucks.

The subjects involved, after regular registration (www.sistri.it), will be equipped with a code by which they will pick up at the Chamber of Commerce territorially competent the technological devices necessary to monitor the movements of waste.

The Concessionaires and the free Collectors of COOU must conform to these provisions equipping themselves with the USB flash drive in their operations units, installing black boxes on their trucks, defining specific internal management procedures and training the personnel involved.

1.3 New European Legislation

Directive 2008/98/EC: the management of waste

The new European directive 2008/98/EC dictates clearer and stricter rules so that any policy of management of waste can pursue the aim of minimizing the negative consequences on human health and on the environment coming from the production and management of waste. Compared with the previous directive on waste (Directive 2006/12/EC, repealed), Directive 2008/98/EC is based on an approach aiming at prevention and re-use of waste as a resource, extending the field of interest to the traceability of the whole life cycle of the products and materials and not only to the end of life phase. According to what is provided for by this Directive, the Member States must take within 24 months (December 2010) the measures necessary to handle waste in accordance with a *hierarchy* applied following a *priority order*:

- Prevention of waste production
- Preparing for re-use
- *Recycling*
- Other recovery, e.g. energy recovery
- Disposal

The Member States have the following obligations:

- To take legislative or other kind of measures aiming at assuring that any individual or legal person who professionally develops, produces, transforms, handles, sells or imports products is responsible for the management of the waste he or she has produced (principle of extended liability of the producer).

- As regards the hazardous waste ensure the control (art. 17), ban on the mixing (art. 18), labelling (art. 19) and take the necessary measures to ensure the separate collection of waste oils and organic waste (art. 21 and 22).

For waste oils, the Member States must take the necessary measures to ensure that they are collected and handled in compliance with articles 4 and 13; particularly, the priority of re-refining on combustion is defined in article 4, which sets recycling as a primary choice before other ways of recovery (e.g. energy recovery).

The Consortium, by means of its network operating nationwide, ensures the promotion of recycling of a hazardous waste through waste oil re-refining, which enables to transform it into a new resource available on the market.

Besides the Consortium System provides for a shared management from the point of view of the cash flows and of responsibilities, the system organization and structure provide for the management costs of waste oil to be charged to the producers and to the subjects putting the oils on the market (associated companies).

2 Used Oil Re-Refining

All mineral used oils, on the one hand, can be adequately treated and recycled. On the other hand, their improper disposal causes environmental pollution.

The re-refining process consists in the elimination, through a suitable treatment, of the carbon residues and of the metal oxides from waste oils to obtain re-usable bases.

This process of disposal represents the priority destination of used oil according to the national legislation: *Italy is the first country in Europe for the quantity of used oil sent to re-refining*. In the last few years, the use of re-refined bases in place of the virgin bases has considerably increased for legislative and economic reasons (re-refined bases have a better quality–price relationship than virgin bases).

2.1 The Statutory Consortium of Waste Oils

Waste lubricants are categorized as hazardous waste and they must be managed accordingly from the administrative, the storage and handling point of view. Used oil collection in Italy is managed by the Statutory Consortium of Waste Oils (COOU), an agency established with DPR nr. 691/1982.

Waste oil is defined in legislation as "*any industrial or lubricating oil, mineral or synthetic, that has become unsuitable for its original purpose, particularly the waste oils of combustion engines and transmission systems as well as the mineral oils for machineries, turbines or hydraulic drives and those contained in the waste filters*".

Besides, legislation states that: "*also the oily mixtures, represented by the fluid or liquid waste compounds only partially made of mineral or synthetic oil, including the tanks oily residues, the water and oil mixtures and the emulsions, are subject to the rules and regulations provided for waste oils*".

Waste oils must be disposed of avoiding any damage to health and to the environment; in fact, the law prohibits:

- Any deposit and/or discharge of waste oil harmful to the soil and any uncontrolled discharge of residues resulting from the processing of waste oils.
- Any processing of waste oils causing air pollution, which exceeds the level prescribed by existing provisions.
- Any discharge of waste oils into internal surface waters, ground water, coastal waters and drainage systems.

According to the legislation in force, the industrial companies producing waste oils and those which during the year hold for any reason a quantity of waste oil higher than 300 L/year are obliged:

- To stock the waste oils properly to avoid any mixture between emulsions and oils in the strict sense of the word or any dispersion or contamination with other substances.
- Not to mix the waste oils with toxic and noxious substances.
- To give and convey all the waste oils held directly to COOU or to the companies authorized to collection and/or disposal, communicating to the cessionary all the data concerning the origin and the previous uses of the waste oils.
- To reimburse to the concessionaire the charges connected with the disposal of the single oily mixtures, of the waste oils that cannot be processed and of the contaminated oils.

The companies putting base and finished lubricating products on the market and the re-refining companies producing lubricating bases from waste oils are members of the Consortium. It is a privately managed structure, paid by the associates (in 2010 the contribution to COOU was 155 €/tonne and it is estimated to fall to 130 €/tonne in 2011), but under the State's policy and control.

The contribution is imposed on the price of the oil sold by the companies, which put lubricating oils on the market, and it is given to COOU, which uses a part of it to support the waste oil collection and another to support the re-refining companies.

The Consortium's obligations are stated in article 11 paragraph 10 of D.Lgs. 95/92. It is obliged to:

- Awaken public opinion on the matters concerning waste oil collection and disposal.
- Ensure and boost waste oil collection picking them up at the holders and at the authorized companies.
- Carry out directly the waste oil collection activities at the holders' which make directly request in the provinces where collection is not made or is insufficient or economically difficult compared with the quantities of lubricating oils put on the market.
- Segregate the waste oils collected for their proper disposal.
- Send the used oils to the authorized companies for their disposal.
- Carry on and boost the study, experimentation and realization of new treatment processes and alternative uses.

- Operate respecting the principles of competition, free movement of goods, economy management as well as health and environment protection from any kind of air, water and soil pollution.
- Write down and elaborate all the technical data concerning waste oil collection and disposal and communicate them annually accompanied by an explanatory report to the Ministers deputed to control.
- Assure to re-refiners, within the limits of the regenerable used oils collected and of the plant production, the quantities of used oils requested at a fair price and in any case not higher than the direct cost of collection.

In Italy, every year about *450 thousand tonnes of lubricating oil* are put on the market. When the oil is no longer suitable for its original purpose, the part which is not consumed during use is the waste oil, which must be collected neutralizing its polluting characteristics.

The polluting power of the oil is underestimated by most of the layman. On the contrary, we must not forget that oils can be, first of all, a resource as they can be re-refined and re-used countless times.

Used oil is re-usable as raw material for the production of re-refined lubricating bases. In quantitative terms, 1 kg of used oil can produce about 0.7 kg of lubricating base, as well as other products (gasoil, fuel, bitumen, etc.).

In conclusion, it is possible to state that recovered oil represents a real economic resource: this is proved by the fact that, in 25 years of COOU's activity, *Italy has saved about 1 billion euro on the National energy bill* thanks to used oil recycling.

3 Used Oil Re-Refining Activity

Among the different used oil treatment processes the one which better exploits the oil potentiality to be reused is re-refining, a process defined both by the national and by the European legislations as the priority destination of used oil.

Italy is the first Country in Europe for the quantity of used oil sent to re-refining. The Italian refineries processing used oil are the following:

- Viscolube SpA Pieve Fissiraga (Lo)
- Viscolube Spa Ceccano (Fr)
- Ra.M.Oil: Casalnuovo (Na)
- Distoms Porto Torres (SS)
- Siral San Vitaliano (Na)
- S.I.R.O.: Soriano di Corbetta (Mi)

Among these, 80% of the used oil is processed by Viscolube S.p.A.

Viscolube SpA, a company established in 1963, is one of the few European companies specializing in used oil re-refining, and today it is one of the few European companies able to produce a lubricating base oil whose characteristics are better than those of the virgin bases coming from crude oil refining currently on the market. Present in Italy with two production facilities (Pieve Fissiraga in the province of Lodi and Ceccano in the province of Frosinone), Viscolube treats every year about 130,000 tonnes of used oil, from which about 90,000 tonnes of re-refined oil are produced.

Since its establishment Viscolube has operated respecting the environment with the philosophy of obtaining more and more qualitatively advanced re-refined oils. In this context, the company has developed, in collaboration with the French company Axens, an advanced technology, which enables to obtain re-refined oils having similar or even better characteristics than virgin oils. This technology, called Revivoil, has already been successfully exported to different countries.

The hydrofinishing unit built in the factory of Pieve Fissiraga enables, through a high pressure treatment with hydrogen, to produce oils having API group II characteristics, that is low sulphur and unsaturates content and very low aromatics content.

4 The Used Oil Collection Step

The first step of the Consortium's activity, that is the used oil collection, is carried out by collecting companies and concessionaires located on the whole Italian territory, which arrange for the used oil collection and for their storage in their depots.

Then these used oils are transferred in the storage depots of the Consortium. The collection process is divided into two steps:

- A primary transport step in which the oil is collected from the holders and carried to the collectors' storage depots
- A secondary transport step in which the oil is carried to the Consortium's depots

Therefore, the whole collection step consists in the oil transport by trucks, both from the holders to the collectors' depots (primary collection) and from the latter to the storage depots (secondary collection).

The qualitative and quantitative results of the used oil collection carried out by the Consortium must be connected with the values of lubricating oil put on the market in the same year

The trend of the oils put into the market is sensibly decreasing (Fig. 1) [1]. The cause of these reductions must be ascribed to the development of new technologies in the industrial and automotive sectors, which imply lower oil consumptions for the same production and/or km covered and to the world economic crisis.

The Consortium uses a collection network made up of 72 companies among which 66 are concessionaires and 6 are collectors located in all the regions of Italy which, through their own vehicles or through sub-collectors, visit the holders to collect the used oils and store them in their depots.

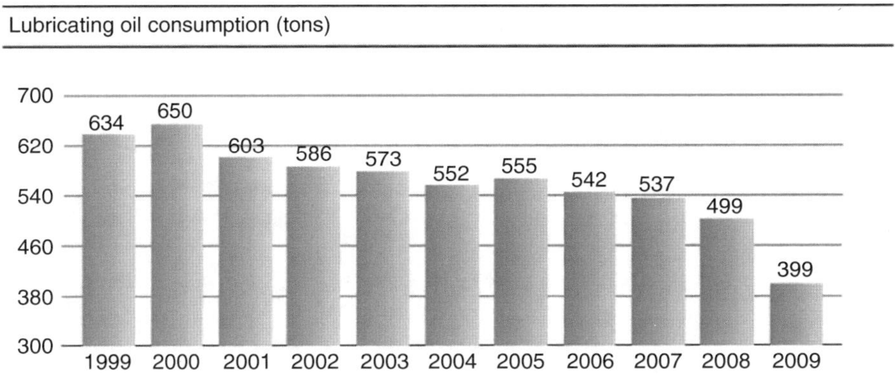

Fig. 1 Lubricating oil consumption

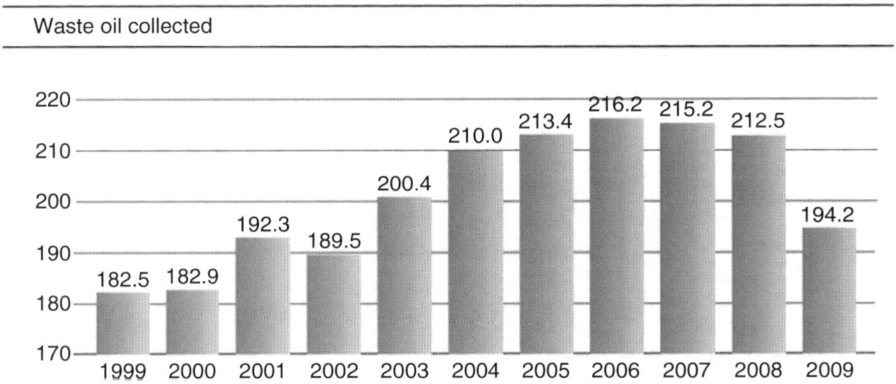

Fig. 2 Collected waste oil

The used oil collection in 2009 settled at the value of 194,209 tonnes, 8.6% lower than 2008 (Fig. 2) [1]. This trend was mainly affected by the strong shrinkage of the lubricants market and by the consequent reduced used oil production.

In spite of this, the ratio between the used oils collected and the lubricants put on the market (real performance indicator of the chain) has increased from 42.6% in 2008 to 48.7% in 2009 (Fig. 3) [1].

Considering a supposed estimate of the actual consumptions of 440,000 tonnes, the ratio between the used oil collected and the lubricants consumption would decrease to about 44% which, in any case, would represent the umpteenth record for COOU.

According to a study carried out in 2004 by COOU and ENI Technologies and General Consult (2004) aiming at evaluating the quantity of used oil potentially

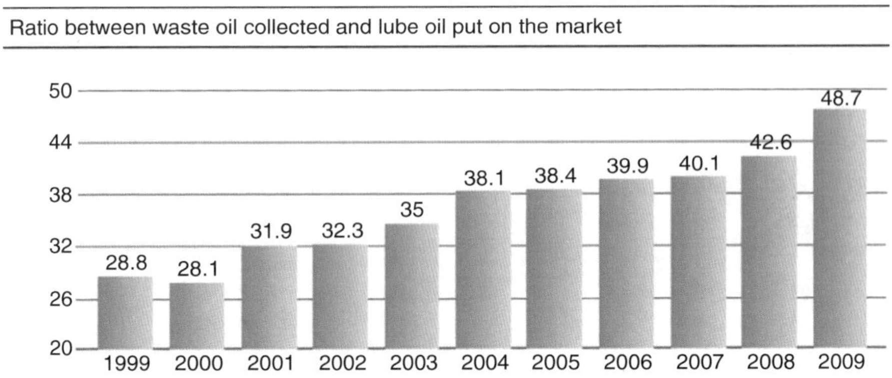

Fig. 3 Ratio waste oil collected/lube oil market

recoverable compared with that put on the market and the consequent possibility of improvement in the collection, it came out that:

- The theoretical level of recovery in the automotive sector can be placed between 56 and 59% of the total oil put on the market in this segment
- As regards, on the contrary, the oil used in the industrial sector the theoretical level of recovery is estimated around 33–35%.

The survey was started on the basis of the lubricants put on the market and, following the applicative cycle of the product typologies, their uses, applications, consumptions and losses were evaluated; in a second phase, the results were validated through a field survey.

On the basis of these considerations, it was estimated that the quantity of used oil collected by COOU is steadily over 95% of the collectible oil; the not intercepted oil is mainly associated with the "do-it-yourself" and with the industrial sector, where the supposed ends are self-consumption (unauthorized combustions) and the use in oily emulsions.

Specific requirements are not requested to collectors except the minimum legal requirements concerning the possession of:

- Authorization to manage a provisional depot, issued by the competent authorities in accordance with the legislation in force.
- Authorization to manage vehicles for used oils collection, which enable to run directly the business.

The minimum requirements requested by COOU to conclude a contract with a concessionaire are:

- The possession of the authorization, issued by the competent authorities in accordance with the legislation in force, to manage a provisional used oils depot (as for the collectors). Besides this, depot must have a total authorized capacity of at least one-thirtieth of the yearly collection and in any case not lower than 90 m^3 divided into at least three tanks of 30 m^3.

- The possession of the authorization to run at least two trucks (just like for the collectors, who nevertheless do not have a minimum limit of vehicles). For consortiums or companies' associations, the number of vehicles must be twice the number of the member companies.
- The achievement, not later than 31st December 2004, of ISO 9000 quality certification and of ISO 14000 certification or, as an alternative to the two aforesaid certifications, the achievement of EMAS II registration (requirement not foreseen for the collectors).

4.1 Used Oil Classification

The used oil treatment phase following collection is storage.

For the used oils storage management, the Consortium enters into service contracts with third parties' depots.

The quantities conveyed for typology of waste oil are shown in the following graph (Fig. 4) [1].

Currently, COOU's depots are 5: Reol, Viscolube Lodi, Monticelli, Viscolube Ceccano and Ra.M.Oil (see Table 1 below).

The oil conveyed by the collectors is submitted by COOU to lab analyses to classify it. The classification is used for specifying the oil typology and consequently

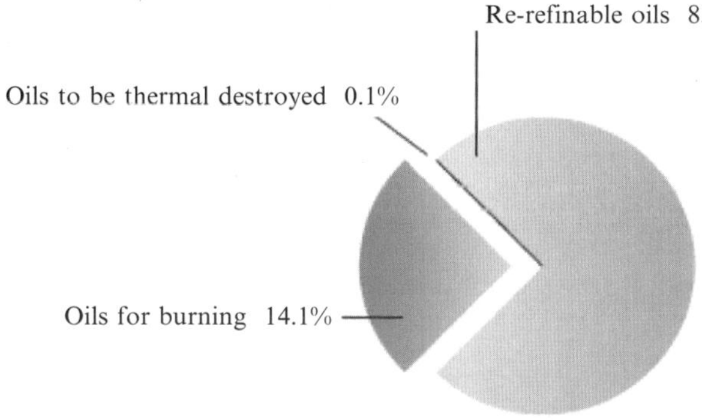

Fig. 4 Waste oil destination in Italy

Table 1 Quantity of 2009 addressed to the tank farms (tonnes)

Tank farm	Quantity entered into the depot
Reol (Soriano di Corbetta, Milano)	7.359
Viscolube Lodi (Pieve Fissiraga, Lodi)	74.237
Monticelli (Mezzanino Po, Pavia)	36.796
Viscolube Ceccano (Ceccano, Frosinone)	48.519
Ra.M.Oil (Casalnuovo, Napoli)	21.598
Total	188.509

Table 2 WASTE OILS for regeneration (Attachment A – D.M. 392/96)

Parameters to be analysed, methods of analysis and minimum/maximum reference values for used oils processing through re-refining

Parameters	Methods		Unit of measurement	Values
Water	NOM 7–70	ASTM D 95	% weight	Max 15
Density at 15°C	NOM 42–83	ASTM D 1298	Kg/I	Max 0,920
Total sediments	NOM 112–71	ASTM D 2273	% weight	Max 3,0
Viscosity	NOM 46–71	ASTM D 445	°E at 50°C	Min 1,8
Polychlorinated biphenyls	CEI 10/19	ASTM D 4059/95	mg/kg	Max 25
Total chlorine	NOM 98–72	ASTM D 1317	% weight	Max 0,5
Sulphur	NOM 97–80	ASTM D 1552	% weight	Max 1,50
Diluents	NOM 39–90	ASTM D 322	% volume	Max 5,0
Lead + Zinc	IRSA 64		mg/kg	Max 4.000
Cadmium + Chrome + Nickel + Vanadium	IRSA 64		mg/kg	Max 50
Neutralization Nr[a]	NOM 86–88	ASTM D 664	mg KOH/g	Max 3,5
Saponification Nr[a]	NOM 81–71	ASTM D 94	mg KOH/g	Max 18,0

[a]Analyses to be made on the sample after proceeding to the loss by heating (ASTM D 3607)

Table 3 WASTE OILS for burning (Attachment A – D.M. 392/96)

Parameters to be analysed, methods of analysis and minimum/maximum reference values for used oils processing through combustion

Parameters	Methods		Unit of measurement	Values
Water	NOM 7–70	ASTM D 95	% weight	Max 15
Density at 15°C	NOM 42–83	ASTM D 1298	Kg/I	Max 0,920
Total sediments	NOM 112–71	ASTM D 2273	% weight	Max 3,0
Flash point (Cleveland)	NOM 83–71	ASTM D 92	°C	Min 90
Polychlorinated biphenyls	CEI 10/19	ASTM D 4059/95	mg/kg	Max 25
Total chlorine	NOM 98–72	ASTM D 1317	% weight	Max 0,5
Sulphur	NOM 97–80	ASTM D 1552	% weight	Max 1,50
Ash	NOM 12–88	ASTM D 482	% volume	Max 1,50
Lead + Zinc	IRSA 64		mg/kg	Max 4.000
Cadmium + Chrome + Nickel + Vanadium	IRSA 64		mg/kg	Max 50
Lead	IRSA 64		mg/kg	Max 2.000
Fluorine[a]	NOM 98–72	ASTM D 1317	mg/kg	Traces

[a]With selective electrode

identifying the best way for its disposal. Then the oil is sent to a storage tank where it is kept until it is sold.

The three processes that used oils undergo for their reuse and/or disposal are re-refining, combustion and incineration.

Storage 2009
Storage Input
Used oils (tonne) 194.616
Storage Output
Re-refining (tonne) 154.118
Combustion (tonne) 39.918
Incineration (tonne) 191
Other uses (tonne) 389

Used oil classification is made in accordance with the legislations in force, summarized in Tables 2 and 3.

5 The Used Oil Re-Refining Process

The re-refining process converts the waste oil into a re-refined base oil re-usable as lubricant (after appropriate additivation).

Re-refining is the best process for a correct and useful use of the waste oil as it enables to remove the carbon residues and the metal oxides from waste oils to obtain re-usable base oils.

In fact, through the most advanced available techniques it is possible to recover from 100 kg of dehydrated waste oil:

- 76 kg of new oil
- 13 kg of additive for bituminous membranes
- 7.6 kg of low sulphur content automotive/combustion gasoil
- 0.4 kg of sulphur 99% pure
- 3 kg of by-products

Moreover from the combustion of the incondensable gases and vapours of light hydrocarbons in the first distillation, it is possible to produce all the energy requirements in terms of steam necessary for the plants running (steam for the ejectors of the vacuum groups, stripper steam and steam for heating in the process exchangers).

Among the most advanced processes for the production of re-refined bases, there is the one adopted by Viscolube in Italy consisting of three steps (Fig. 5) [2]:

- *Step I – Preflash*: the used oil is heated up to 140°C and then distilled in a vacuum column, where the water and light hydrocarbons are separated.
- *Step II – Thermal De-Asphalting TDA*: the dehydrated oil is distilled at 360°C in a vacuum de-asphalting column (TDA), where the asphaltic and bituminous products remain at the bottom and, at the same time, three side cuts of different viscosity are extracted. An intermediate gas oil is extracted at the top of the column. The three side fractions and the gas oil are sent to storage to be subsequently hydrofinished in batches in a high pressure (105 bar partial pressure) catalytic plant.

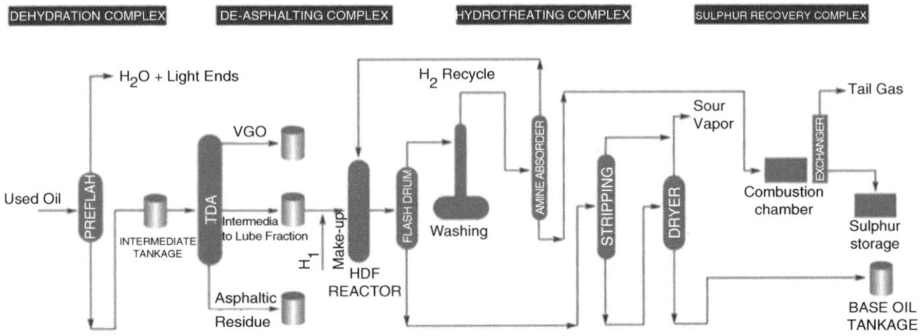

Fig. 5 Viscolube re-refining process – Revivoil

- *Step III – Hydrofinishing*: the Hydrofinishing process starts in a fired heater, where the oil and hydrogen are heated up to 300°C. They are then sent to a reactor containing a catalyst favouring hydrogen reaction with the unsaturate compounds, sulphur and nitrogen. The reactor effluent is then separated into two phases, the vapour phase and the liquid phase; the first one is washed with water to remove the chlorine and sulphur compounds, the second one is stripped with steam to eliminate the most volatile compounds and restore the Flash Point.

The water contained in the oil after stripping is then removed in a vacuum dryer. The streams containing sulphur are sent to an amine plant, where the hydrogen sulphide is separated from the other compounds. Then they go to a Claus plant, where H_2S is transformed into pure liquid sulphur. The final result is a clear oil with very low sulphur and polynuclear aromatics (PNAs) content. This lubricating base oil has many advantages: it is beneficial to health and the environment as well as demonstrating excellent performance on the lubrication circuits where it is used.

Through this plant, it is possible to obtain finished products having advanced environmental and health characteristics:

- Very low sulphur content both in the lubricants and in the gas oil
- No polynuclear aromatics
- Saturation of the aromatic compounds until the achievement of API Group II specifications (Table 4)
- Elimination of by-products and wastes such as waste clay, sludge acid and other substances harmful to the environment
- Up-grading of the finished lubricating product in terms of performances/quality
- Greater oxidation stability
- Better colour
- Lower volatility
- No odour
- Higher yield in lubricant

Table 4 API definition for base oil

Group	Saturates		Sulphur	Viscosity index	Manufacturing method
I	<90%	And/or	>0.03%	≥80 and <120	Solvent refining
II	≥90%	And	≥0.03%	≥80 and <120	Hydroprocessing
III	≥90%	And	≥0.03%	≥120	Severe hydroprocessing
IV	Polyalphaolefins (PAOs)				Chemical reaction
V	All other basestocks not included in groups I, II, III, or IV				

The chemical reactions involved in the hydrofinishing process are of two types:

The *hydrotreating reactions:*

- Elimination of sulphur
- Elimination of nitrogen and oxygen
- Elimination of metals
- Saturation of olefins and diolefins

And the *hydrocraking reactions.*

5.1 *Hydrotreating Reactions [3]*

A. Hydrogenation of sulphur compounds

- Mercaptans, sulphides and disulphides react easily leading to the corresponding saturated or aromatic compounds.
- Sulphur combined into cycles of aromatic structures, such as thiophene, is more difficult to eliminate.
- These reactions are exothermic, they produce hydrogen sulphide and consume hydrogen.

Examples:

- Mercaptans

$R - SH + H_2 \rightarrow R - H + H_2S$

- Sulphides

$R - S - R + 2\,H_2 \rightarrow 2\,R - H + H_2S$

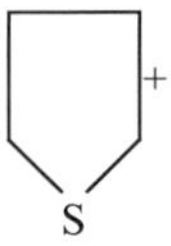

$+ 2\,H_2 \rightarrow C_4\,H10 + H_2S$

- Thiophene

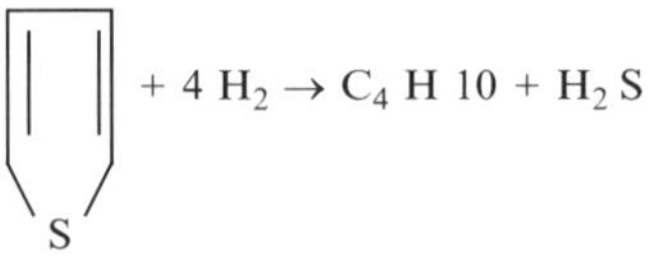

- Dibenzothiophene

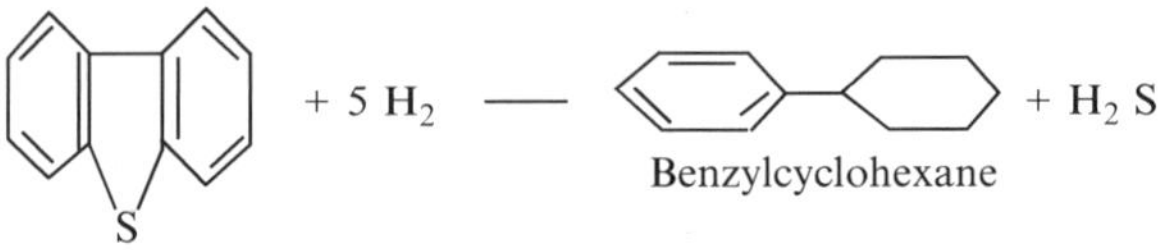

B. Hydrogenation of the nitrogen compounds

- The rate is lower than for the desulphurization reactions.
- These reactions lead to ammonia formation.
- These reactions are also exothermic.

Examples:

- Pyridine

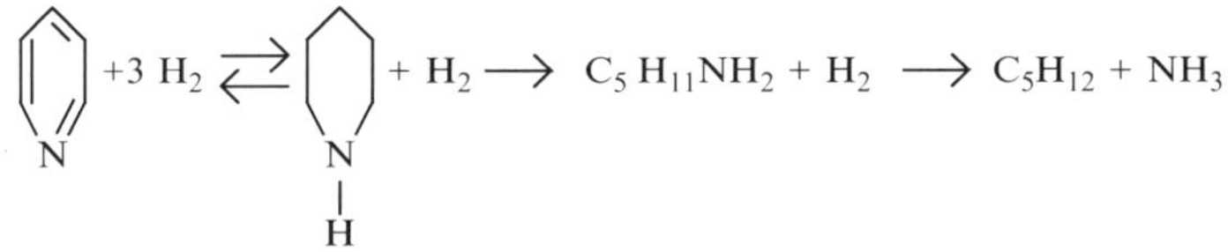

Pyridine Piperidine

- Amine

$R - NH_2 + H_2 \rightarrow RH + NH_3$

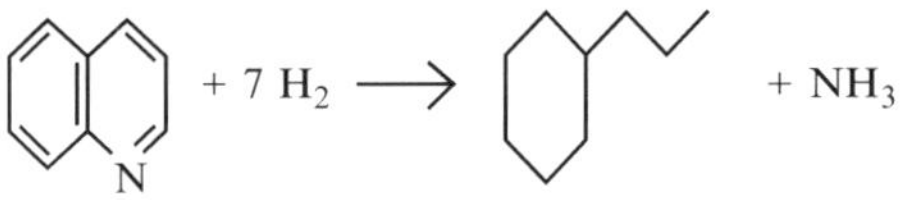

C. Hydrogenation of oxygenated compounds

Hydrogenation of bond C–O

- Alcohols and phenols

$R - OH + H_2 \rightarrow R - H + H_2O$

- Acids

$$\mathrm{R-\underset{\underset{OH}{|}}{C}=O + H_2 \rightarrow R-\underset{\underset{H}{|}}{\overset{\overset{O}{||}}{C}} + H_2O}$$

Hydrogenation of bond C=O

- Aldehydes

$$\mathrm{R-\underset{\underset{H}{|}}{C}=O + 2H_2 \rightarrow R-CH_3 + H_2O}$$

D. Hydrogenation of olefinic compounds

- These reactions are highly exothermic. Olefins and diolefins are converted to saturated compounds.

The hydrogenation rate of olefins and diolefins is higher than the hydrodesulphurization rate.

Examples:

- Olefins

$$\mathrm{R-CH=CH_2+H_2 \rightarrow R-CH_2-CH_3}$$

E. Hydrogenation of aromatic compounds

The hydrogenation of aromatics has thermodynamic limitations. These reactions are exothermic and the number of molecules decreases. Therefore, they are favoured by low temperature and high pressure.

For a given pressure, when the temperature increases the hydrogenation rate increases first, to reach a maximum, and then decreases as the temperature continues to increase.

For a given temperature, the hydrogenation rate increases rapidly with the pressure.

Examples:

- Benzene

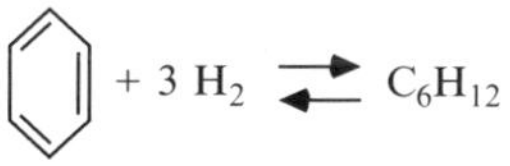

- Naphthalene

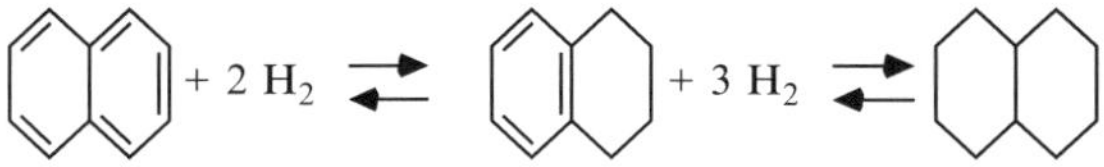

Naphtalene Tetraline Dicyclohexane

F. **Demetallization**

- The organometal compounds (containing As, Pb, Cu, Ni, Va...) are cracked and the metals are trapped on the catalyst.

Hydrocracking Reactions

It is an undesirable reaction, which has to be minimized, because it consumes hydrogen, reduces the product yield and the hydrogen purity of the recycle gas.

It is limited by the selection of catalysts with low hydrocracking capacity and working at low temperature.

(a) Hydroisomerization and then cracking into lighter isoparaffins.

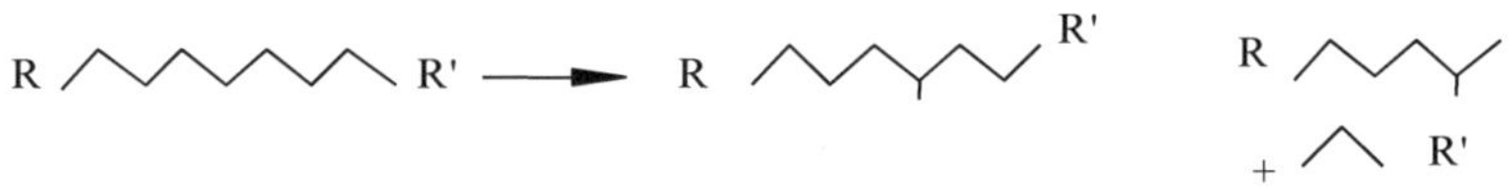

(b) Monocyclic naphthenes are dealkylated and then isomerized into light isoparaffins and lower molecular weight naphthenes.

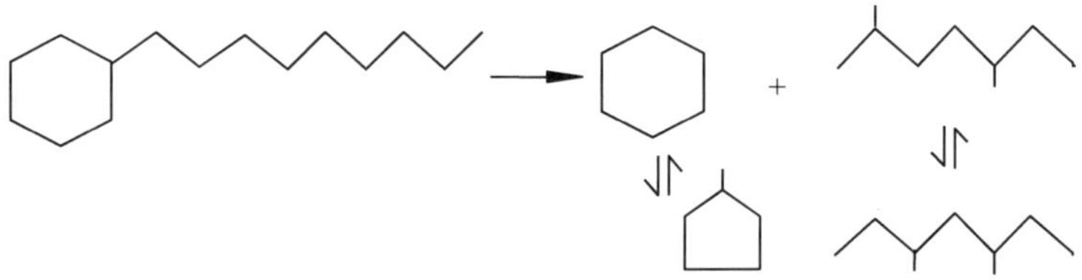

(c) Bicyclic naphthenes are reacting to open one ring to form alkyl-substituted monocyclic naphthenes, which follow the path described in b.
(d) Alkylbenzenes are dealkylated to form aromatics and isoparaffins. They are also hydrogenated to form monocyclic naphthenes, which follow the reactions described in b.
(e) Benzonaphthenes react primarily by opening of naphthene rings to produce alkylbenzenes, which react as described in d. They may also undergo hydrogenation of the benzene ring to form bicyclic naphthenes, which follow the reaction path described in c.
(f) Polyaromatics first undergo hydrogenation of one ring to form benzonaphthenes. The benzonaphthenes then follow the reaction path described in e.

Coking

This phenomenon is favoured at higher than design catalyst bed temperatures and lower than design hydrogen pressures. Coke deposits are the major contributor to catalyst deactivation.

Excessive coking rates will significantly diminish the cycle length, causing loss of revenue and higher than normal maintenance costs.

6 Benefits of Recycling Used Oil

The purpose of recycling used oil is to restore its chemical composition so that it can be re-used to produce fresh lube oil products (i.e. motor oil) over and over again. Because of its indefinite life cycle, substantial benefits from recycling used oil are achieved with respect to energy, environmental impact and resource conservation.

These benefits affect the consumer at the local level. However, when viewed from a national perspective, these benefits support a national agenda for closed-loop recycling and the re-refining of used oil. Re-refining contributes to save energy, reduces air and water pollution, conserves oil, and thus reduces its dependence on foreign oil. Each of these three benefits is summarized below:

- *Energy Savings* – recycling used oil provides greater benefits from a total energy resource conservation than burning used oil.
- *Environmental Benefits* – recycling used oil extracts toxic metals (zinc, cadmium, chromium, and lead, among others) that are emitted into the atmosphere during combustion.
- *Resource Conservation* – Burning used oil destroys this resource, which is particularly disadvantageous after expending substantial additional resources to develop it. Recycling reverses this pattern and extends the life cycle of this non-renewable natural resource so that it can be used indefinitely. For example, 1 kg of recycled lubricating oil can be produced from 1.5 kg of used oil, while 63 kg of crude oil are necessary to produce this same amount.

6.1 *Energy Savings*

Of the estimated 945 million gallons of used oil collected each year in the United States, 83% is burned and only 17% is recycled. In 1995, the U.S. Department of Energy (DOE) conducted a study to evaluate the energy requirements for both burning used oil and recycling used oil. The results of that study showed that recycling *presents greater benefits from a total energy resource conservation standpoint than burning used oil* (see Table 5 below). As noted in the table below, the energy comparison made to used oil burning favours recycling over using used oil as a fuel.

The analysis concluded shows that the total net theoretical energy savings for recycling compared to burning is 8.1% of 780 million gallons of the estimated volume of used oil that is recovered and burned per year. This equates to an annual energy savings of up to 63 million gallons of fuel ($63 million at a fuel oil price of $1.00/gallon).

Furthermore, the EPA stated that production of base oil from used oil (recycling) requires *50–85% less energy than refining crude oil into lubricating oil*. This is due in large part to the low quantity of base oil yielded from crude oil: approximately 42 gallons of crude oil is required to yield 1 gallon of base oil.

Table 5 Energy saving for burning vs. re-refining used oil

Energy balance Btu/Bbl of waste oil)	Burning	Re-refining	Variance	%
Transportation	−144	−198	−54	0.90
Processing consumed	−294	−742	−448	−7.60
Processing saved	745	1,722	977	16.60
Energy recovered	5,564	5,564	0	0.00
Net energy recovered	5,871	6,346	474	***8.10***

Source: Department of Energy, Argonne National Laboratory, August 1995

Table 6 Listing of common additives in used oils

Antiwear	Zinc dithiophosphates, acid phosphates, organic sulphur and chlorine compounds. Sulphurized fats, sulphides and disulphides
Detergent	Metallo-organic compounds of sodium, calcium and magnesium phenolates. Phosphonates and sulphonates
Anticorrosion	Zinc dithiophosphates, metal phenolates, fatty acids and amines
Dispersant	Alkylsuccinimides, alkylsuccinic esters
Friction modifier	Organic fatty acids. Lard oil. Phosphorous-based compounds
Pour point depressant	Alkylated naphthalene and phenolic polymers, polymethacrylates
Seal swell agent	Organic phosphates aromatic hydrocarbons
Viscosity modifier	Polymers of olefins, methacrylates, dienes or alkylated styrenes
Antifoamant	Silicone polymers, organic copolymers
Antioxidant	Zinc dithiophosphates, hindered phenols. Aromatic amines, sulphurized phenols
Metal deactivator	Organic complexes containing nitrogen and sulphur amines, sulphides

Source: "UK Waste Oils Market 2001", by David Fitzsimons, Nicholas Morley, Peter Lee, Oakdene Hollins LTD

6.2 *Environmental Impact*

The additive content of lubricants, especially in motor oils, raises concerns to the environmental concerns involved in the combustion of used oils.

Up to 20–25% of typical motor oil blends are commonly made up of additives used to improve the quality, stability, and longevity of motor oils in combustion engine applications. Table 6 summarizes some of the common additives and their chemical makeup.

A 1995 analysis conducted by the DOE compared the environmental impact of pollutants released into the atmosphere by burning used oil versus the recycling alternative. The key conclusion with regards to recycling is that the noxious compounds are solidified and stabilized in the form of asphalt, one of the main by-products of recycling and pose minimal environmental impacts.

Alternatively, combustion of used oil results in air emissions, the magnitude of which is dependent on the quality of the air pollution control equipment. Space heaters, for instance, which commonly use oil, have higher emissions than cement kilns, industrial burners, and hot mix asphalt plants because they do not feature any emissions control equipment.

Table 7 Environmental impact characteristics for used oil re-use options

Lead	50+% to air, balance to ash deposits	50%+ to air (90+% less with pollution control equipment), balance to ash	100% to asphalt
Cadmium	50+% to air; balance to ash	50%+ to air (90+% less with pollution control equipment), balance to ash	100% to asphalt
Chromium	<50% to air balance to ash	50%+ to air (90+% less with pollution control equipment), balance to ash	100% to asphalt
Zinc	50% to air, balance to ash	50%+ to air (30+% less with pollution control equipment), balance to ash	100% to asphalt
Sulphur	SOX to air	SOX to air, possibly scrubbed to form neutral salt	Burned to SOX scrubbed with caustic to form neutral salt
Nitrogen	NOX to air	NOX to air	NOx to air or pollution control equipment
Polynuclear hydrocarbons	COx to air, ash/soot	COX to air, ash/soot	Removed by hydrotreatment
Chlorinated hydrocarbons	HCl to air minimized in feeds	HCl to air, possibly scrubbed to form neutral salt	HCl scrubbed with caustic for form neutral salt
Phenols	COx to air	COX to air, fraction to wastewater	To fuel by-product or to wastewater treatment
Waste streams	Ash deposits	Wastewater, filtration solids, oily sludges, tank bottoms, ash	Wastewater, tank bottoms, ash

Source: Department of Energy, Argonne National Laboratory, August 1995

The environmental benefits clearly favour recycling because the toxic heavy metals (zinc, cadmium, chromium, and lead, among others) are extracted from the used oil, rather than emitted into the atmosphere during combustion as summarized in Table 7.

A peer-reviewed study conducted by the California EPA and University of California (Berkeley) compared the life cycle environmental impacts of combusting used oil and recycling used oils to produce base oils. This study was initiated as no comparable past studies could be found in the United States and because other studies had not fully assessed the impacts on human health or the environment. The results showed that heavy metal emissions from uncontrolled combustion of used oil fuels could be hundreds of times higher than the comparable recycling process or from the combustion of virgin fuel oil.

The data from uncontrolled combustion of used oil and from recycling were then categorized into life cycle impact characteristics. A comparison was made using a ratio of the results for combustion divided by the recycling results. Values of more than one indicate that uncontrolled burning results in more environmental impacts than recycling.

The results indicate that for most of the studied environmental parameters, the impact of burning used oil (assuming no pollution controls) produces more adverse environmental impacts than the alternative of re-refining used oil. Additionally, the heavy metal emissions may cause up to 150 times the eco-toxicity impacts compared to re-refining if air pollution control technology is not used. To put this into perspective, the California EPA study estimated that the zinc emissions from space heaters alone amount to approximately 7% of the total U.S. zinc air emissions. This assessment showed that combustors need more than 99% air pollution control efficiency before the toxicity impacts become comparable to re-refining. In addition, air pollution control equipment and subsequent ash management from the equipment use was not included in this analysis and may lead to even more significant impacts from used oil burning.

A study made by IFEU Institute in Germany (Heidelberg) in February 2005 and commissioned by G.E.I.R. the European Group of Re-refining Industries belonging to the U.E.I.L. the European Union of Lubricating oils Industries, shows that regeneration (by means used oil re-refining) scores better than primary virgin production of lubricants (from crude oil) for all the environmental impact categories considered (Fig. 6) [4], reducing the environmental drawbacks, in particular:

- *Global warming*: 2 times
- *Nutrification*: 3–4 times
- *Acidification and fine particulates (PM10):* >5 times
- *Carcinogenic risk potential*: 10–20 times
- *Fossil resources*: >30 times

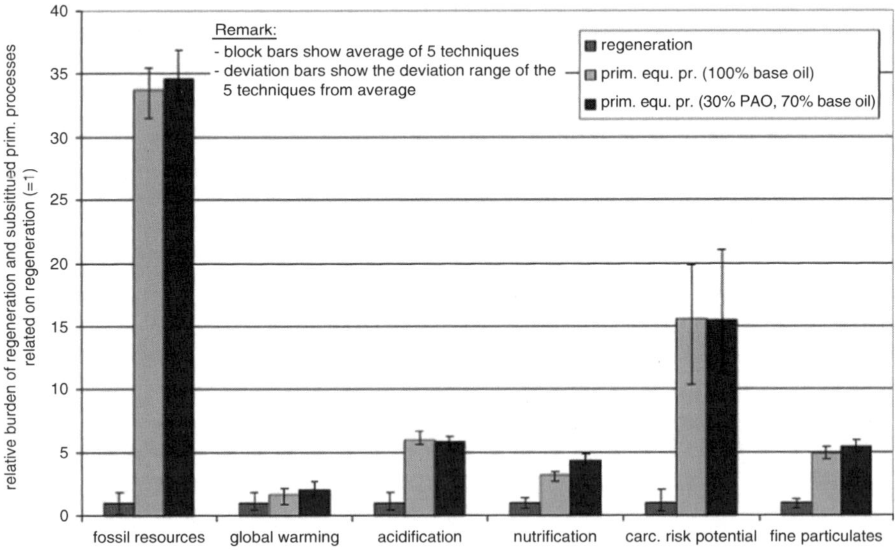

Fig. 6 IFEU – L.C.A.: comparison between waste oil regeneration and virgin lube oil production

6.3 Resource Conservation

Oil is a precious natural resource that is non-renewable. Burning used oil destroys this resource, which is particularly disadvantageous after expending substantial additional resources to develop it. Recycling reverses this pattern and extends the life cycle of this non-renewable natural resource so that it can be used indefinitely. For example, 2.5 quarts of recycled lubricating oil can be produced from 1 gallon of used oil, while 42 gallons of crude oil are necessary to produce this same amount.

In addition, the national security and economic prosperity of the U.S. is threatened by oil dependence, and the reliance on oil imports infringes on American foreign policy. Buying recycled oil helps reduce this dependence on imported oil, which currently accounts for approximately two-thirds of the U.S. demand.

The key steps to expanding the reuse of used oil relate to two key objectives:

- Increasing the effectiveness of used oil recovery programs to minimize the potential for disposal of used oils into landfills or other improper outlets.
- Increasing the volume of used oil available for re-refining since this process offers the maximum energy conservation and environmental benefits.

The waste management hierarchy detailed below is expressed in the waste management policy sections of "The Resource Conservation and Recovery Act and Pollution Prevention Act" in USA. The "Considered Action Step" describes the application of this used oil hierarchy from the most preferred environmental option to the least preferred environmental option.

Resource Conservation Hierarchy

Waste Management Ranking (from Environmental Perspective)	*Option*	*Considered Action Step*
Most Preferable	*Prevent the waste in the first place*	*Source reduction (e.g., extended oil drain intervals)*
↓	*Reuse and reclaim the product*	***Re-refine used oil***
	Recover energy by burning	*Combust used oil for heating value recovery*
Least Preferable	*Dispose of the waste by land filling or incineration*	*Recover and collect used oil for proper disposal*

Source: Department of Energy, Used Oil Re-refining Study to Address Energy Policy Act of 2005, Section 1838

Recycling used oil is the most preferable option over burning. The same type of hierarchy has been adopted by the European Commission in 2008 with a mandatory application by the end of 2010 from all Member States. The amount of use oil available for recycling is diminished by extended drain intervals and the fact that the volume of used oil that is currently burned is more than four times the volume that is currently recycled. This provides substantial room for the development of more recycling capacity. More importantly, as extended oil drain intervals become more prevalent, it will significantly increase the incentive to recycle. As the lubricating industry continues to produce higher quality products, the trend will favour recycling in as it will be economically disadvantageous to burn and destroy these high quality lubricants.

7 Conclusions

The Italian system is today a successful model substantially because:

- It rigorously strenghtens the legislation in force and it supports the principle of *sustainability*.
- It enforces the EC principle of *Life Cycle Thinking*.
- It strictly enforces the *priority order* of the EC hierarchy established by EC Directive 98/2008, where recycling comes before combustion.
- It enforces the concept that "*the polluter pays*" laying emphasis on the "*producer responsibility*".
- It supports waste oil collection and recycling with *economic incentives and voluntary agreements* as provided for by the EC Directive.
- It enables the companies to invest into *advanced technologies* to be able to compete with the virgin product, putting on the market base oils having the same qualities and performances as those produced by crude oil refineries and advanced environmental and toxicological characteristics (low sulphur and aromatics content).
- It enables to comply with *REACh Regulation* on the chemical substances and it enables re-refiners to produce lubricating oils, which can be registered in compliance with what is provided for by REACh thanks to the *sameness* of the recycled substance with the original one.
- It enables re-refiners to use the instruments made available by the Directive on *Green Public Procurement* and to join the Recycled Materials Repertory to supply the Public Administrations and Authorities and achieve the target of 30%. In fact, D.M. 203 of 8th May 2003 decrees that Regions shall adopt the necessary measures, addressed to the public administrations and to the companies with mostly public equity capital, also in the services management, to guarantee that articles and goods produced with recycled materials should cover at least *thirty per cent of their annual purchases*.
- It enables the producers of recycled oils to benefit by the *Ecolabel* brand. There is already an example in Spain, where the Catalan region has instituted an ecolabel for regenerated lubricants that requires lubricants to contain at least 25% of regenerated oils.
- It contributes, as stated by the European Commission in 2008 in the climate-energy package, to achieve the *20–20–20 targets*, namely to reduce by 20% the CO_2 emissions and to use at least 20% of renewable energies within 2020.

All that thanks to:

- Legislation
- Technology
- Organization of the collection system

It is now proven that in the countries where re-refining has been developed thanks to the political will to support it, waste oil collection and proper disposal have been implemented and important targets have been attained.

The *Italy System* has already become an example for some countries such as Poland and Venezuela, which were instructed by COOU to build a collection system similar to the Italian one and by Viscolube to benefit from an advanced technology producing top quality lubricating oils and abreast with the times performances.

It would be sufficient that, within Europe 27, also the other countries take the Italian system as an example so that a potential of *2.5 million tonnes/year* of waste oil would be sent to re-refining (and in Europe the plants capacity is available) to produce about *1.7 million tonnes/year* of recycled oil against the current 500,000 tonnes.

Moreover, Europe could benefit from a harmonized system at European level supported by laws and regulations equal for all avoiding dangerous cross-border transports of an hazardous waste such as waste oil, which nowadays is moved from a Member State to another on roads, railways, seas and rivers with the risk of accidents and pollution, increasing CO_2 emissions due to long distance transport and with sometimes dubious and illegal destinations or favouring combustion instead of recycling, making disappear a valuable fossil raw material, which is running out such as lubricating oil which, unlike many other waste, can be recycled countless times keeping unchanged its rheological and performance characteristics.

Unfortunately, the lobbies of those who have an interest in combustion (cement factories, steelworks, power stations) and, somctimcs, the lobbies of the oil companies, which see that the quantity of recycled product is increasing to the detriment of the virgin one in a climate of more and more reduced consumptions, are stronger than those supporting recycling and the environment protection.

References

1. Sustainability Report 2009 – Statutory consortium of used oil
2. Information from Viscolube. www.viscolube.it
3. Axens – Hydrofinishing Unit – Operating instruction manual 2009
4. IFEU – Ecological and energetical assessment of used oil re-refining – 2005

Emissions of Additives from Plastics in the Societal Material Stock: A Case Study for Sweden

Tomas Rydberg, Jenny Westerdahl, Elisabet Hallberg, Andreas Öman, Patrik L. Andersson, Peter Haglund, Tomas Holmgren, Filippa Fuhrman, Sverker Molander and Johan Tivander

Abstract Estimating the size of the problems related to release, fate, exposure and effects from the human use of chemical substances in materials and consumer products is daunting. More than 100,000 chemical substances are in commercial use and a reasonable description of their existence in, and release from, plastic polymers, glues, paints, fibres, lubricants etc. comprise a big challenge. Here we report the initial results from a generic emission model that has been developed and applied to estimate emissions of a set of organic chemicals from products. The scope of the study was to estimate emissions from products containing plastic materials during their average lifetime within the geographical boundaries of Sweden. The results show that approximately 2% of the plastic additives are emitted annually. Plasticisers, flame retardants, organic pigments and stabilisers are the use categories of additives that are emitted in the largest quantities. Until now, the method has only been used to estimate emissions of additives from plastic materials, but it is believed to also be applicable to other materials.

Keywords Additives, Diffuse emissions, Diffusion, Emission model, Material stock, Plastic additives, Plastics

---

T. Rydberg (✉), J. Westerdahl, E. Hallberg, and A. Öman
IVL Swedish Environmental Research Institute, Box 5302, 400 14 Göteborg, Sweden
e-mail: tomas.rydberg@ivl.se

P.L. Andersson, P. Haglund, and T. Holmgren
Department of Chemistry, Umeå University, 901 87 Umeå, Sweden

F. Fuhrman, S. Molander, and J. Tivander
Environmental Systems Analysis, Chalmers University of Technology, 412 96 Göteborg, Sweden

B. Bilitewski et al. (eds.), *Global Risk-Based Management of Chemical Additives I: Production, Usage and Environmental Occurrence*, Hdb Env Chem (2012) 18: 253–264,
DOI 10.1007/698_2011_107, © Springer-Verlag Berlin Heidelberg 2011,
Published online: 21 June 2011

Contents

1 Introduction

This chapter will present a methodology for estimating emissions of organic substances from the societal stock of products. It will also present the resulting estimated emissions of plastic additives from the Swedish stock of products. The data presented here are interim results from ongoing research in the Swedish research programme ChEmiTecs – Organic Chemicals Emitted from Technosphere Articles, which has been funded by the Swedish Environmental Protection Agency (EPA). The goal of the ChEmiTecs research programme has been aiming at an improvement of the understanding of emissions of organic substances from articles and to clarify and determine the magnitude of them.

In society, a vast number of different products and materials are accumulated. When calculating emissions from the entire stock of products and materials in society, the sheer number of products and materials does not make it feasible to describe the exact chemical composition of specific materials and products in detail and estimate the emissions based on that [1]. The methodology presented here is instead aimed at a simplified initial approximation of emissions of a set of organic chemicals from products containing plastic materials.

The emissions from the considered plastic articles during the year 2006 within the geographical boundaries of Sweden are the scope of the emissions approximation. This includes products manufactured in Sweden as well as imported into Sweden. Even though emissions from processes outside Sweden occur, no emissions outside Sweden are covered here.

2 Method

The methodology used comprises four steps. Firstly, the accumulated stock of selected product categories was estimated using trade statistics. Secondly, material composition data were compiled for the product categories with the largest stock.

Information on additive compositions for a number of plastic materials contained in the selected product categories was gathered and mapped to the product categories. Finally, the emissions from the stock of products were estimated using a simple emissions model. In the following sections, this methodology will be described in more detail.

2.1 *Quantification of Accumulated Stock of Goods*

In order to quantify the emissions of organic chemicals from technosphere articles, the accumulated stock of articles in society needs to be estimated. A previous study by the Swedish EPA and the Swedish Chemicals Bureau [2] described that the accumulated stock of articles in society could be estimated based on the annual net supply of articles to society in combination with the average lifetime of the articles according to the following relation:

$$\text{Accumulated stock} = \text{Net supply} \times \text{Average lifetime} \quad (1)$$

This relation is only valid if the net supply has reached steady state (that is the net supply does not change over time). In this study, the net supply was estimated based on national trade statistics covering manufacturing of goods in Sweden, together with import to and export from Sweden. The net supply can be obtained by:

$$\text{Net supply} = \text{Domestic manufacturing} + \text{Import} - \text{Export} \quad (2)$$

Trade statistics are compiled by the Statistics Sweden and are reported according to the customs Combined Nomenclature. The Combined Nomenclature is comprised of the so called Harmonised System nomenclature with further Community subdivisions [3]. In the Combined Nomenclature, statistics are reported for different groups of articles, called CN-product categories. All these 99 product categories are hierarchically divided into different subcategories, called CN2, CN4, CN6, and CN8, where the ingoing articles are divided into smaller subgroups with increasing tier. To give an example, product category 39 (CN2-level) is described as plastics and articles thereof. Within this category, there are 26 subcategories of articles at the CN4-level, for example, '3901 Polymers of ethylene in primary forms'. Within each CN4 category, there are additional subgroups, CN6-categories and so on.

The net supply of articles was compiled for the year 2006, reported as tonnes/year for all CN4 product categories [4]. A portion of all products is also transferred over the national boundary, but not exported in the statistical meaning, and therefore not accounted for in national customs statistics. This includes private import and export and illegal smuggling. These product flows are therefore not included.

Here, the average lifetime regard all episodes of how products are handled, actively or passively. The lifetime is defined as the period from the point in time when products are either domestically manufactured or when they enter Sweden as imported goods, until all points in time when the products are incorporated as components of other products or are discarded and enter the waste stream. The average lifetime of the different product categories was obtained from the Swedish EPA and the Swedish Chemicals Bureau [2]. There, the average lifetime was assigned as 1, 5, 25 or 50 years for the different CN4 categories. For all CN4 categories, the accumulated stock was calculated according to (1). Since the ChEmiTecs research programme focuses on emissions of organic chemicals from technosphere articles, product categories that were assumed to not contain any organic chemicals were excluded. This means that, for example, sand and other inorganic ore-based products were excluded. The product categories with an accumulated stock larger than 100,000 tonnes were selected for further investigation, for which material compositions were to be described.

2.2 Specifications of Materials in Product Categories

For the selected CN4 product categories, the contents of materials were specified aiming at a description of the type of material and the amount in each product category. In Table 1 below, the main plastic materials that were identified in the studied product categories are listed.

The specifications of what product categories contained certain amounts of the plastic materials were based on information of articles included in the different product categories, in combination with information from life cycle assessments and building product declarations, covering the material content of specific articles. The specifications resulted in mass percentages of the materials contained in a product category. Furthermore, the average thickness of the ingoing materials was

Table 1 Plastic materials included in this study

Plastic materials	Abbreviation
Acetals	
Acrylics	
Acrylonitrile butadiene styrene	ABS
Epoxy resin	
Phenolic resin	
Polyamide	PA
Polycarbonate	PC
Polyester	
Polyethylene	PE
Polypropylene	PP
Polystyrene	PS
Polyurethane	PUR
Polyvinyl chloride	PVC

Table 2 Typical thicknesses of different materials

Example of materials	Thickness
Varnish and paint layers, thin plastic films	0,00001–0,0001 m
Textiles, foils and paper	0,0001–0,001 m
Plastic components, carpets	0,001–0,01 m
Wood, tyres, soles	0,01–0,1 m
Concrete, building materials	0,1–1 m

estimated in order to calculate the exposed surface area of the ingoing materials, which in turn was needed for the emission calculation of additives. Surface areas were estimated based on the shape of the products. The areas were modelled as being constant over the entire product lifetime. The surface areas were calculated according to (3).

$$\text{Area} = \text{Mass}/(\text{Thickness} \times \text{Density}) \tag{3}$$

The thickness of a specific material in a product category can vary, depending on specific application, producer etc. To cover the likely thicknesses of a material used in a product category, a thickness interval was assigned for each material in a specific product category. When estimating the thickness, product specific information was preferred, but such information was not available for all product categories. For those product categories where no such information was known, a list of different materials and their typical thicknesses was developed and used instead, see Table 2. If it was known that a product category contained a material with very different thicknesses, two or more intervals were assigned to cover the range of thicknesses of that specific material. In order to calculate the area from known masses the density of all the plastic materials was set to 1,000 kg/m^3.

2.3 *Specification of Chemicals in Materials*

For the identified plastic materials in the selected product categories, more detailed chemical specifications were made. Lists of additives commonly used in different plastic materials were compiled, based on a thesis by Jansson [5] and information from Zweifel et al. [6]. Out of all the identified chemicals, the high production volume chemicals and the low production volume chemicals were selected for further study at this stage.

All additives were divided into use categories, for example, antioxidants, flame retardants, fillers etc. For each additive category and each plastic material, a minimum and maximum percentage of content was defined. Within each use category, there can be several different specific additive compounds possible to use to obtain a certain function in the plastic. Often only one or a few of these additives is used in a single product. Since no information was available regarding what specific additive that was used in a certain article, it was instead assumed that

Table 3 Assumed use of different plastic additives in plastic products

Additive	Amount of plastics containing a specific additive
Antifogging additive	10%
Antioxidant	100%
Antistatic additive	10%
Flame retardant	10%
Lubricant	10%
Pigment	50%
Plasticiser	50%
Slip additive	10%
Stabiliser	100%
UV stabiliser	10%
Whitening agent	10%

a mixture of all listed additives in a use category was used in all the materials To avoid overestimation, the minimum and maximum amounts of each additive were divided by the number of additives within each use category before summation.

All products do not contain all use categories of additives, and to account for this fact, an estimation of how commonly the different categories of additives are used in products were developed, see Table 3. These correction factors were applied to additives in plastic products to avoid overestimation of the stock of additives.

2.4 Estimation of Emissions of Organic Chemicals from Plastic Materials

The emissions of organic chemicals were estimated based on mass, area and additive composition of all plastic materials in the selected CN4 product categories. The method used is based on Fick's Second Law of diffusion [7] and the emissions can be calculated according to (4) and (5).

$$N_{\mathrm{add}} = 0.02 \times A_{\mathrm{polymer}} \times F_{\mathrm{add,polymer}} \times \rho_{\mathrm{polymer}} \times \left(\frac{D_{\mathrm{add}} \times t}{\pi}\right)^{0,5} \tag{4}$$

$$D_{\mathrm{add}} = \frac{10^{(-7{,}83-0{,}0062\,\times\,\mathrm{Mw})}}{10{,}000} \tag{5}$$

N_{add} is the emitted amount of an additive (kg), A_{polymer} is the area of a certain polymer (m^2), $F_{\mathrm{add,\ polymer}}$ is the mass percentage of additive in the polymer and ρ_{polymer} is the density of the polymer which here is assumed to be 1,000 kg/m^3 for all plastics and t is the time period for the calculation (s). D_{add} is the diffusion coefficient (m^2/s) of the specific plastic additive, which is estimated based on its molecular weight (M_w) at 40°C according to (5). By using (4) and (5), the emission of plastic additives from the polymer surface as a result of passive diffusion in the polymer is obtained. In the model, it is assumed that the additives are uniformly distributed within the polymer and that the additives are not chemically bound to the polymer. It is also assumed that the polymer is not subject to physical or biological degradation [7]. Finally, the model does not consider how the properties (e.g. temperature) of the surrounding media could affect the diffusion.

The time period of interest in this study is 1 year. When using the above (4) and (5), it is predicted that the rate of diffusion will slow over time. When calculating the emissions during 1 year for the accumulated stock, it is not appropriate to calculate this based on the first year of a products lifetime, since the products in the stock are of varying age and doing so would overestimate the emission. Here, the emission from the stock during 1 year was instead calculated as the average emission during the entire lifetime. That is

$$\frac{\text{Emission}}{\text{Year}} = \frac{N_{\text{add}}(t = \text{average lifetime})}{\text{Average lifetime}} \tag{6}$$

This assumption is valid if the accumulated stock has reached steady state.

These calculations were performed for all additives in all materials and product categories, resulting in emissions that can be traced back to the different product categories.

3 Results

In the following sections, the obtained results from the different calculation steps will be presented, i.e. information regarding the accumulated stock of plastic materials and additives in the Swedish societal stock of products as well as estimated emissions of plastic additives from these products for the year of 2006.

3.1 Accumulated Stock of Plastic Materials and Additives in the Swedish Society

In total, the stock calculations showed that approximately 100 CN4 categories had a stock larger than 100,000 tonnes. These CN4 categories constitute approximately 99% of the total stock of articles in Swedish society. In this study, the focus was on plastic-containing products excluding e.g. wood or concrete from this study.

Many of the product categories with a stock larger than 100,000 tonnes have a relatively long average lifetime, 25 years or longer. If the average lifetime is short, the products do not have the ability to accumulate to the same extent as products with long average lifetime, thereby obtaining a smaller accumulated stock in society.

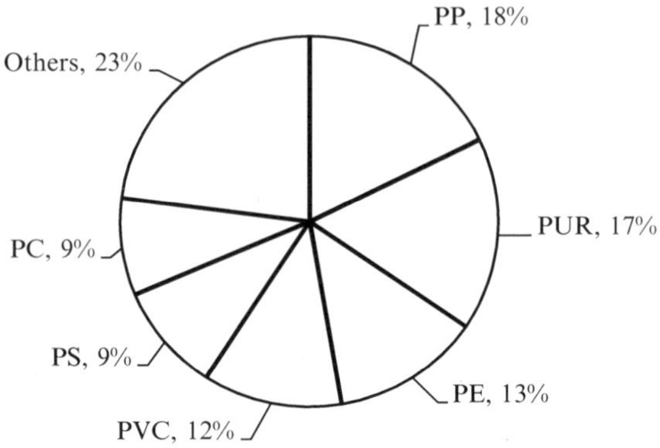

Fig. 1 Distribution of plastic materials in the Swedish stock of products

In total, 61 CN4 product categories were identified as containing organic chemicals and were studied in more detail. The Swedish accumulated material stock of these product categories was estimated to be 130 million tonnes in the year of 2006, out of which approximately one third, or 43 million tonnes, was assumed to be plastic materials. Figure 1 shows the relative distribution of different polymers in the accumulated stock. In van Oers et al. [8], statistics on the demand of different plastics materials per polymer in the EU were presented. Compared to these statistics, the fractions of PE and PET are lower in this study. Here, packaging materials covering finished products are not included since the trade statistics used in this study covers data on products excluding their packaging. Since two major packaging materials are PE and PET, the fraction of these materials therefore becomes lower in this study. However, the accumulated stock of packaging materials sold as packaging materials is included in this study. The fractions of the stock that is constituted by PP, PVC and PS correspond well with the sales data presented in van Oers et al. [8], whereas for PUR and PC, this study suggests higher fractions.

The plastic materials in the stock contain a number of additives. The results from this study show that the amount of additives in the plastic material stock studied here corresponds to approximately 6% (w/w) or 2.7 million tonnes of additives.

In Table 4, the net inflow and stock of additives in the Swedish society is presented for different additives use categories. Plasticisers are the additives that have the largest stock in Swedish society, for which the main application is softened PVC. Flame retardants, organic pigments and stabilisers are other categories of additives with large stocks.

Comparing the net inflow obtained here with the sales data for plastic additives presented in van Oers et al. [8], it can be concluded that the Swedish net inflow of additives represents approximately 1–4% of the global annual sales of additives. This estimated net inflow seems reasonable as the Swedish economy constitutes approximately 0.5% of the world economy [9].

Table 4 Estimated net inflow of additives and amount of additives in the accumulated stock of plastic products in the Swedish society in 2006

Additive	Net inflow (1,000 tonnes/year)	Stock (1,000 tonnes)
Antifogging additives	0.98	10
Antioxidants	8.2	140
Antistatic additives	1.4	17
Flame retardants	36	450
Bromine based compounds	*31*	*350*
Phosphorous based compounds	*4.0*	*80*
Other flame retardants	*1.0*	*20*
Lubricants	6.1	80
Organic pigments	38	480
Plasticisers	66	1,100
Phthalate plasticisers	33	550
Other plasticisers	33	550
Slip additives	0.13	1.3
Stabilisers	25	370
UV stabilisers	1.2	18
Whitening agents	1.1	20
Total	180	2,700

3.2 *Emissions of Plastic Additives*

The estimated emissions of different types of additives are presented in Table 5. In total, approximately 2% of the additives in the stock are estimated to be emitted through molecular emissions during 1 year.

The surface areas of the materials as well as the molecular weight of the additives are the two main parameters that influence the calculated emissions. Large emissions correspond to large areas of plastic materials and low molecular weights of the additives. The emissions are also influenced by the stock of additives in the products, where higher concentrations also yield higher emissions.

Since the emitting area is one of the major parameters that influence the emission of additives, the product categories with the largest areas will give rise to the largest emissions. Of the product categories included in this study, the CN2-category with the largest stock and emitting area are CN39 plastics and articles thereof. Other product categories that give rise to large emissions of plastic additives are CN85 which consist of electrical machinery and equipment and parts thereof such as engines and televisions, CN87 which consist of vehicles and vehicle parts and CN 94 which partly consist of prefabricated buildings. It should be noted that all subcategories within these CN2-categories have not been included, e.g. for CN94, 'CN9406 Prefabricated buildings' is the only CN4 category included.

Table 5 Emissions of plastic additives from the accumulated stock of plastic products in the Swedish society in the year of 2006

Additive	Emission (1,000 tonnes/year)
Antifogging additives	0.26
Antioxidants	0.66
Antistatic additives	0.078
Flame retardants	5.6
Bromine based compounds	*3.7*
Phosphorous based compounds	*1.6*
Other flame retardants	*0.3*
Lubricants	1.4
Organic pigments	6.9
Plasticisers	24
Phthalate plasticisers	*13*
Other plasticisers	*11*
Slip additives	0.042
Stabilisers	8.0
UV stabilisers	0.36
Whitening agents	0.026
Total	47

4 Uncertainties and Improvements

Within these calculations, there are different uncertainties which affect the results. Some of them are linked to the model, some to the object of study and others to the data quality. As well as for net inflow of additives, it is believed that the emissions of additives are overestimated. The origin of these uncertainties as well as their effect on the end results is discussed below.

The emissions of plastic additives are based on the accumulated stock of additives found in plastic products in Swedish society. When calculating the accumulated stock of products, net inflow of the selected product categories for the year 2006 in combination with the average lifetime of the product categories was used. The net inflow of a product category can vary from year to year. Using trade statistics from a single year, as in this study, yields uncertainties. If trade statistics covering more than 1 year are used instead, the uncertainty within the stock estimations can be reduced.

The area of the stock that was obtained by using (3) also has a large impact on the emissions of additives. In (3), one uncertainty lies within the estimation of the thickness of the material. A product category can be quite diverse and it can therefore be difficult to estimate the thickness of the material, which causes uncertainties in the obtained emissions.

Additionally, the emission model itself also cause uncertainties in the estimated emissions. The emission model used here estimates emissions of additives from plastic materials to air through molecular diffusion in the plastic material, where it is assumed that the additives are not chemically bound to the matrix. Since chemical interactions between the additives and the polymer do occur, this assumption causes an overestimation of the emissions. Another factor that causes uncertainties is the temperature of the material and the surroundings. In this model the diffusion coefficients were, due to the empirical equation, calculated at 40°C. As the average temperature in Sweden is lower than 40°C and the diffusion in plastic materials is affected by the temperature, this assumption will give rise to an overestimation. More information regarding important assumptions and uncertainties within the emission model can be found in the Emission Scenario Document on Plastic Additives by the OECD [7]. To improve the quality of the obtained emissions and to reduce the uncertainties connected to the emission model, a new emission model is being developed [10].

As a result of these uncertainties, the final emissions are judged as being overestimated.

5 Conclusions

By using information from national trade statistics and a rough emission model, it is possible to estimate emissions of additives from the accumulated stock of products in society during 1 year of the product's average lifetime. Thus far, the method has only been used to estimate emissions of additives from plastic materials, but it is believed to also be applicable to other materials.

The obtained results show that the accumulated stock of products containing plastic materials is approximately 130 million tonnes and that out of the stock approximately 43 million tonnes consist of plastics. These plastic materials contain almost 3 million tonnes of additives, of which plasticisers, flame retardants, organic pigments and stabilisers constitute the largest fraction. According to model calculations, almost 2% of these additives are emitted to the environment yearly. The additive categories that are emitted in the largest amount correspond well to the categories with the largest stock, i.e. plasticisers, flame retardants, organic pigments and stabilisers.

The results contain uncertainties, which are caused by both model uncertainties and uncertainties in the input data. The most influential factors are believed to be the net inflow, the average life time, the thickness of the material and the emission model used. By using trade statistics covering more than 1 year and by using a more detailed emission model, a large fraction of the uncertainties can be reduced.

The methodology and models presented here will be further improved to minimise the identified uncertainties and thereby improve the emission estimates.

Acknowledgements The authors gratefully acknowledge the financial support from the Swedish EPA given to the ChEmiTecs programme (2008–2012).

References

1. Molander S, Tivander J, Fuhrman F, Rydberg T, Andersson PL, Haglund P, Holmgren T (2011) Conceptualizing emissions of chemicals from products – foundations of a modelling approach. Extended Abstract submitted 2010-11-30 to SETAC Europe Annual Meeting, 23–26 May 2011, Milano, Italy

2. Swedish EPA, Swedish Chemicals Agency (1999) Att finna farliga flöden: kemikalier i samhället: rapport från ett regeringsuppdrag. Naturvårdsverket (in Swedish)
3. European Commission (2010) Taxation and customs union – the combined nomenclature. Available at http://ec.europa.eu/taxation_customs/customs/customs_duties/tariff_aspects/combined_nomenclature/index_en.htm. Accessed 31 May 2010
4. Westerdahl J, Rydberg T, Molander S, Tivander J, Fuhrman F, Andersson PL, Haglund P, Holmgren T (2011) Emissions of chemicals from the economy-wide stock of plastic material – first iteration based on Sweden. Extended Abstract submitted 2010-11-30 to SETAC Europe Annual Meeting, 23–26 May 2011, Milano, Italy
5. Jansson E (2008) Kartläggning av kemiska ämnens förekomst i materialslagen plast, gummi och textil. Självständigt arbete 15hp, 2008:1, Institutionen för Miljökemi, Stockholms universitet, (KEMI) (in Swedish)
6. Zweifel H, Maier RD, Schiller M (2009) Plastics additives handbook, 6th edn. Carl Hanser Verlag, Munich
7. OECD (2009) OECD Series on Emission Scenario Documents, Number 3 – Emission Scenario Documents on Plastic Additives. OECD Environment Health and Safety Publications, Report ENV/JM/MONO(2004)8/REV1
8. van Oers L, van der Voet E, Grundmann V (2011) Additives in the plastics industry. In: Bilitewski B, Darbra RM, Barceló D (eds) The handbook of environmental chemistry – global risk-based management of chemical additives I: production, usage and environmental occurrence. Springer, Heidelberg
9. CIA (2010) Publications – The World Factbook. Available at https://www.cia.gov/library/publications/the-world-factbook/index.html. Accessed 14 Mar 2011
10. Holmgren T, Persson L, Olofsson U, Andersson P, Haglund P (2010) Predictive emission model for organic compounds added to materials in consumer goods. Extended Abstract submitted 2010-11-30 to SETAC Europe Annual Meeting, 23–26 May 2011, Milano, Italy

Recycling of Waste Electrical and Electronic Equipment, a Case Study: Brazil

M. Araujo, A. Magrini, and C.F. Mahler

Abstract Electrical and electronic equipment is one of the most dynamic sectors of the economy, comprising 4.5% of the Brazilian gross domestic product (GDP) [1]. This market is growing faster than the country's GDP. Although business volumes are high, there is no formal specific structure for treatment of waste electrical and electronic equipment (WEEE) in Brazil. A general law on solid waste was approved by the Congress and signed by the president in August 2010, but further enabling regulations still need to be issued.

Keywords Recycling, WEEE

Contents

M. Araujo (✉) and A. Magrini
Energy Planning Department – COPPE/UFRJ – Ilha do Fundao, s/no., Rio de Janeiro, Brazil
e-mail: marcel_g@uol.com.br; ale@ppe.ufrj.br

C. Mahler
GETRES - Group of Studies and Technology of Waste – COPPE/UFRJ – Ilha do Fundao, s/no., Rio de Janeiro, Brazil
e-mail: cfmahler@acd.ufrj.br

B. Bilitewski et al. (eds.), *Global Risk-Based Management of Chemical Additives I: Production, Usage and Environmental Occurrence*, Hdb Env Chem (2012) 18: 265–278,
DOI 10.1007/698_2011_111, © Springer-Verlag Berlin Heidelberg 2011,
Published online: 2 August 2011

Abbreviations

ABINEE	Brazilian Electrical and Electronic Producers Association
CRT	Cathode ray tubes
GDP	Gross domestic product
LCD	Liquid crystal display
Pb	Lead
PC	Personal computer
WEEE	Waste of electrical and electronic equipments

1 Introduction

Brazil has an increasing population with the means to consume electrical and electronic equipment. This trend is very likely to continue, resulting in a huge increase in WEEE and the related environmental impacts [1]. Economic data for the sector are presented in Sect. 2.1, obtained from the Brazilian Electrical and Electronic Producers Association – ABINEE [1]. In Brazil, 85% of the population lives in urban areas, which facilitates a high penetration of electrical and electronic equipment. Sales to consumers are well developed with a concentration in large retail chains and few manufacturers.

Bandini [2] from ABINEE estimated a 2% WEEE recycling rate for the entire country. Other materials are recycled to a large extent in Brazil, particularly aluminum cans, where the recycling rate was 96% in 2005, one of the highest rates in the world [3]. This activity developed completely independent of government actions, starting in the early 1990s. Other materials that are also recycled in Brazil are cardboard boxes and paper. Estimates are that more than a million people live entirely or supplement their income by scavenging for various materials in the country. These people commonly sell their pickings to junkyards and small general recycling companies or sort it themselves and sell it to more specialized recyclers (paper/cardboard, plastics, aluminum, and so on).

There is a good opportunity to start a formal WEEE treatment structure from the beginning with the support of scavengers and recyclers for improved dismantling to achieve higher recovery of metals and appropriate treatment of hazardous substances. Nonetheless, the feasibility of such a structure depends on government incentives, through tax advantages and financial support with lower interest rates for investments. Also, the secondary materials market has an important role in the financial equation. This is particularly true in Brazil where primary metals prices are low, since the country is a big global supplier of several metals.

This report is a preliminary analysis of the WEEE situation in Brazil since the release of the National Solid Waste Policy is recent and will have a substantial effect on the WEEE system that may develop.

A brief description of the market for electronic and electrical equipment is given in Sect. 2. Section 3 presents generated amounts of WEEE, with more detailed data on flows of computers, monitors and cell phones for the period 2000–2009. Section 4 provides a preliminary picture of the current recycling system for WEEE in Brazil. Section 5 discusses the lack of information on the health effects caused by WEEE treatment. Section 6 presents the current status of policies related to WEEE and solid waste management. Finally, Sect. 7 presents the conclusions, emphasizing the critical sucess factors for good WEEE management in Brazil.

2 Problem Definition

2.1 Electrical and Electronic Goods Market

Brazil has an increasing population with the means to consume electrical and electronic equipment. Increasing purchasing power, particularly among lower and lower-middle class people, due to an expanding economy and federal income transfer programs, is boosting the consumption of these products. This trend is very likely to continue, resulting in a huge increase in WEEE and the related environmental impacts [1]. Economic data for the sector are presented in Table 1, obtained from the Brazilian Electrical and Electronic Producers Association [1], including total revenue of the sector in Brazil for 2007–2009 and estimates for 2010.

Sales to consumers are well developed with a concentration in large retail chains. For computers, multinational companies like Dell and Toshiba and local companies like Itautec, Positivo, and CCE dominate the market for preassembled computers. There is also a large "gray market" of desktop computers assembled from off-the-shelf parts by small shops. Brazil was a big manufacturer of cell phones, including for exportation, but in recent years local producers have lost market share to imported devices.

The Brazilian Institute of Geography and Statistics [4] conducts an annual household survey called the PNAD. Among the data collected in this survey is the number of households that own various electronic devices by type of device, see Table 2.

The PNAD data only shows the number of households that possess the equipment, not if a household has two or more devices. Also, it does not include office equipment. Therefore, the PNAD numbers are conservative for equipment like

Table 1 Economic indicators for the electronic and electrical goods sector in Brazil [1]

	2007	2008	2009	2010[a]
Revenues (US$ million)	57,338	66,989	56,459	73,477
Exports (US$ million)	9,300	9,891	7,200	7,200
Imports (US$ million)	24,050	32,033	24,000	26,800
No. of employees (1000)	156	162	160	163

[a]Estimate

Table 2 Brazilian national household survey data for 2008 [4]

Devices	Million units	% Hh
Refrigerators	53.0	92%
Freezers	9.2	16%
Telephones	47.0	82%
Washing machine	23.9	42%
Audio system	51.2	89%
Computers	20.3	35%
Cell Phones	24.1	42%
Televisions	54.8	95%

Table 3 Computer sales and stock in use, 2000–2009 (million units)

	Sales [1] million units	Stock [6] million units
2000	2.9	10.0
2001	3.1	13.0
2002	3.1	16.0
2003	3.2	19.0
2004	4.1	23.0
2005	5.6	28.0
2006	8.2	34.0
2007	10.0	41.5
2008	12.0	50.0
2009	12.0	60.0

Table 4 Cell phone sales and stock in use (million units)

	Sales [1] million units	Stock [5] million units
2000	2.6	23.2
2001	5.2	28.7
2002	11.3	34.9
2003	16.4	46.4
2004	33.3	65.6
2005	36.6	86.2
2006	37.1	99.9
2007	49.7	121.0
2008	55.1	150.6
2009	50.0	174.0

computers and cell phones that have a characteristic or personal use and are used by both households and businesses.

For cell phones and computers, other data sources are also available (Tables 3 and 4). Stock data for Cell phones can be based on the number of active cell phone subscriptions, obtained from the local operators [5]. For personal computers, an annual survey of households and offices is conducted by Meireles [6], estimating the stock of computers "in use". Sales data for both devices were obtained from ABINEE [1].

2.2 *Waste of Electrical and Electronic Equipment (WEEE)*

The Swiss Federal Laboratories for Materials Science and Technology (EMPA) prepared an estimation of WEEE generation in Brazil, based on the consumption and use methodology [7], in which generation is calculated by dividing the stock by a constant average lifetime. The model considers data related to televisions, washing machines, refrigerators, telephones, computers, and audio systems. The lifetimes applied in the model are similar to those used in EMPA surveys for other Latin American countries [8–10]. The authors' estimate of computer waste including monitors in Brazil was 0.3 kg/cap year or 56,270 tonnes in entire Brazil for 2008, considering a computer and monitor lifetime of seven years and total weight of 29.26 kg.

A recent report by UNEP [11] presents estimates for WEEE flow for several countries. For 2005, Brazil's WEEE generation arising from computers (including monitors) was calculated to be over 0.5 kg/cap year, which gives a total amount of computer waste of 96,800 tonnes in 2005, considering a lifetime of five years and total weight of 25 kg.

Comparing these numbers one can see that there is a big difference between estimates, although both use the same methodology. This is mainly due to the estimation of the lifetime parameter, which should consider the time the device is used for all users. In Brazil, cascade use is quite common, with donation or re-sale of equipments for lower power acquisition groups, resulting in a longer lifetime, with regional variances. Available information shows small importation of used computers for reuse as donation to poor communities. Another factor that contributes to the uncertainty of the result is the limited data on volumes.

Considering the discrepancies of WEEE data, a different methodology has been used in this study for estimating the generated amounts of waste consisting of computers, monitors and cell phones.

3 Generation of WEEE in Brazil

This section presents an estimate of the generation of waste of computers (desktop PCs), cellular phones and monitors is presented for the period of 2000–2009. Import and export of WEEE to/from the country is assumed to be very small.

The time-step method was used for this estimation [see (1)] as recommended by the European Environmental Agency [12]. The stock in use and sales quantities are presented in Tables 3 and 4 above.

$$\text{Generation of WEEE} = \text{sales - stock variation} \tag{1}$$

The model's assumptions are:

1. There is no accumulation of stock of equipment at retail points and producers.
2. There is no importation or exportation of WEEE.

3. When desktop computers are discarded, the monitor is also discarded.
4. All computers that were discarded in 2008 were desktops. ABINEE [4] sales data also show the split between desktops and notebooks, where it can be seen that the percentage of notebooks was very low before 2008.
5. Discarded monitors are CRTs (cathode ray tube), since LCD (liquid crystal display) sales only started to increase after 2007.
6. The average weight of a desktop computer is 13 kg, monitors 17 kg, and cell phones 170 g according to USEPA [13].
7. There is only one phone per active cell phone subscription.

The time-step method does not need a lifetime input. It is adequate for markets where the products' lifetimes are variable. Markets that are growing and/or are subject to technological changes do not need a lifetime input. This is the case of personal computers and cell phones, which have undergone continuous technological changes, such as desktops to notebooks, CRT to LCD monitors, or the three generations of cell phone equipment. With this method, guessing the lifetime can be bypassed and cascade use is considered within the estimation of stock in use.

The results obtained by using the time step method for Brazil are shown in Tables 5–7.

Figure 1 shows the sales and waste generation estimated by the model for cell phones in Brazil. By comparing these curves it can be seen there is no constant discrepancy between the two lines. For 2008, cell phone lifetime is around 4.5 years in Brazil. The amount of WEEE generated decreased between 2007 and 2008, probably due to the small increase in equipment sales between 2004 and 2006.

The same analysis for computers is shown Fig 2, where a lifetime of five years is obtained from the curves at the point of 2008 waste.

Both curves show a growing sales trend after 2004. The waste curve has a corresponding increase, but not as well defined. Statistical treatment would allow a better interpretation of the results. Nonetheless, it can be seen that the generation of waste for the devices only recently has been increasing.

Table 5 Cell phone waste generation estimates

	Stock variation	Cell phone waste	
	million units	million units	tonnes
2000	8.2	–	–
2001	5.5	–	–
2002	6.2	5.1	867
2003	11.5	4.9	833
2004	19.2	14.1	2,397
2005	20.6	16.0	2,720
2006	13.7	23.4	3,978
2007	21.1	28.6	4,862
2008	29.6	25.5	4,335
2009	23.4	26.6	4,522

Table 6 Desktop computer waste generation estimates

	Stock variation million units	Computer waste million units	tonnes
2000	2.7	0.3	3,250
2001	3.0	–	650
2002	3.0	0.1	1,300
2003	3.0	0.2	2,600
2004	4.0	0.1	962
2005	5.0	0.6	8,255
2006	6.0	2.2	28,925
2007	7.5	2.5	32,279
2008	8.5	3.5	45,500
2009	10.0	2.0	26,000

Table 7 Monitor waste generation estimates

	Stock variation million units	Monitor waste million units	tonnes
2000	2.7	0.3	4,250
2001	3.0	–	850
2002	3.0	0.1	1,700
2003	3.0	0.2	3,400
2004	4.0	0.1	1,258
2005	5.0	0.6	10,795
2006	6.0	2.2	37,825
2007	7.5	2.5	42,211
2008	8.5	3,5	59,500
2009	10.0	2,0	34,000

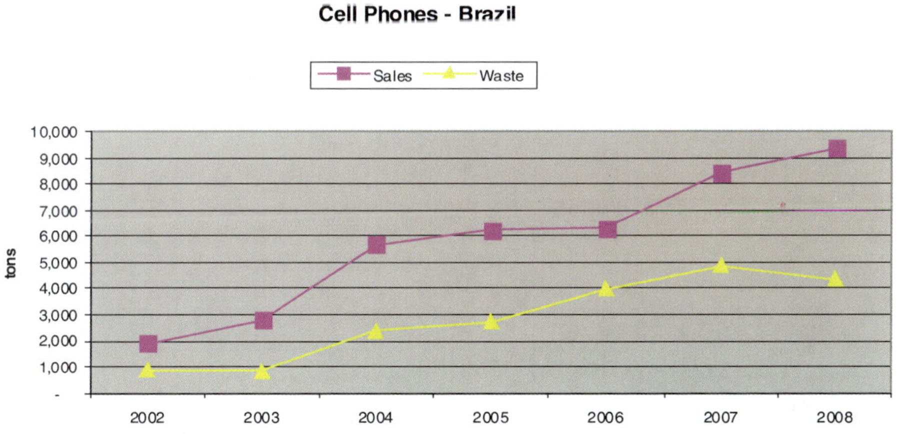

Fig. 1 Sales and waste curves for cell phones in Brazil

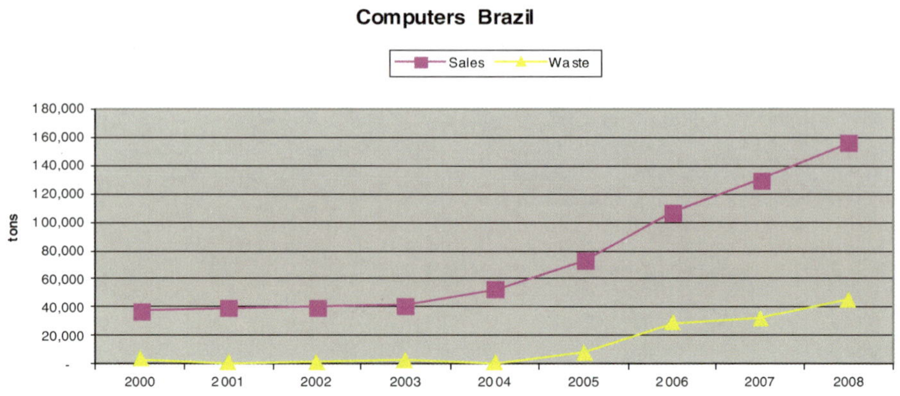

Fig. 2 Sales and waste curves for computers in Brazil

It should be emphasized that most of this waste has not reached conventional disposal sites and is probably stored in households.

4 Methods to Recycle the WEEE

In general, reliable information is scarce and based on assumptions. WEEE in Brazil was analyzed by Rodrigues [14]. The author conducted a field survey and found there is no structured WEEE management in Brazil. She stated there are some recycling companies acting in Brazil, but they are more focused on waste generated by the electronic manufacturing sector (obsolescent items, production line rejects). Usually, the devices are dismantled, separated according to materials and the material fractions are then shredded. The final material is sent to industries for reuse as raw material or is disposed off at industrial landfills. One prominent exception is circuit boards, which are sent to Europe for precious metals recovery. Some of WEEE Recycling Companies in Brazil are shown in Appendix A.

The collection of WEEE in Brazil has been done essentially by scavengers and it is only recently that companies have become interested in the WEEE sector. The majority of these companies are located in the Southeast region of the country, mainly in Sáo Paulo state. This is not surprising, since this region is the most developed part of the country which is led by São Paulo, the country's most developed state.

5 Health and Ecosystem Effects in the Country Based on WEEE Recycling Methods

Very little is known about the WEEE treatment in Brazil, and even less in known about health or ecological effects caused by WEEE treatment. No information about these effects has been found for Brazil. There are only few formal recycling facilities for WEEE treatment and these facilities must be submitted to environmental licensing requirements. Battery recycling was a problem in Brazil, because many scavengers destroyed the batteries using a hammer without any protection to get Pb. However, no records have been collected of disease provoked by this behavior.

6 Current Policies to Deal with These Problems

The NBR 10.004 (Norma Brasileira de Referência – Reference Brazilian Standard) standard from the Brazilian Association of Technical Standards [15] establishes a classification of solid wastes: dangerous, inert, and non-inert wastes. A waste is classified as dangerous according to the criteria inflammability, corrosiveness, reactivity, and pathogenicity.

The National Environmental Council (CONAMA) Resolution 23 [16] is the local regulation resulting from the Basel Convention. It aims to control the international trade of hazardous substances in Brazil. It defines what is considered as dangerous waste, and prohibits the importation of such material. It explicitly cites as dangerous wastes those that contain substances such as halogen, copper, mercury, cadmium, and beryllium.

CONAMA Resolution 313 [17] specifies controls of the generation, storage, transport, disposal, and treatment of industrial solid wastes. They must be handled according to the NBR 10.004 classification. As the name indicates, it only covers industrial solid wastes.

These laws and resolutions are not focused on WEEE and they only prohibit the import and export of hazardous materials and regulate the treatment of industrial solid waste. Appendixes B and C present the most important legal instruments and standards of solid waste in Brazil.

The Brazilian National Solid Waste Management Policy (PNRS) is defined in the recently enacted Law 12,305/2010 [18] and is based on the concepts of shared product responsibility, life cycle of products, and mandatory reverse logistics for six specific product streams: agricultural chemicals, batteries, tires, lubricant oils, fluorescent lamps, and electrical and electronic waste. Producers and governmental authorities must prepare plans that provide diagnosis of situations, scenarios and goals for reduction, reuse and recycling, and for decreasing disposal in landfills. Producers and retailers must establish a reverse logistics chain for these six products. Consumers have the obligation to package these waste products adequately for the reverse logistics chain. Finally, producers must provide an environmentally adequate destination for waste products and their packaging.

The Brazilian Congress took about 19 years to prepare and approve the PNRS. Unfortunately, the final text is too vague and the opportunity to create firm rules and contribute to a modern solution of the waste problem was partially lost, though some ground can be reclaimed depending on the stringency of the enabling regulations that are eventually issued by the competent authorities.

As indicated, follow-on regulation of the National Solid Waste Policy is still pending. At this stage, it is the main gap in the legal and regulatory framework.

7 Conclusion

The development of an informal market in Asian countries was fostered by the need for metals [19]. This is not the case for Brazil. Brazil is one of the biggest international suppliers of ores and metals. This is likely one of the reasons why the informal market of WEEE recycling has not developed in the country, with exception of alumina cans and cardboard/paper. The existence of a huge primary raw materials market can be a disadvantage for the WEEE recycling companies, because the demand for secondary material is fundamental for the success of recycling facilities.

The efficiency of a WEEE policy in the country will depend on several critical factors:

1. Development of regulations that are adequate for the different regional market conditions.
2. Effective consumer education initiatives regarding selective collection. It is not feasible to impose fines on consumers, since the country has a large area, diversified regional characteristics and weak legal enforcement. In addition, many areas, particularly shanty towns, are not adequately served by even non-selective trash pick-up. The shared product responsibility concept in the new Solid Waste Law is not clear about penalties on consumers who do not meet their responsibility, and in the final analysis consumer compliance is the key to effective WEEE management.
3. Structuring of a reverse logistics stream. Reverse logistics channels for each product type must be competitive and environmentally and technologically well structured.
4. Development of treatment technologies and exchange of experiences with research organizations from developed countries.
5. Fostering the reuse, refurbishment and secondary recycled material market.
6. Incentives for the creation of scavenger associations and recycling facilities, by tax breaks and low-interest loans to small and medium enterprises.
7. Establishment of effective and safe controls.
8. Establishment of a specific legislation on WEEE.

Finally, many Brazilian specialists, entrepreneurs, and authorities argue that selective collection is too expensive. This shortsighted attitude ignores the costs that will inevitably be borne by future generations to solve the environmental problems they will face. By ignoring the medium and long-term costs of failing to take action now, and only looking at the immediate bottom line, they misrepresent the economics of adequate WEEE treatment, recycling and disposal. We hope this situation changes as the Brazilian population becomes more aware of the issues involved.

Appendix

Appendix A: Recycling Companies In Brazil

UMICORE – Guarulhos, São Paulo.
SUZAQUIM – Suzano, São Paulo.
Essencis – Nordeste.
TCG Recycling – Americana, São Paulo.
Cimélia – Campinas, São Paulo.
Baterias Pioneiro – Treze Tílias, Santa Catarina
Pioneiro Ecometais – Água Doce, Santa Catarina.
Suzaquim/Faarte – São Paulo, SP
SIR Company – São Paulo, SP
Interamerican Ltda – São Bernardo do Campo, SP
Lorene Importação e Exportação Ltda. – São Paulo, SP
Oxil - Manufatura Reversa – Paulínia, SP
Sanlien Exportação Ltda. – São Paulo, SP
Sucata Eletrônica – São Paulo, SP
Target Trading S.A – São Paulo, SP
Xerox Comércio e Indústria Ltda. – São Paulo, SP

Appendix B: Important Solid Waste Legislation In Brazil

MINTER Edict 053 of March 1, 1979
Approval by a governmental authority is mandatory for projects for treatment and disposal of solid waste.
CONAMA Resolution 1-A of January 1986
Regulation of transport of hazardous substances.
CONAMA Resolution 05 of August 5, 1993
Regulation of management of solid waste at ports, airports, railroad, and trucking terminals.
CONAMA Resolution 024 of December 7, 1994
Requires authorization of the National Nuclear Energy Commission for importation or exportation of radioactive material.
CONAMA Resolution 023 of December 12, 1996
Regulates the importation and use of hazardous substances and classifies solid waste.
CONAMA Resolution 273 of November 29, 2000
Establishes rules on prevention and control of pollution at service stations.

CONAMA Resolution 307 of July 5, 2002
Establish directives, criteria and procedures for the management of solid waste from construction projects.
CONAMA Resolution 313 of October 29, 2002
National Inventory for Industrial Solid Waste.
CONAMA Resolution 316 of October 29, 2002
Establises procedures and criteria for thermal treatment for solid waste.
CONAMA Resolution 348 August 16, 2004
Includes asbestos as a hazardous substance in CONAMA Resolution 307/02.
CONAMA Resolution 358 of April 29, 2005
Treatment and final disposition of solid waste from hospitals and other health care facilities.
CONAMA Resolution 375 of August 29, 2006
Establishes criteria and procedures for the agriculture use of sewage from treatment plants.
CONAMA Resolution 380 of October 31, 2006
Corrections in CONAMA Resolution 375/06
CONAMA Resolution 401 of November 4, 2008
Establishes maximum limits for lead, cadmium, and mercury for batteries and presents criteria and procedures for sound management.
CONAMA Resolution 404 of November 11, 2008
Establishes criteria and directives for the environmental licensing of small-scale landfills.
CONAMA Resolution 416 of October 1, 2009
Prevention of environmental harm from unserviceable tires and their final disposal.
CONAMA Resolution 420 of December 28, 2009
Establishes directives for environmental management of contaminated soil, criteria, and chemical levels for soil quality.
Law 12,305 of August 2, 2010
Establishes the National Solid Waste Policy and alters Law 9,605/1998

Appendix C: Standards For Solid Waste In Brazil – Brazilian Association of Technical Standards (ABNT)

ABNT Standard For Solid Waste	
NBR 10.004	Solid waste – classification
NBR 10.005	Waste leachate
NBR 10.006	Solubilization of waste
NBR 10.007	Sampling of solid waste
NBR 10.703	Soil degradation – terminology
NBR 12.988	Free líquids – checking of waste samples

(continued)

ABNT standards for landfills	
NBR 8418	Presentation of projects of hazardous industrial waste landfills
NBR 8419	Presentation of projects of urban solid waste landfills
NBR 10.157	Landfills of hazardous waste – criteria for projects, construction and operation
NBR 13.896	Landfills for non-hazardous waste – criteria for projects, construction and operation
Standard ABNT about treatment, storage, and transportation of waste	
NBR 11.174	Solid waste storage
NBR 11.175	Incineration of hazardous solid waste – performance criteria NB 1265
NBR 13.894	Landfarming
NBR 98	Storage and use of flamable combustibles liquids
NBR 7.505	Storage of petroleum and liquids products and ethanol
NBR 12.235	Storage of harzadous solid waste (old NB 1183)
NBR 11.174	Storage of non inerts and inerts waste (old NB 1264)
NBR 13.221	Waste transportation
NBR 7.500	Symbols of risk and use for treatment and storage of materials
NBR 7.501	Transportation of dangerous material – terminology
NBR 7.502	Transportation of dangerous material – classification
NBR 7.503	Emergency sheet for transportation of dangerous materials
Standards ABNT about characterístics and dimensions	
NBR 7.504	Envelope for transportation of dangerous material– dimensions
NBR 13.786	Selection of equipments and systems for underground instalations of combustible for service stations
NBR 13.784	Detection of leaching in service station
Standards ABNT about health solid waste	
NBR 12.807	Health solid waste – terminology
NBR 12.808	Health solid waste – classification
NBR 12.809	Use of health solid waste – procedure
NBR 12.810	Colection of health solid waste – procedure

References

1. Associação Brasileira da Industria Elétrica e Eletrônica – ABINEE. (Brazilian Electrical and Electronic Producers Association) 2009. Desempenho Setorial. (Sector Performance). São Paulo, Brasil. Avaiable at www.abinee.org.br. Accessed on January 10, 2010
2. Bandini M (2009) Seminário Internacional de Resíduos Eletroeletrônicos (International seminar on electrical and electronic waste). Minas Gerais, Brasil. Available at http://www.seminarioree.com.br/. Accessed on December 12th, 2009
3. Gatti JB, de Castilho QG, Corrêa Garcia EE (2008) Recycling of aluminum can in terms of life cycle inventory (LCI). Int J LCA 13(3):219–225

4. IBGE – Instituto Brasileiro de Geografia e Estatística (Brazilian Institute of Geography and Statistics) 2009. Pesquisa Nacional por Amostra de Domicílios 2008 (National Household Survey 2008). Rio de Janeiro (in portuguese).
5. Associação Brasileira de Telecomunicações - TELEBRASIL. 2010. O Desempenho do Setor de Telecomunicações no Brasil Series Temporais 2 T2010 (Performance of the Brazilian Telecommunication Sector – 2 T2010). Rio de Janeiro, August 2010
6. Meireles F (2010) Tecnologia de Informação 20ª Pesquisa Anual – 2009 (Information Technology Annual Survey – 2009). Fundação Getúlio Vargas. São Paulo. Available at www.fgv.br. Accessed on January 11th, 2010
7. Rocha HTR, Gomes FVB (2009) Diagnóstico da Geração de Resíduos Eletroeletrônicos no Estado de Minas Gerais (Diagnosis of the generation of electrical and electronic waste in minas gerais state). Fundaçao Estadual do Meio Ambiente (FEAM), Swiss Federal Laboratories for Materials Science and Technology (EMPA) (in portugese)
8. Steubing B, Boni H, Schluep M, Silva U, Ludwig C (2010) Assessing computer waste generation in Chile using material flow analysis. Waste Manag 30:473–482. doi:10.1016/j.wasman.2009.09.007
9. Protomastro, G.F. 2007. La cadena de valor de los RAEE: Estudio sobre los circuitos formales e informales de gestión de Residuos de Aparatos Eléctricos y Electrónicos en Argentina (The value chain of WEEE: A study of the formal and informal circuit of WEEE management in Argentina). http://www.residuoselectronicos.net (in spanish).
10. Ott D (2008) Gestión de resíduos electrónicos en Colombia: Diagnóstico de computadores y telefónos celulares (Management of electronic waste in Colombia: diagnosis of computers and cell phones). Swiss Federal Laboratories for Masterials Science and Technology (EMPA), Secretaría Federal de Asuntos Econômicos, Universidad de los Andes, Cámara Colombiana de Informática y Telecomunicaciones (in spanish)
11. UNEP (2009) Recycling – from e-waste to resources: sustainable innovation and technology. UNEP transfer industrial sector studies. July 2009
12. Environmental European Agency (EEA) (2003) Waste from Electrical and Electronic Equipments WEEE – quantities, dangerous substances and treatment methods. European Topic Centre on Waste, Copenhagen
13. United States Environment Protection Agency - USEPA (2007) Management of electronic waste in the United States: approach two. Draft final report. EPA530-R-07-004b
14. Rodrigues AC (2007) Impactos socioambientais dos resíduos de equipamentos elétricos e eletrônicos: estudo da cadeia pós-consumo no Brasil. (Environmental and social impacts of electrical and electronic waste: survey on the post-consumer chain in Brasil – dissertation) - Universidade Metodista de Piracicaba, Faculdade de Engenharia, Arquitetura e Urbanismo, Programa de Pós-Graduação em Engenharia de Produção. São Paulo
15. Associação Brasileira de Normas Técnicas - ABNT (Brazilian Association of Technical Standards). 1987. NBR 10.004 – Resíduos Sólidos. (Solid Waste Standard) September, 1987
16. Conselho Nacional do Meio Ambiente CONAMA (National Environmental Council) 1996. Resolução no. 23 de 12 de dezembro de 1996 - (Resolution 023 of December 12, 1996). Brasília, Brazil
17. Conselho Nacional do Meio Ambiente CONAMA (National Environmental Council) 2002. Resolução no. 313 de 29 de outubro de 2002 (Resolution 313 of October 29, 2002). Brasília, Brasil
18. Republica Federativa do Brasil 2010. Política Nacional de Resíduos Sólidos Lei 12.305 (National Solid Waste Policy - Law 12,305). Brazil. August, 2010
19. Zoeteman BCJ, Krikke HR, Venselaar J (2010) Handling WEEE waste flows: on the effectiveness of producer responsibility in a globalizing world. Int Jr Adv Manuf Technol 47:415–436

Part III
Conclusions

Conclusions

D. Barceló, R.M. Darbra, and B. Bilitewski

Abstract The main outcomes of the book are summarized in this chapter. The importance of additives in the product final characteristics is an essential element of the chapter. Each of the additives gives a set of properties that improve the final product. These additives can be applied during the manufacture of the product or at the finishing process. The difficulty in gathering data on emissions, fate, and exposure of the additives has been highlighted in some occasions, especially at the end of the product's life. A review of the main conclusions for each sector is also presented in this chapter. The highlights of each case study are also presented.

Keywords Chemical additives, Environmental and Human Impact, Lack of data, Risk

After reviewing the additives used in the different sectors selected by Riskcycle project, a set of conclusions can be presented, being the main one that there exists a very wide variety of chemical compounds used for improving the properties of the products. These substances provide better characteristics to the final product but during the life of the product and at the end of this one, they can be released into the environment affecting the humans and the ecosystems.

D. Barceló (✉)
Department of Environmental Chemistry, IDAEA-CSIC, Jordi Girona, 18-26 08034, Barcelona, Spain
e-mail: dbcqam@cid.csic.es

R.M. Darbra
CERTEC, Department of Chemical Engineering, Universitat Politècnica de Catalunya, ETSEIB, Diagonal 647, pav. G. 08028, Barcelona, Spain

B. Bilitewski
Institut für Abfallwirtschaft und Altlasten, TU Dresden, Pratzchwitzer Str. 15, 01796 Pirna, Germany

B. Bilitewski et al. (eds.), *Global Risk-Based Management of Chemical Additives I: Production, Usage and Environmental Occurrence*, Hdb Env Chem (2012) 18: 281–286,
DOI 10.1007/698_2011_121,
Published online: 20 October 2011

The additives can be applied in mainly two different moments: (a) throughout the manufacturing process and (b) during the finishing process to confer the desired quality properties to the final product. Some of these substances are hazardous to both human health and the environment. These additives can be emitted from the products during all the life cycle of the equipment, i.e., during the production, use phase, and/or end-of-life treatment. For many of them, information regarding emission, fate, exposure, and toxicity is scarce, especially at the end of the product's life. To be able to better assess the risks connected to emissions of additives from the different products and to suggest appropriate management actions to reduce the risks, more data on releases and toxic effects caused by these substances are needed.

Historically, from an environmental perspective the paper industry has attracted particular interest due to the known large amounts of organic compounds released to the environment by this sector. However, additives have deserved much less attention. With the review carried out in this book, information on the current state-of-the-art additives used in the paper and board industry has been presented. Additives to the paper are as follows: sizing agents, fillers, wet strength agents, coatings, optical brighteners, biocides, and dyes. The main environmental issues identified are emissions to water and air, sludge buildup, and energy consumption. It is the pollution of water bodies, however, which is of major concern due to the large volumes of wastewaters generated during the paper manufacturing, depending on both the raw material and the type of paper produced.

Even though the tanning process of leather is known since long ago, it is still today an active industrial sector subjected to continuous evolution through the use of new techniques and chemicals. During the last years, the leather industry has become an important commercial value for some developing and emerging countries like China, India, or Brazil, all being major producers of raw leather that is subsequently distributed worldwide for further manufacturing of final articles (i.e., shoes, clothing, or other goods). Therefore, leather production and trade have reached a global dimension, and their environmental outcomes should be considered at that scale. Leather industry has been typically characterized by its environmental impact due to the high loads of chemicals released, which include among others heavy metals such as trivalent chromium, acids, and bases, all of them used as major reagents throughout the leather production process. Furthermore, there is a rising concern about the environmental fate and effects of additives, since most of the additives still remain on the final product. For example, alkylphenols used as surfactants (i.e., nonylphenol) are known to cause endocrine disruption in some aquatic species, and are included in the list of priority compounds in Europe under the provisions of the Water Framework Directive (Directive 2008/105/EC).

In order to give the everyday improving electric and electronic equipment, the desired properties, additives are added to the materials. As mentioned before, some of these substances are hazardous to both human health and the environment. End-of-life treatment has been identified as a life cycle phase where high emissions of additives can occur. The amount of emitted additives from the end-of-life treatment is to a large extent affected by how the e-waste is treated. If a controlled treatment is applied, the emissions can be minimized and so will the risks to human health and the environment. When the e-waste is treated in the uncontrolled system, for example in Asia or Africa, there is a large risk that it will be dismantled manually and burned in open fires by people with insufficient protection and appropriate equipment. Growing awareness of risks involved in trading waste globally has led to the Basel Convention on the Control of Transboundary Movements of Hazardous Wastes and Their Disposal, usually known as the Basel Convention, which entered into force in 1992. This international treaty aims at preventing the shipping of hazardous waste, including e-waste, from developed to less developed countries, which have ratified the Basel Convention and the Basel Ban Amendment, adopted in 1995. Also initiatives like "Solving the E-waste problem" (StEP), founded by various UN organizations, address the problem of exporting of e-waste to countries where informal recycling techniques pose a risk to humans and the environment. Nevertheless, large volumes of e-waste are still shipped to poorer countries by being falsely declared as "reusable."

Additives used in the textile sector are mainly applied during the finishing process and have many different uses, such as improving the antistatic properties of the textile, protecting it from microbial agents or fire, etc. Most of these additives may pose an environmental and health damage during the textile production, the product's use, or at the end of its life. For instance, PFCs are extremely persistent, bioaccumulative, and toxic compounds. Triclosan is highly toxic to aquatic organisms and HBCD, one of the most extended flame retardants, is not readily biodegradable since it has a very high potential to adsorb to soils and sediments. The assessment of the damage of these additives has been estimated in some cases during the production and use. However, difficulties arise when trying to characterize their risk at the end of the product's life. Data about the exposure levels to these substances during their last lifecycle stages (disposal or recovery) are scarce. On the contrary, it would be very interesting to be able to carry out research at this stage. Since most of the textile products end in landfills, all the additives present in clothing or textiles are going to end up mainly on the soil or groundwater, originating the associated pollution and being able to affect the environment and the population.

The most important application of lubricants is in the automotive sector, where lubes are used to avoid damages to the engines. They also play a key role in the reduction of environmental burden by saving important resources, avoiding leakage losses, and reducing exhaust gas emissions. Most ecological effects are induced by the mineral base oils because additives are normally insoluble in water. Focusing on used oils, it is possible to say that the toxicity of these products derives from their improper disposal. Heavy metals, PCBs, dioxins, and PAHs can be released into the environmental as a result of lube combustion. Modern lubricants do not contain halogenated additives; so very small amounts of halogens may be present as contaminants from chemical processing. Moreover, modern re-refining techniques, like thermal deasphaltation, reduce the environmental impact and the toxicity of the used oils that otherwise could be disposed using incorrect ways with higher environmental load.

A plastic, randomly chosen, will contain about 20 additives. These additives have different functions in providing the plastic material with the desired characteristics. Some types of additives, like colorants or flame retardants, are used in all polymers. Others are more specific, such as heat stabilizers or plasticizers. The additives make out a significant share of the material, in some cases up to 60% of the weight. Statistical data on the production of plastic additives are scarce as well as on their emissions and effects. Life Cycle Inventory data are not available for plastic additives, especially for emission and waste information. This makes very difficult to obtain a picture of plastics-related emission of additives to the environment. China can be appointed as one of the fastest growing markets for plastics and thus plastic additives. Environment-friendly products such as nonhalogenated flame retardants, organic heat stabilizers, conductive compounds, wood composites, nano composites, and fiber-reinforced thermoplastics represent some of the potential growth areas for global additives. The market for plastic additives such as flame retardants, antioxidants, and organic heat stabilizers, particularly for vinyl, is strong as Europe banned lead-based heat stabilizers.

Pulp and paper production of Vietnam increases over years, but not in a sustainable way. The unbalance among Recovered Paper Utilization, Utilization Rate, and Collection Rate is a problem that affects the paper recycling industry. The paper recycling activity in craft villages has contributed as a remarkable sign in the recycling sector. However, there is still a problem in harmonization between production and environmental protection. The recycled paper products with low quality, especially for hygienic paper delivered throughout the country, may cause health problems for customers.

The survey of chemicals which are potential candidates for substitution within the Danish printing industry resulted in about 200 substances/substance groups. About 60 of these substances fulfill one or more of the criteria for the EU REACH Annex XIV candidate list. Some of these, like the phthalates and the lead chromate pigments, may be relevant when looking at the potential hazard of globally recycled paper based on printed matter. The Danish printing industry will work to substitute these substances for other less pollutant.

The textile industry plays a significant role in the Indian economy. The textile effluent from dyeing and printing units gets treated in the Common Effluent Treatment Plant (CETP), which leads to generation of chemical sludge in voluminous quantities. This sludge is considered as hazardous waste as per the Indian Hazardous Waste Regulation. Presently, the only option available for disposal of this sludge is secure landfill, which is a costly option for small-scale units. The characterization study revealed the concentration of heavy metals (such as Cr, Cu, Ni, Zn, Cd, and Pb) in the sludge, and leachate was found to be less than prescribed limits (Indian Hazardous Waste Rules and UESPA limits for leachate). The presence of a significant amount of calcium, magnesium, silica, and aluminum oxides in the sludge indicated that it had the potential to be used as construction material.

The leather industry is using a high number of chemicals in the production, and it is not uncommon that tanneries use 300–400 different chemicals in their production. A careful and accurate management of chemicals is therefore necessary for the industry to minimize the health and safety aspects by being a chemically intensive industry. An example of a modern and well-developed tannery in Sweden is presented in this book to show how reducing the environmental impact from leather production is possible. The main incentives for the tannery to implement this chemical management are the environmental policies of the company, environmental legislation, and customer demands. The difficulty in getting information on chemicals and the lack of resources are the two main barriers for the tanneries to improve their environmental management.

In Italy, used oil re-refining practice is common since the 1950s, and it is today a reality, thanks to an excellent organization system and thanks to the priority given to the used oil recycling vs. burning. This practice has now been strengthened by the European Union with the recent Waste Framework Directive which imposes a priority order hierarchy for waste management, where recycling is at an upper level than burning with energy recovery. Italy, within the past EU15, was the only Member State which did not receive an infringement procedure by the Court of Justice of Luxembourg for not having respected the priority of regeneration according to Directive 75/439/EEC. Legislative framework, collection system, incentives, good communication policies, and technology are the key points that made Italy the most advanced European Country in the used oil re-refining sector.

Through the use of information from national trade statistics and a rough emission model, a Swedish case study has shown how it is possible to estimate emissions of additives from the accumulated stock of products in society during 1 year of the product's average lifetime. The obtained results show that the accumulated stock of products containing plastic materials is approximately 130 million tonnes, and that of the stock approximately 43 million tonnes consist of plastics. These plastic materials contain almost three million tonnes of additives, of which plasticizers, flame retardants, organic pigments, and stabilizers constitute the largest fraction. According to the model calculations, almost 2% of these additives are emitted to the environment yearly. The additive categories that are emitted in the largest amount correspond well to the categories with the largest stock, i.e., plasticizers, flame retardants, organic pigments, and stabilizers.

In Brazil, electrical and electronic equipment is one of the most dynamic sectors of the economy, comprising 4.5% of the Brazilian gross domestic product (GDP). This market is growing faster than the country's GDP. Although business volumes are high, there is no formal specific structure for treatment of waste electrical and electronic equipment (WEEE) in Brazil. A general law on solid waste was approved by the Congress and signed by the president in August 2010, but further enabling regulations still need to be issued. Many Brazilian specialists, entrepreneurs, and authorities argue that selective collection is too expensive. This shortsighted attitude ignores the costs that will inevitably be borne by future generations to solve the environmental problems they will face. By ignoring the medium and long-term costs of failing to take action now, and only looking at the immediate bottom line, they misrepresent the economics of adequate WEEE treatment, recycling, and disposal.

As it can be seen, the risk assessment of chemicals at global scale is not an easy subject. In many cases, information is not available for production and use of chemicals, and also for the emissions, fate, and exposure of them. However, assessing the environmental risks associated with additives along the whole product's life cycle at global scale is an issue of paramount relevance, which requires the development of proper methodologies and effort to gather information. This is the main objective to be achieved by the RISKCYCLE project. Hopefully, by the end of the project some answers to this question will be given, and they will be captured in the second volume of this book.

Index

B. Bilitewski et al. (eds.), *Global Risk-Based Management of Chemical Additives I: Production, Usage and Environmental Occurrence,* Hdb Env Chem (2012) 18: 287–290,
DOI 10.1007/698_2011,